Alternative Fuels and Petroleum Technology

Alternative Fuels and Petroleum Technology

Edited by **Kurt Marcel**

SYRAWOOD
PUBLISHING HOUSE

New York

Published by Syrawood Publishing House,
750 Third Avenue, 9th Floor,
New York, NY 10017, USA
www.syrawoodpublishinghouse.com

Alternative Fuels and Petroleum Technology
Edited by Kurt Marcel

International Standard Book Number: 978-1-68286-118-9 (Hardback)

Contents

Permissions

List of Contributors

Preface

This book has been a concerted effort by a group of academicians, researchers and scientists, who have contributed their research works for the realization of the book. This book has materialized in the wake of emerging advancements and innovations in this field. Therefore, the need of the hour was to compile all the required researches and disseminate the knowledge to a broad spectrum of people comprising of students, researchers and specialists of the field.

Petroleum technology is a discipline that integrates concepts of several fields such as geology, engineering and geophysics; it primarily covers exploration and extraction of crude oils. However, depletion of conventional fuel sources and their extensive use has resulted into deterioration of environment and increased demand for exploration and consumption of alternative fuels. This book traces the availability of alternate fuels and highlights some of the key concepts and applications of these fuels. The topics included in this book on petrophysics, thermodynamics, biofuels are of utmost significance and bound to provide incredible insights to readers. This book includes contributions of experts and scientists which will provide a multidisciplinary perspective into the fields of alternative energy resources and petroleum technology.

At the end of the preface, I would like to thank the authors for their brilliant chapters and the publisher for guiding us all-through the making of the book till its final stage. Also, I would like to thank my family for providing the support and encouragement throughout my academic career and research projects.

Editor

Performance test of palm fatty acid biodiesel on compression ignition engine

Praveen K. S. Yadav[1]*, Onkar Singh[2] and R. P. Singh[1]

[1]Department of Oil and Paint Technology, Harcourt Butler Technological Institute, Kanpur, India.
[2]Department of Mechanical Engineering, Harcourt Butler Technological Institute, Kanpur, India.

Vegetable oil causes problem when used as fuel in compression ignition engines. This problem is due to high viscosity and low volatility of vegetable oils, which can be minimized by the process of trans-esterification. The relatively high cost of refined vegetable oils render the resulting fuels unable to compete with petroleum derived fuel. To reduce the cost of biodiesel a relatively low cost palm fatty acid which is the by- product of palm oil refinery was chosen as feed stock. A two step acid catalyzed methanolysis process is employed for efficient conversion of palm fatty acid into palm fatty acid methyl esters (FAME). The conversion of palm fatty acid into palm FAME was done on pilot plant. This paper presents the results of performance of compression ignition engine fueled with different blends of high speed diesel with biodiesel obtained from conversion of palm fatty acid. Parameters like torque, brake specific fuel consumption and brake thermal efficiency were calculated at different loads for pure diesel and different combinations of dual fuel. The results indicate that, in case of palm fatty acid biodiesel, the dual fuel combination of B40 can be used in the diesel engines without making any engine modifications.

Key words: Trans-esterification, methanolysis, palm fatty acid, fatty acid methyl esters (FAME), compression ignition engine, triglyceride (TG).

INTRODUCTION

Energy is an essential factor for economic growth. Building a strong base of energy resources is necessary for sustainable economic and social development of any country. In view of limited fossil fuel reserves and environmental concerns the efforts are made to search a potential alternate. Moreover, the exhaust of petroleum diesel is the main reason for environmental pollution. Under these circumstances it becomes necessary to look for some self sustainable, biodegradable and environment friendly alternate fuels. Biodiesel, which is fatty acid methyl ester (FAME) is biodegradable and substantially reduces the exhaust emissions, when used in different blend ratios with petroleum diesel. Biodiesel does not increase the level of carbon dioxide in the atmosphere and therefore helps in minimizing green house effect. There is a growing interest in biodiesel because of its similarities in properties with fossil fuel

(Prafulla et al., 2009; Ramadhas et al., 2004). Biodiesel will mitigate the vulnerability and the adverse effects of use of fossil fuels. Several countries have introduced policies encouraging the use of biodiesel to replace part of their fossil fuel use and also prevent environmental degradation. However, production cost of biodiesel is not economically competitive with fossil fuel due to higher cost of lipid feedstock. The production of biodiesel can be made economical by using low grade lipid feedstocks containing high amount of free fatty acids (FFA). Waste cooking oil and non conventional oil seed oils viz. Tobacco seed oil with high free fatty acid content can be used as substitute of conventional diesel (Wang et al., 2006; Zlatica and Predojevic, 2008; Veljkovic et al., 2006). Lipid feedstock which are non edible in nature like karanja oil and Mahua oil can also be used for production of biodiesel (Stalin and Prabhu, 2007; Ramadhas et al., 2004; Shashikant and Raheman, 2005). However, the limitation with non edible feedstock is its low availability. In this study palm fatty acid (PFA) which is the by-product of refining of palm oil is used as feedstock for producing biodiesel. Here the study has been carried out on PFA

*Corresponding author. E-mail: pkyhbti@yahoo.co.in.

feedstock with 93 wt% FFA content. The use of PFA as feedstock for production of biodiesel has the following advantages:

i.) It does not compete with edible grade oil bearing seeds of food market as it is the by-product of refining of palm oil and is non-edible.
ii.) It is easily available.
iii.) Generally, the high cost of biodiesel is the major obstacle for its commercialization, as the biodiesel produced from vegetable oil or animal fat is usually more expensive than the petroleum diesel. In view of low cost of PFA, the biodiesel produced will be economically competitive.

A two step process is used for the conversion of PFA into FAME. The first step of the process is to reduce FFA content by esterification with methanol in presence of acid catalyst. In the second step, triglyceride (TG) portion of the lipid reacts with methanol in presence of base catalyst to form methyl ester and glycerol (Meher et al., 2006; Siti Zullaikah et al., 2005; Fangrui et al., 1999; Vicente et al., 2004). The acid catalyst is generally sulfuric acid and base catalyst is usually sodium or potassium hydroxide (Vicente et al., 2004; Zheng et al., 2006; Di Serio et al., 2005; Pramanik, 2003). Product from the reaction is separated into two phases under gravity. The FAME portion is washed with water and dried to meet the biodiesel fuel standards. The biodiesel produced from low grade lipid feedstock and its blends with fossil fuel can be used in diesel engines without any modifications of the engine. The use of biodiesel reduces the gas emissions, improves lubricity and thus increases the life of engine ((Pramanik, 2003; Ramadhas et al., 2005; Mustafa Canakei et al., 2006).

MATERIALS AND METHODS

PFA was procured from M/s Kanpur Edibles Pvt. Ltd. Kanpur, India. The composition of the palm fatty acid was determined with the help of gas liquid chromatography (Figure 1). The palm fatty acid contains 0.2% Lauric acid, 1.2% Myristic acid, 42.7% Palmitic acid, 0.1 % Palmitoleic acid, 4.5% Stearic acid, 38.9% Oleic acid, 11.5% Linoleic acid, 0.7% Linolenic acid, 0.1% Arachdic acid and 0.1 % traces. All chemicals including methanol, sulfuric acid and sodium hydroxide were of analytical grade. The acid value, saponification value and iodine value of PFA were 184, 198 and 52, respectively.

EXPERIMENT

The process of transesterfication was used to reduce the high viscosity of triglycerides. The transesterification reaction is represented by the general equation in Figure 2. A two step process was used for conversion of PFA into palm FAME. The first step of the process was to convert PFA into FAME by esterification with methanol (MeOH) in presence of acid catalyst. In the second step the PFA content was further reduced by reaction with MeOH in presence of base catalyst (Stalin and Prabhu, 2007).

Production of biodiesel

The operating conditions for production of biodiesel were first optimized at laboratory scale by studying the effect of variation of molar ratio of PFA to MeOH from 1:1 to 1:12 keeping the wt% of catalyst and temperature constant. The amount of catalyst was decided by varying the wt% of H_2SO_4 catalyst from 0.2 to 1.2% for first step and wt% of NaOH catalyst from 0.5 to 2.5% for second step keeping the molar ratio of PFA to MeOH and temperature constant. The effect of temperature was studied by varying the temperature from 40 to 65 °C keeping other parameters constant. The production of biodiesel in bulk was done at the optimized conditions of molar ratio 1:10 (PFA: MeOH), 1 wt% of H_2SO_4 and 2 wt% of NaOH at 65°C using the pilot plant in the Oil and Paint Technology Department, Harcourt Butler Technological Institute, Kanpur, India. The plant mainly consists of three continuous stirred tank reactor, a dosing unit, a condenser, a boiler, storage tanks for raw material and finished product and pumps. A schematic flow diagram and pilot plant used for biodiesel production are shown in Figure 3a and b respectively. Preheated PFA and catalyst (H_2SO_4) in methyl alcohol solution were fed in continuous stirred tank reactor. After the completion of the reaction the reaction product was transferred to separator for the separation of methyl esters under gravity. The top phase containing FAME also contained excess methanol and water formed during the reaction. Excess methanol was separated and purified and reused as starting material. The FAME phase was taken off at the bottom and passed into evaporator to remove and collect the traces of methanol. The FAME still had residual PFA requiring further purification. The PFA was neutralized with sodium hydroxide in methanol (NaOH–MeOH) solution. Both neutralization and transesterification reactions took place at the same time. The product obtained from the purification step was settled in a separator. The FAME phase was separated and washed with hot water to remove impurities. The final product obtained was characterized and its comparison with petroleum diesel is given in Table 1.

Analysis of palm fatty acid methyl ester

The composition of biodiesel was confirmed by FT-IR spectrum (Figure 4). Sharp band in the range of 2925.13 cm^{-1} is due to C-H stretching vibrations of methyl and methylene groups. A sharp band at 1743.67 cm^{-1} is attributed to C=O stretching frequency. Absorption at 1437 and 1463.25 cm^{-1} is assigned to asymmetric -CH$_3$ or -CH$_2$ bending vibrations. Bands at 1246.56, 1196.93 and 1171.19 cm^{-1} are due to C-O stretching of ester. The bands obtained at 1117.90, 1017.31 and 880.31 cm^{-1} are due to C-C stretching.

Calorific value of biodiesel

Different blends of biodiesel and conventional diesel were prepared and the calorific value of each sample was determined with the help of Bomb Calorimeter and the values are given in Table 2. The blending of FAME with petroleum diesel was done by intense mixing of both with help of electric stirrer at room temperature on v/v basis. For example, for making a sample B20, first 20% by volume of FAME was taken in the beaker and further 80% by volume of petroleum diesel was added and then it was vigorously mixed with help of electric stirrer at room temperature.

Engine test

The performance of the prepared palm FAME was studied and compared with the conventional diesel fuel. The tests were performed at Mechanical Engineering Department, Harcourt Butler

PEAK REPORT (ANRF)

Method Name	: Default
Sample Name	: PFA Methyl Ester
Run No	: 20
Analysis Time	: 9/6/09 4:05:41 PM

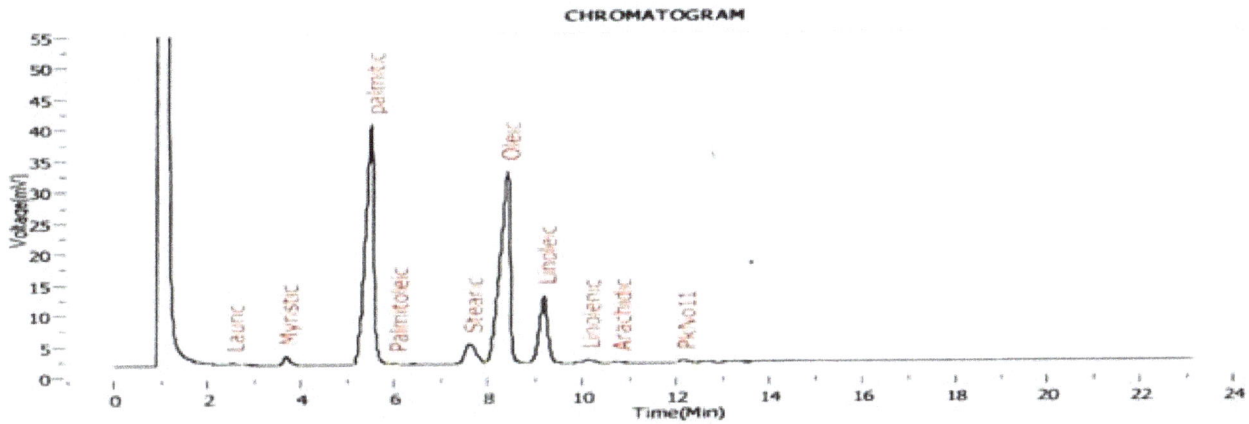

CHROMATOGRAM

Peak No	RT (Min:Sec)	Area (mV-Sec)	Height (mVolt)	RF -	Amount (MI)	Amount% -	Component Name
1	2:32	1.8299	0.2837	1	1.8299	0.178	Lauric
2	3:41	12.3957	1.2991	1	12.3957	1.169	Myristic
3	5:30	491.4392	38.6281	1	491.4392	42.664	palmitic
4	6:01	0.819	0.0733	1	0.819	0.102	Palmitoleic
5	7:37	50.7973	3.1879	1	50.7973	4.493	Stearic
6	8:24	457.3748	30.9331	1	457.3748	38.997	Oleic
7	9:10	133.9003	10.9324	1	133.9003	11.443	Linoleic
8	10:08	7.8221	0.4835	1	7.8221	0.704	Linolenic
9	10:46	1.128	0.1066	1	1.128	0.129	Arachidic
10	12:10	3.795	0.3384	1	3.795	0.053	PKNo11

Number of Peaks : 10
Total Peak Area : 1161.301 mV-Sec

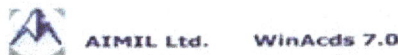

AIMIL Ltd.   WinAcds 7.0

Figure 1. Chromatogram of palm fatty acid methyl ester.

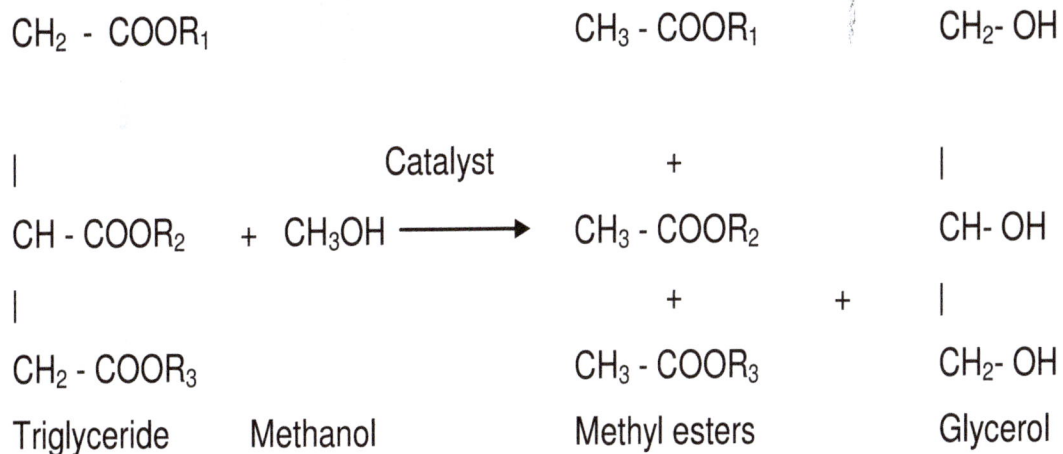

Figure 2. General equation of transesterification.

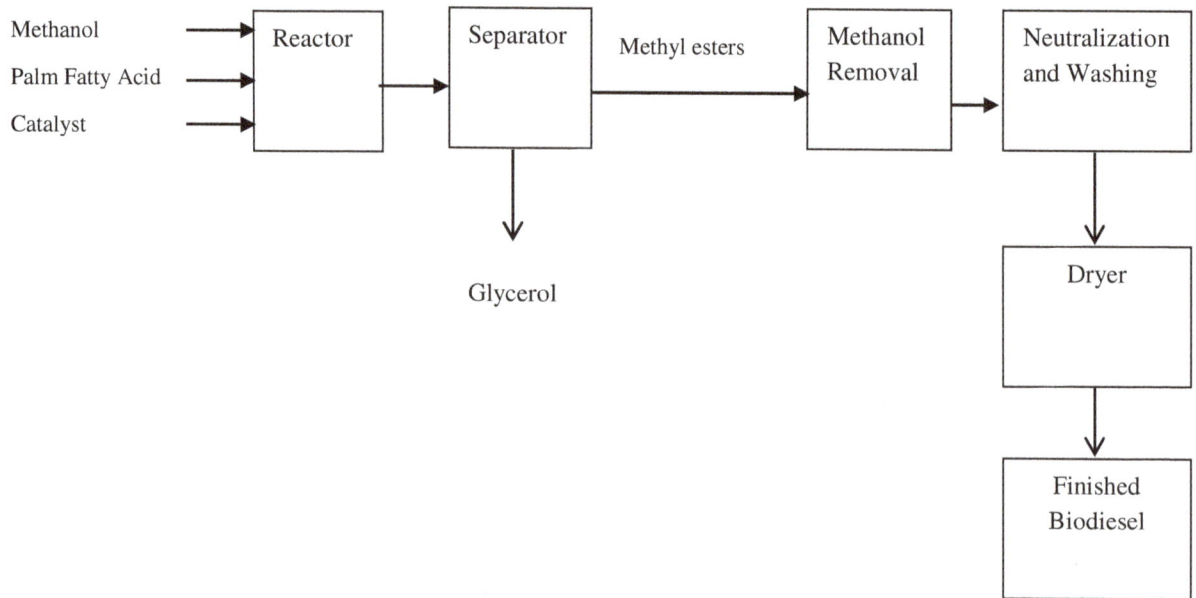

Figure 3a. Schematic diagram for biodiesel production.

Figure 3b. Photograph of Pilot Plant used for Biodiesel production.

Table 1. Chemical and physical properties of petroleum diesel and PFA methyl ester.

Properties	Petroleum diesel	PFA methyl ester
Density gm/cc	0.831	0.897
Kinematic Viscosity cSt (40°C)	4.223	4.951
Calorific value MJ/Kg	43.79	38.05
Flash Point (°C)	67	165
Acid Value mg of KOH/g	-	0.42

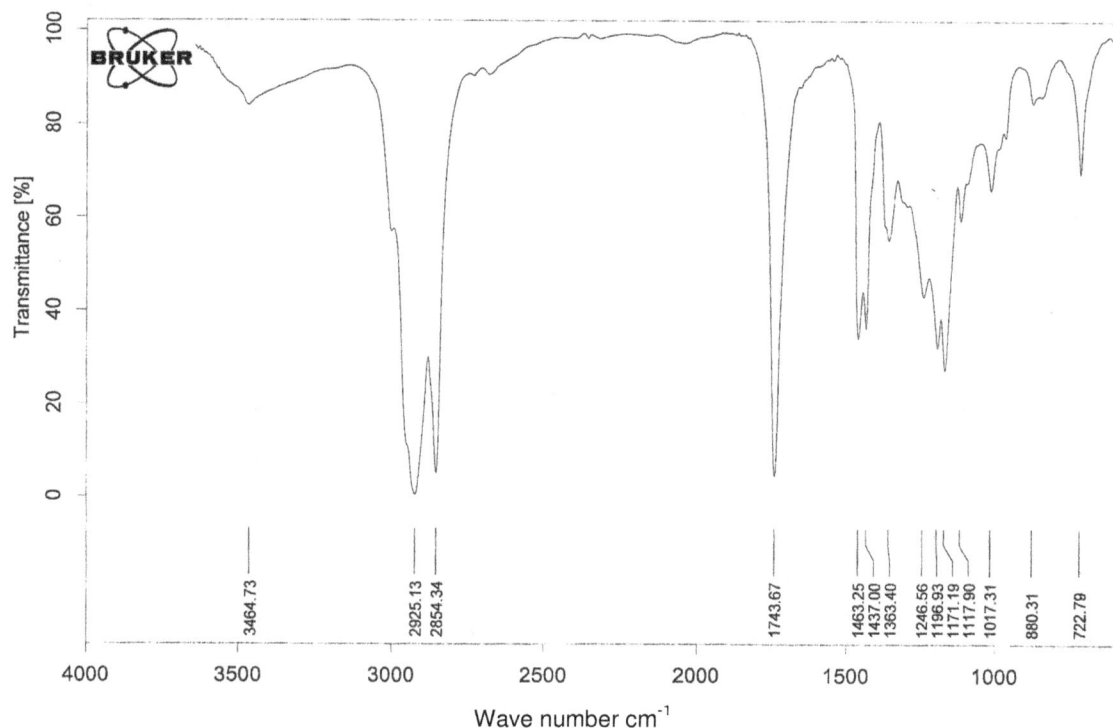

Figure 4. FT-IR spectrum of produced biodiesel.

Table 2. Calorific value of different blends of biodiesel and conventional diesel.

Fuel	M2 (g)	M2-M1 (g)	Δ T (°C)	Cv(MJ/kg)	% Decrease
B0	0.4944	0.3996	1.868	43.79	-
B5	0.4959	0.4011	1.854	43.26	1.2
B10	0.4964	0.4016	1.838	42.8	2.3
B20	0.4967	0.4019	1.820	42.27	3.5
B30	0.5012	0.4064	1.811	41.57	5.1
B40	0.5028	0.4080	1.796	41.01	6.3
B50	0.5048	0.4100	1.772	40.19	8.2
B60	0.5066	0.4118	1.755	39.58	9.6
B70	0.5085	0.4137	1.748	39.22	10.4
B80	0.5102	0.4155	1.736	38.74	11.5
B90	0.5116	0.4168	1.728	38.42	12.3
B100	0.5122	0.4174	1.716	38.05	13.1

Table 3. Engine specifications.

Make: Kirloskar India Limited
Number of cylinder: 1
Number of strokes: 4
Rated power: 3.7 KW at 1500 rpm
Loading device: Eddy Current
Fuel oil : High Speed Diesel, Biodiesel
Bore and stroke: 80 x 110 mm

Technological Institute, Kanpur, India. The engine used for this study is a single cylinder, four strokes, water cooled vertical diesel engine manufactured by Kirloskar India Limited. The technical details of the engine are given in the Table 3. Engine tests were performed at the estimated speed of 1500 rpm at varying load. The tests were performed with pure diesel and different blends of dual fuel for the study of Torque, brake specific Fuel efficiency, brake power and brake thermal efficiency. All the observations were performed thrice to obtain the average value of each observation so as to minimize the influence of measurement and observation errors. A three way hand operated control valve and two fuel

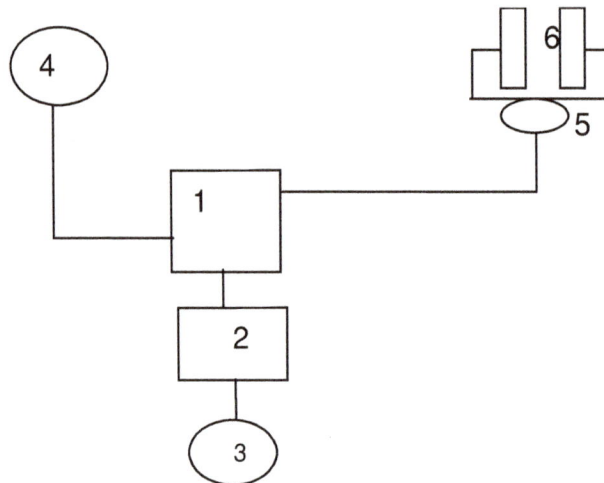

Figure 5. Schematic diagram of engine test set up. (1) Engine (2) Dynamometer (3) Control unit (4) Cooling system (5) Three way valve (6) Fuel meters.

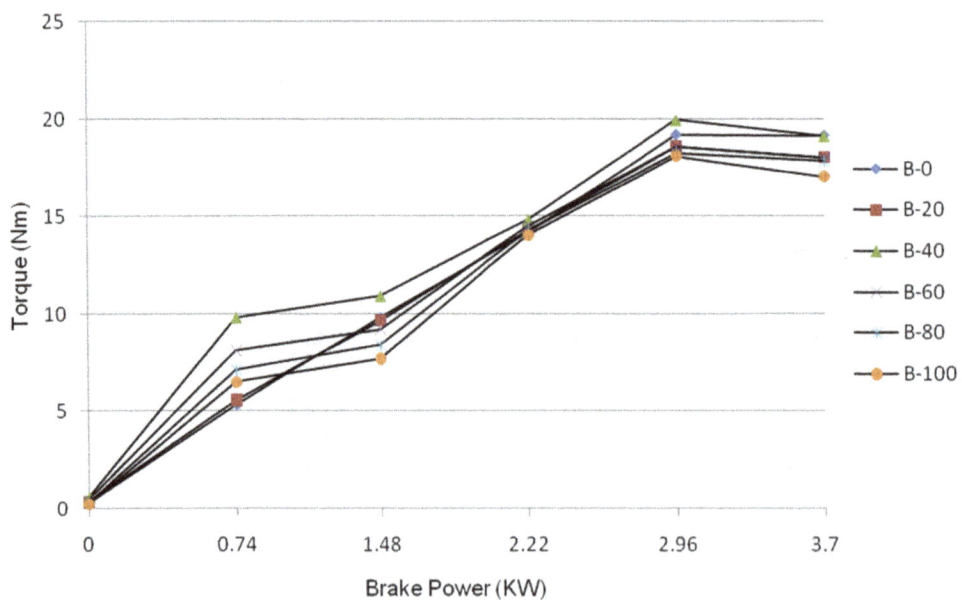

Figure 6. Comparison of torque Vs brake power for different proportions of biodiesel in petroleum diesel fuel.

meters were added to modify the engine fuel system as per the requirement. The schematic diagram of the experimental set up is shown in Figure 5.

Effect of blending on torque

The variation of brake power output and the torque for different fuel combination is depicted in Figure 6. It is observed that for all fuels as the brake power increases, torque increases to the maximum at 80% load and then decreases for all the fuel samples. The graphical representation reveals that the torque increases for B-0 to B-40 fuel samples and then decreases. A dual fuel B-40 means

40% of biodiesel is blended with conventional diesel. This could be because of higher cetane number of biodiesel and complete burning of fuel. In the case of dual fuel mixtures with higher proportions of biodiesel, the torque produced is less due to lesser energy released. This is attributed to the low calorific value of the palm FAME.

Effect of blending on brake specific fuel consumption

Figure 7 shows the effect of blending on brake specific fuel consumption for various fuel combinations. The brake specific fuel consumption is observed to decrease sharply for all fuels at higher

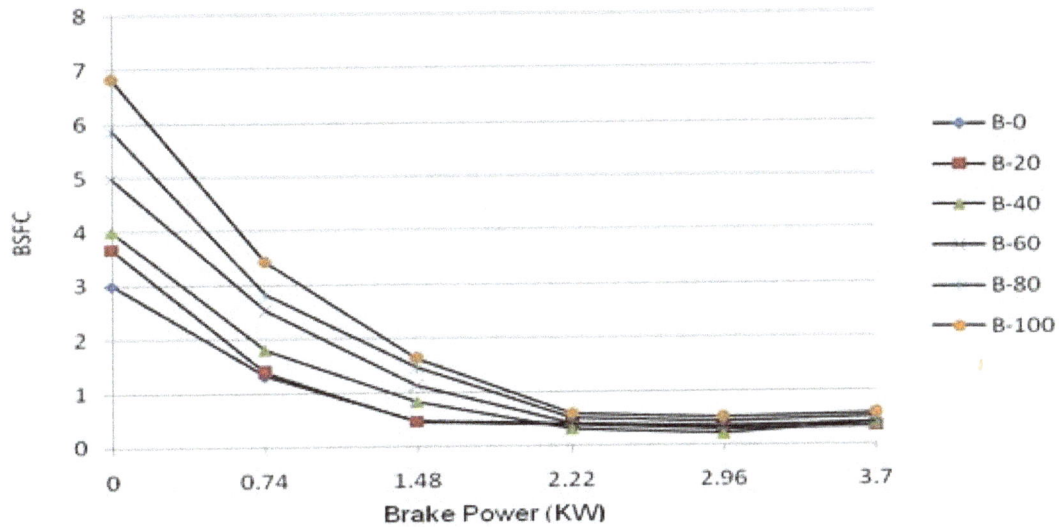

Figure 7. Comparison of Brake specific fuel consumption Vs Brake power for different proportions of biodiesel in petroleum diesel fuel.

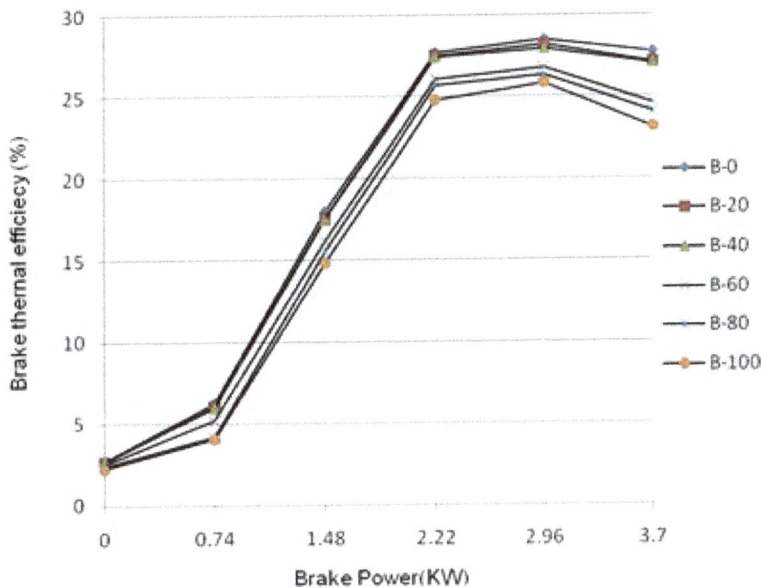

Figure 8. Comparison of Brake thermal efficiency Vs Brake power for different proportions of biodiesel in petroleum diesel fuel.

load. The main reason for this could be that percent increase in fuel required to operate the engine is less than the percent increase in brake power due to relatively less portion of the heat losses at higher load. This indicates that the compression ignition engines run efficiently at higher load than at part load. Graphical representation shows that for certain brake power the specific fuel consumption is found to be lowest in case of pure diesel (B-0) and it increases as blending is increased from B-0 to B-100. However, at higher load the specific fuel consumption for B-40 is lowest. The lower fuel consumption of B-40 as compared to B-0 could be because of possible synergistic effect of biodiesel with diesel as the oxygen present in biodiesel might have helped in improved combustion of the blend. However, on increasing the biodiesel

proportion in the blend further this effect was negated due to the reduced calorific value of these blends.

Effect of blending on brake thermal efficiency

The effect of blending on brake thermal efficiency for various fuel combinations is depicted in Figure 8. The brake thermal efficiency of engine was low at part load as compared to the engine running at higher load. This is due to relatively less portion of the power being lost with increasing load. The variations in brake thermal efficiency between various blends of fuel at higher load was less than that at part load and is in accordance with the trend observed

Emissions, NOx

(a)

Emissions, CO

(b)

Emissions, HC

(c)

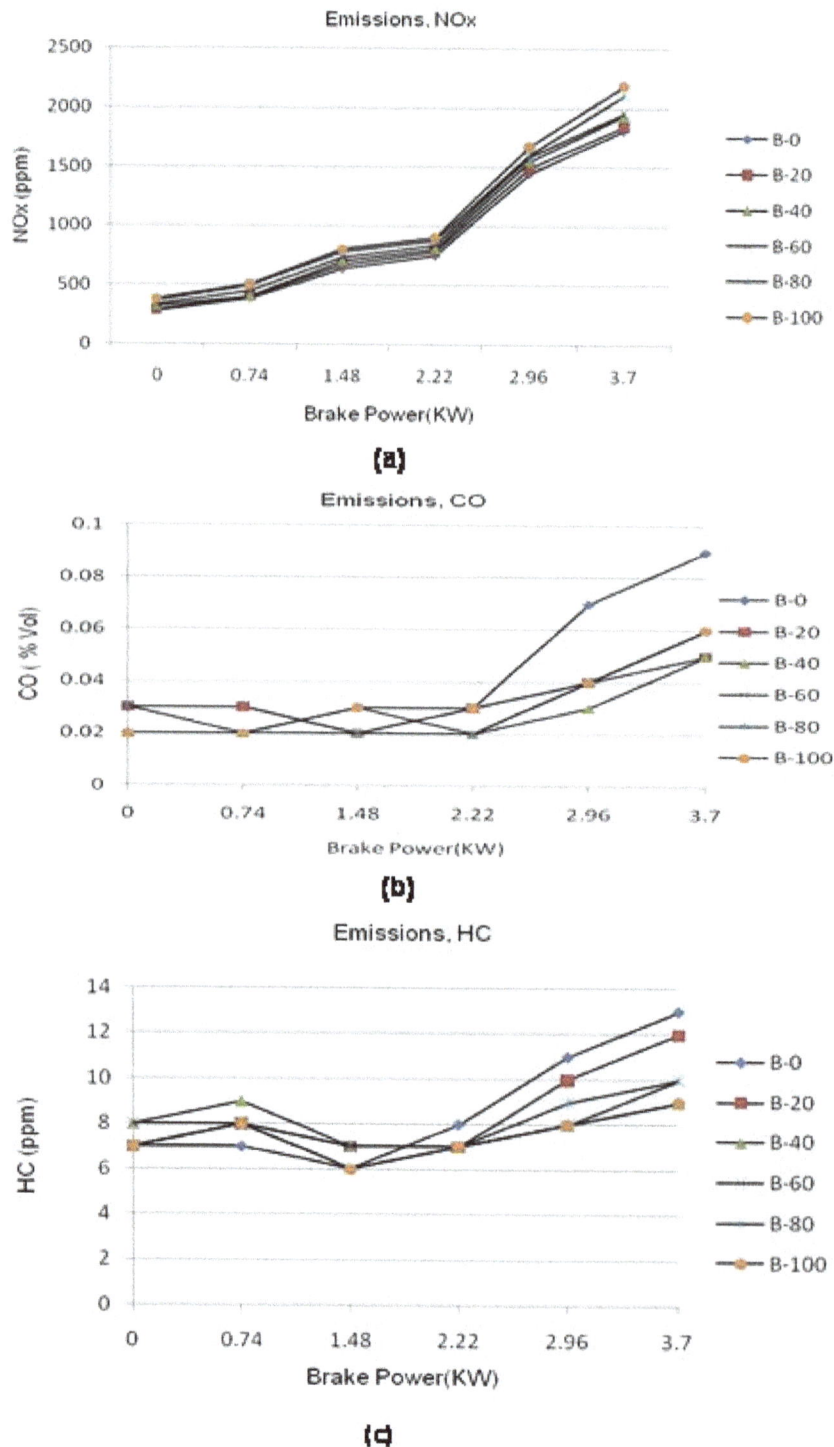

Figure 9. (a) Comparison of nitrogen oxides Vs brake power (b) Comparison of carbon monoxide Vs brake power. (c) Comparison of hydrocarbon Vs brake power.

for brake specific fuel consumption. At higher load the brake thermal efficiency for B-40 was nearly the same as that of B-0. It is because of increased temperature inside the cylinder due to more amount of fuel burning at higher load, resulting in proper atomization and ease of vaporization, which in turn forms a better oil-air mixture in the combustion chamber and yields better combustion.

Emission characteristics

Figure 9a shows the comparison of nitrogen oxide (NOx) emissions of different fuel combinations with brake power. Graphical representation shows that there is an increase in NOx emission with blends of biodiesel fuel as compared to petroleum diesel. This occurs due to high temperature generated because of fairly complete combustion process on account of adequate availability of oxygen in fuel and hence, the fuel NOx is also found along with thermal NOx.

Figure 9b shows the variation of carbon monoxide (CO) emissions of different fuel combinations with respect to brake power. The graphical representation shows that with the increase of biodiesel content in the fuel the CO emissions decrease. At low load the difference between the CO emission on pure diesel and B-40 fuel are low, which significantly increases at high load. This is due to the fact that fuel B-40 containing oxygen atoms leads to almost complete combustion and therefore, results in lower CO emission.

The variation of hydrocarbon (HC) emission of different fuel combinations with brake power is shown in Figure 9c. It is observed from the graphical representation that the HC emission is reduced with the increase of biodiesel content in the fuel at higher load. This could be due to almost complete combustion of fuel because of presence of oxygen atom.

Conclusion

In the present study, biodiesel is produced from low cost PFA with high FFA using two step trans-esterification process. Various parameters viz. molar ratio, catalyst concentration and temperature were optimized and then fuel quality biodiesel was produced from PFA at pilot scale using the optimized process conditions of molar ratio 1:10 (PFA: MeOH), 1 wt% of H_2SO_4 and 2 wt% of NaOH at 65°C. The final product is a light yellowish material meeting the ASTM standards for biodiesel. A high quality product is produced in continuous stirred tank reactor with low residence time, and at lower temperature which is considered to be a more economic solution. It is concluded that the biodiesel produced from low used in diesel engines. The results obtained in the study of torque output, brake specific fuel consumption and brake thermal efficiency indicate that dual fuel combination of B-40 can be used in diesel engines without any modifications in the engines.

ACKNOWLEDGEMENT

Authors are thankful to all India Council for Technical Education, New Delhi, India, for providing the financial support for pursuing this research work at H.B.T.I. Kanpur, U. P., India.

REFERENCES

Prafulla D, Patil, Shuguang Deng (2009). Optimization of biodiesel production from edible and non edible vegetable oils. Fuel, 88: 1302-1306.
Ramadhas AS, Jayaraj S, Muraleedharan C (2004). Use of vegetable oils as I.C. engine fuel - a review. Renewable Energy, 29: 727-742.
Stalin N, Prabhu HJ (2007). Performance test of IC engine using karanja biodiesel blending with diesel. ARPN J. Eng. Appl. Sci., 2: 32-34.
Ramadhas AS, Jayaraj S, Muraleedharan C (2004). Biodiesel production from high FFA rubber seed oil. Fuel, 84: 335-340.
Shashikant VG, Raheman H (2005). Biodiesel production from mahua oil having high free fatty acids. Biomass and Bioenergy, 28(6): 601-605.
Wang Y, Ou S, Liu P, Xue F, Tang S (2006). Comparision of two different processes to synthesis biodiesel by waste cooking oil. J. Mol. Catalysis A: Chemical; 252: 107-112.
Zlatica PJ (2008). The production of biodiesel from waste frying oils: A comparison of different purification steps. Fuel, 87: 3522-3528.
Veljkovic VB, Lakicevic SH, Stamenkovic OS, Todorovi ZB, Lazic ML (2006). Biodiesel production from tobacco seed oil with a high content of free fatty acids. Fuel, 85: 2671-2675.
Meher LC, Vidyasagar D, Naik SN (2006). Technical aspects of biodiesel production by transesterification - a review. Renewable Sustain. Energy Rev., 10: 248-268.
Siti Zullaikah, Chao- Chin Lai, Shaik Ramjan Vali, Yi- Hsu Ju (2005). A two- step acid - catalyzed process for the production of biodiesel from rice bran oil. Biores. Technol., 96; 1889-1896.
Fangrui Ma, Milford A, Hanna (1999). Biodiesel production: A review. Biores. Technol., 70: 1-15.
Vicente G, Martinez M, Aracil J (2004). Integrated biodiesel production: a comparison of different homogeneous catalysts systems. Biores. Technol., 92; 297-305.
Zheng S, Kates M, Dube MA, McLean DD (2006). Acid-catalyzed production of biodiesel from waste frying oil. Biomass Bioenergy, 30: 267-272.
Di Serio M, Tesser R, Dimiccoli M, Cammarota F, Nastasi M, Santacesaria E (2005). Synthesis of biodiesel via homogeneous Lewis acid catalyst. J. Mol. Catal. A: Chem., 239 (102): 111-115.
Pramanik K (2003). Properties and use of jatropha curcas oil and diesel fuel blends in compression ignition engine. Renewable Energy, 28: 239-248.
Ramadhas AS, Jayaraj S, Muraleedharan C (2005). Performance and emission evaluation of a diesel engine fueled with methyl esters of rubber seed oil. Renewable Energy, 30:1789-1800.
Mustafa Canakei, Ahmet Erdil, Erol Arcaklioglu (2006). Performance and exhaust emissions of a biodiesel engine. J. Appl. Energy, 83: 594-605.

2

Performance and emission characteristics of karanja methyl esters: Diesel blends in a direct injection compression-ignition (CI) engine

Nanthagopal K.*, Thundil Karuppa Raj R. and Vijayakumar T.

Automotive Research Centre, SMBS, VIT University, Vellore-14, Tamilnadu, India.

Biodiesel is a fatty acid alkyl ester which is renewable, biodegradable and non toxic fuel which can be derived from any vegetable oil by transesterification process. One of the popularly used non edible biodiesel in India is karanja oil methyl ester (KOME). In the present investigation, karanja oil based methyl ester (biodiesel) blended with diesel were tested for their use as a substitute fuel for diesel engine. Experiments were conducted to study the performance and emission characteristics of a direct injection diesel engine using blends of karanja methyl esters with diesel on a 10, 20, 30, 40 and 50% volume basis, respectively. The acquired data were analysed for various parameters, such as brake thermal efficiency, carbon monoxide (CO), hydrocarbon (HC), Smoke and nitrogen oxide (NOx) emissions. Engine performance showed that brake thermal efficiency for 50% biodiesel blend is 21.69% poorer than diesel. Emission characteristics of CO, HC and smoke were found lower with biodiesel blends. However, oxides of nitrogen were 26% higher for 50% biodiesel blend as compared to diesel.

Key words: Biodiesel, transesterification, karanja oil, methyl ester, emission.

INTRODUCTION

Invention of the internal combustion engine has tremendously increased our day-to-day energy demand. This has resulted in the wide spread exploitation of the petroleum reserves, which are getting depleted at a rapid rate. Moreover, the combustion of these fuels has polluted the environment to alarming levels. In search of the alternate fuels, people have used various fuels like propane (LPG), methane (CNG), hydrogen (H_2), alcohols, vegetable oils, etc., (Nanthagopal and Rayapati, 2009). It has been found that vegetable oil has special promise in this regard, since they can be produced from plants grounded in rural areas (Babu et al., 2003). In this contest, many varieties of vegetable oils have been evaluated in many parts of world, but only very few oil such as jatropha oil and Karanja oil can be considered to be economically affordable to nations like India in particular. The use of straight vegetable oils is restricted by some unfavorable characteristics, like carbon deposit buildup, poor durability, poor fuel atomization and also poor thermal efficiency. The problems associated with straight vegetable oils can be solved by any one of the four processes via pyrolysis, micro emulsification, dilution or transesterification of oil to produce biodiesel (Sahoo and Das, 2009; Ramadhas et al., 2004). It has been reported that transesterification process is the best effective method of biodiesel production and an important role in the viscosity reduction of vegetable oil. Biodiesel is referred to as the mono-alkyl-esters of long chain fatty acids derived from renewable lipid sources. It is the name for a variety of eater based oxygenated fuel from renewable biological sources. In the present investigation, the transesterified non edible oils of karanja have been considered as a potential alternative fuels for diesel engines.

A large number of experiments were carried out with vegetable oils as a replacement of compression-ignition (CI) engine fuel by researchers from various part of the world. The use of karanja methyl ester in diesel engine has been reported by (Rehmann and Phadatare, 2004). It

---

*Corresponding author. E-mail: nanthu_gopal@rediffmail.com.

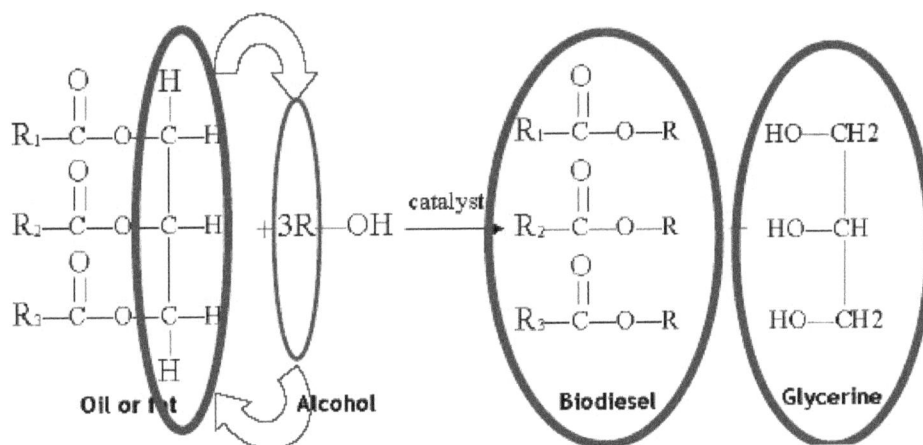

Figure 1. Transesterification process of biodiesel.

Table 1. Properties of karanja (KB) methyl ester and diesel.

Property	Diesel	Karanja oil methyl ester
Density (kg/m^3)	850	883
Kinematic viscosity at 40°C (Cst)	2.87	4.37
Flash point (°C)	76	163
Calorific value (kJ/kg)	44000	42133

was reported that blends of karanja methyl ester with diesel reduces emissions, such as carbon monoxide (CO), smoke density and nitrogen oxide (NOx) on an average of 80, 50 and 26%, respectively. It was found out that brake power output increases on an average 6% up to B40 biodiesel blend and decreases further with increase in biodiesel blends.

This study targets at making a comparison of the methyl esters of karanja oil in a diesel engine vis-à-vis diesel fuel. The performances and emissions of the engine characteristics were evaluated using different proportions and were compared using diesel.

MATERIALS AND METHODS

Transesterification

Transesterification is also known as alcoholysis. It is a chemical process in which triglycerides in vegetable oils are converted to mono alkyl esters of the same fatty acid, which are called biodiesel. It is the reaction of fat or oil with an alcohol form esters and glycerin. A catalyst is used to improve the reaction rate of the transesterification process (Figure 1). Among the alcohols, methanol and ethanol are used commercially because of their low cost and their physical and chemical advantages. To complete a transesterification process, 3:1 molar ratio of alcohol is needed. Enzymes, alkalis or acids can catalyze the reaction, that is, lipases, KOH, NaOH and sulphuric acid, respectively. Among these, alkali transesterification is faster, hence, it is used commercially (Forson et al., 2004; Reed et al., 1992; Murugesan et al., 2009).

Transesterified karanja was blended with diesel oil in varying

proportions with the intention of reducing its viscosity close to that of the diesel fuel. The blends prepared were stable under normal conditions. The fuel properties of karanja methyl esters and diesel are summarized in a tabular form as shown in Table 1.

Experimental set up and test procedure

A 4.4 kW, 1500 rpm, single cylinder, four-stroke, naturally aspirated, air-cooled, direct injection vertical diesel engine coupled with an eddy current dynamometer is used in this analysis (Figure 2). Specifications of the engine are presented in Table 2. Engine torque was measured with a strain gauge type load cell mounted between the stator and the base frame of dynamometer. The major pollutants in the exhaust of a diesel engine are smoke. AVL smoke meter was used to measure the smoke density of the exhaust from diesel engine. QRO-TECH five gas analyzer was used for the measurement of CO, hydrocarbon (HC) and NOx emissions. The engine was operated on diesel first and then on methyl esters of karanja and its blends. The different fuel blends and mineral diesel were subjected to performance and emission tests on the engine. The performance data were then analyzed from the graphs regarding thermal efficiency, brake-specific fuel consumption and smoke density of all fuels. The brake-specific fuel consumption is not a very reliable parameter to compare different fuels, as the calorific values and the densities are different.

RESULTS AND DISCUSSION

A series of exhaustive engine tests were carried out using diesel and different biodiesel blends ranging from B10 to B50. The engine performance and emissions data

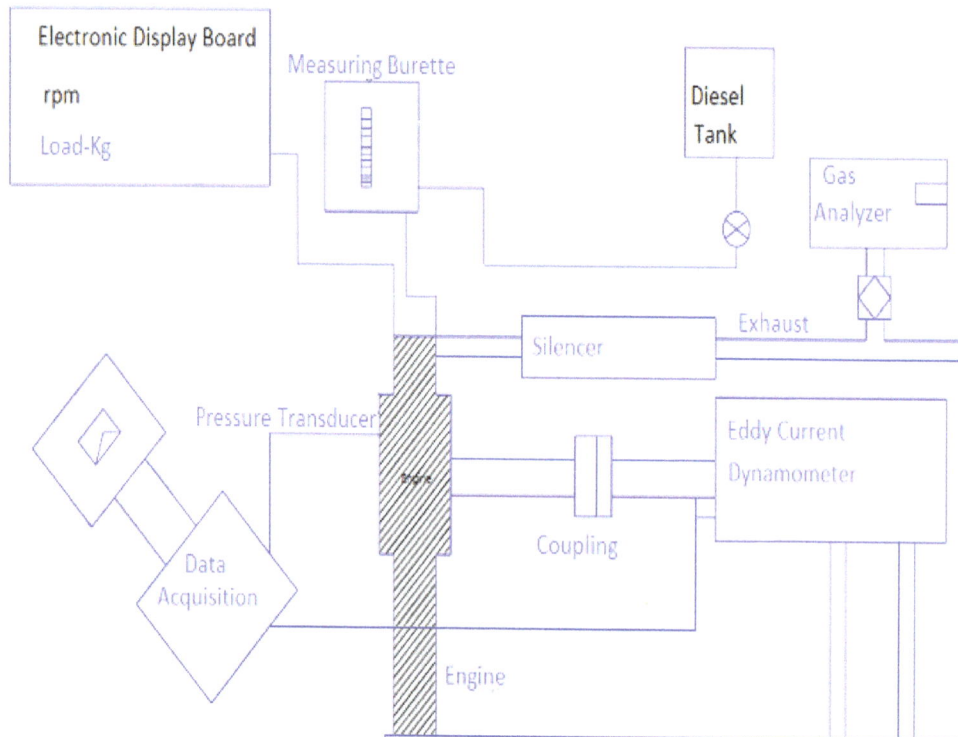

Figure 2. Schematic diagram of the experimental set up.

Table 2. Engine specifications.

Model	Kirloskar TAF1
Type	Single cylinder, four stroke. Direct injection, bowl-in-piston combustion chamber
Bore and stroke	87.5 × 110 mm
Compression ratio	17.5 :1
Rated power	4.4 kW at 1500 rpm
Injector opening pressure	205 bar
Injection timing	23°bTDC
Dynamometer	Eddy current

obtained for biodiesel blends were compared with baseline data for diesel. Characteristics curves for brake thermal efficiency, CO, HC, smoke and NOx emissions were drawn for different biodiesel blends and diesel. These curves are shown in Figures 3 to 7.

Brake thermal efficiency

Brake thermal efficiency (BTE) is one the main performance parameters which indicates the percentage of energy present in the fuel that is converted into useful work. The comparison of BTE of the various blends of karanja with diesel (KB10, KB20, KB30, KB40 and KB50) and clean diesel is as shown in Figure 3. The BTE of karanja blends were lower than diesel for the entire load.

The decreasing trend in efficiency with increase in concentration of biodiesel in diesel may be because of lower calorific value of methyl esters than that of diesel. It may also be caused by its poor atomization due to its high viscosity.

Smoke opacity

The variation of smoke emission with brake power for biodiesel blends and diesel is as shown in Figure 4. The smoke that is formed due to incomplete combustion is much lower for biodiesel and its diesel blends as compared to neat diesel. This is because of complete combustion of methyl esters as compared to diesel fuel. However, with increase in percentage of biodiesel blends,

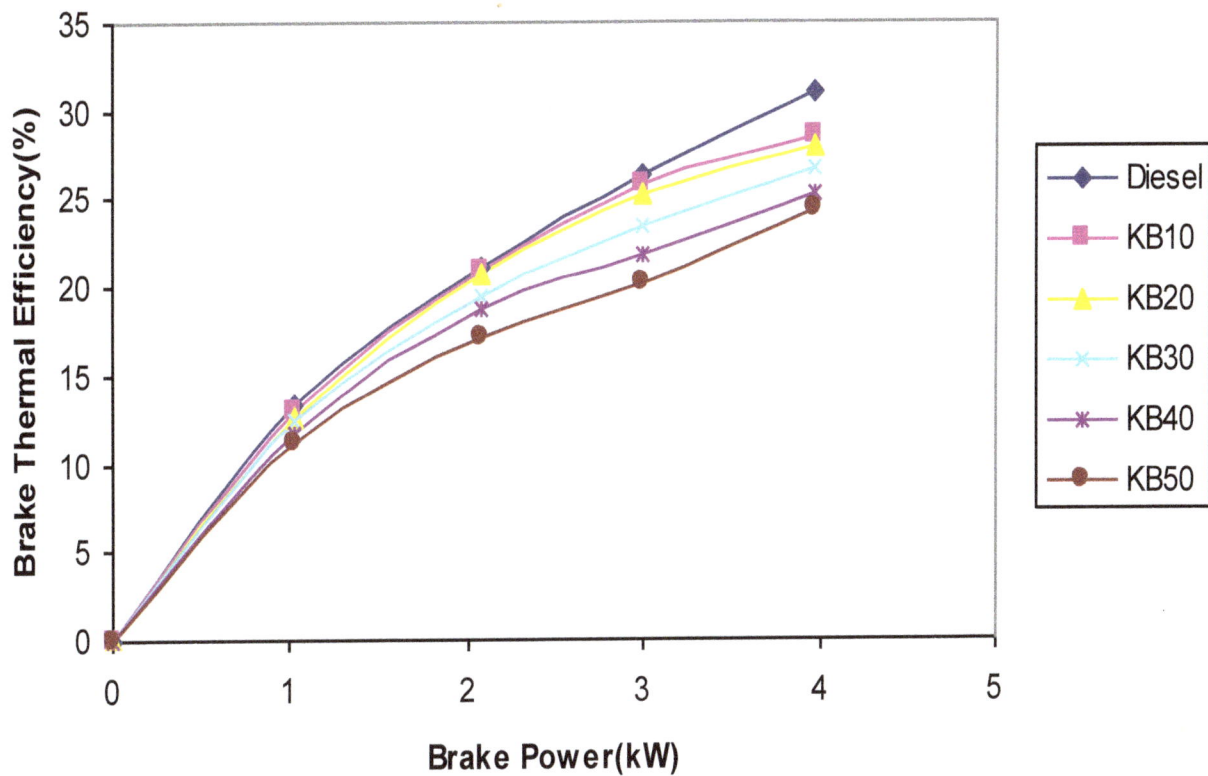

Figure 3. Variation of BTE with brake power.

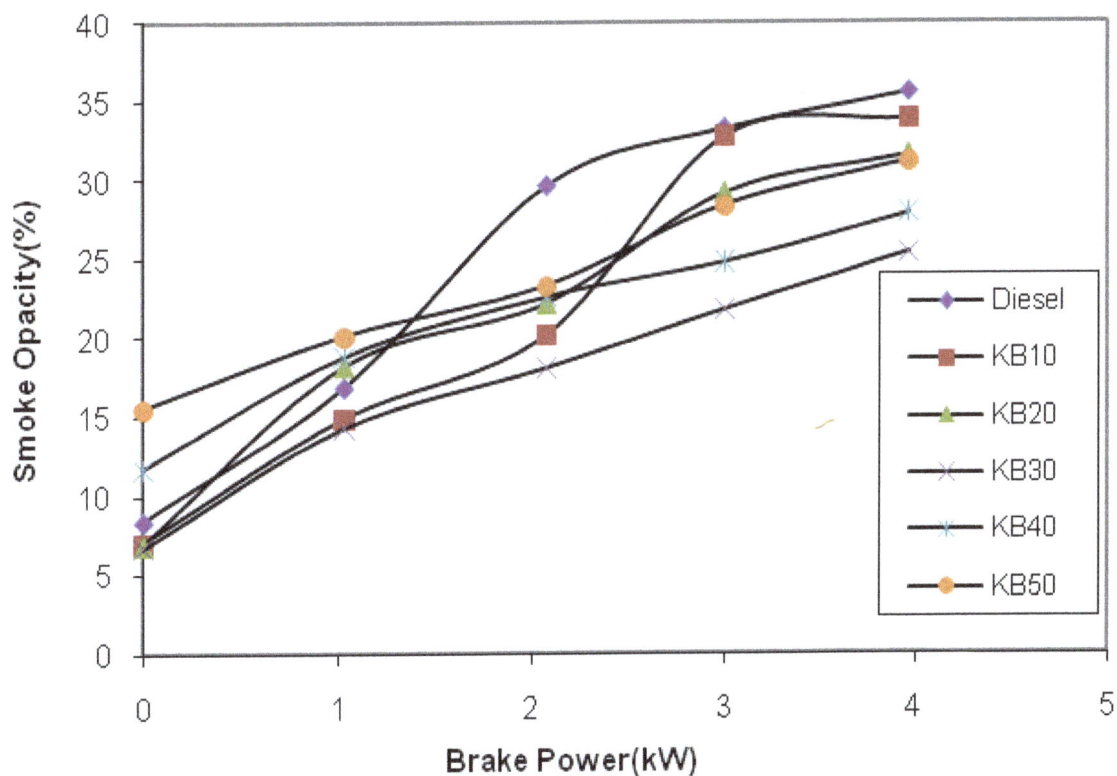

Figure 4. Variation of smoke opacity of different fuels.

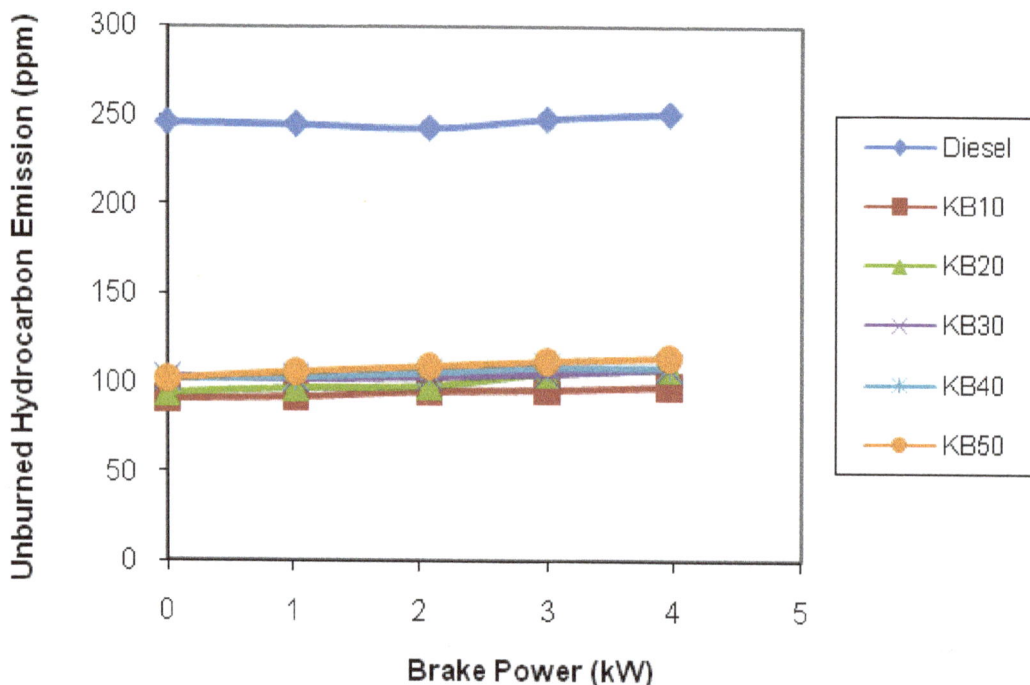

Figure 5. Variation of HC emissions of different fuels.

smoke density decreases.

Unburned HC emission

Unburned HC is also an important parameter for determining emission behaviour of the engines. Figure 5 shows the variation of HC emission with brake power for different fuels. It is observed that HC emission of the various blends was lower at partial load, but increased at higher engine load. This is due to the availability of less oxygen for the reaction when more fuel is injected into the engine cylinder at higher engine load. It is also observed from the Figure 5 that biodiesel blends give relatively lower HC as compared to the diesel. This is because of better combustion of the biodiesel inside the combustion chamber due to the availability of excess content of oxygen in the biodiesel blends as compared to clean diesel.

Emission of NOx

NOx emissions are extremely undesirable. NOx is one of the main emissions in diesel engine. NOx is more likely to cause respiratory problems such as asthma, coughing, etc. Three conditions which favor NOx formation are higher combustion temperature, more oxygen content and faster reaction rate. The above conditions are attained in biodiesel combustion very rapidly as

compared to neat diesel. Hence, NOx formations for biodiesel blends are always greater than neat diesel. It can be observed from Figure 6 that at higher power output conditions, due to higher peak temperatures, the NOx values are likely higher for both biodiesel and clean diesel fuel.

Carbon monoxide (CO) emission

Carbon monoxide emissions occur due to the incomplete combustion of fuel. The emissions of carbon monoxide are toxic. The comparative analysis is as shown in Figure 7. Biodiesel blends give less carbon monoxide as compared to diesel due to complete combustion. When the percentage of blend of biodiesel increases, carbon monoxide decreases. This is due to more amount of oxygen content of biodiesels that result in complete combustion of the fuel and supplies the necessary oxygen to convert CO to CO_2.

Conclusion

Conclusively, the performance and emissions characteristics blends of the biodiesel from karanja are analysed and compared with that of the diesel. It was found out that biodiesel blends have lower brake thermal efficiency than diesel. Emissions of NOx are found to be high for biodiesel blends. CO, HC and smoke emissions are lower for biodiesel blends as compared to diesel.

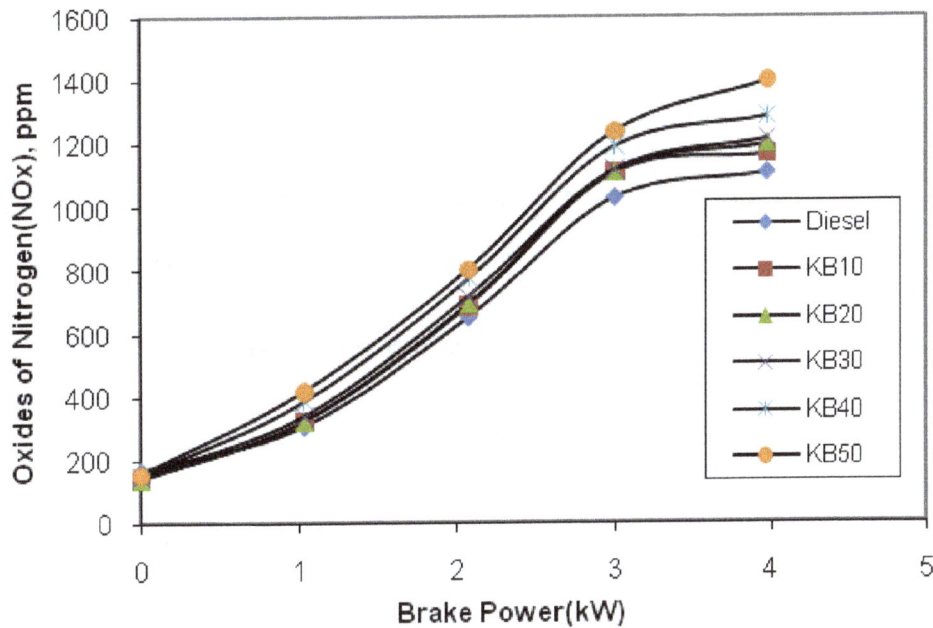

Figure 6. Variation of NO$_x$ emissions of different fuels.

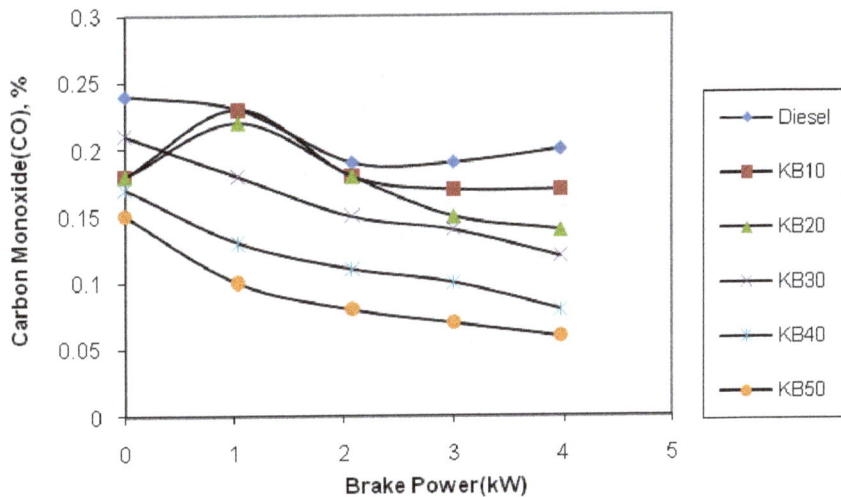

Figure 7. Variation of CO emissions of different fuels.

REFERENCES

Babu AK, Devarajane G (2003). Vegetable oils and their derivatives as fuels for CI engines: An overview. SAE Technical Paper. 2003-01-0767.

Forson FK, Oduro EK, Hammond–Donkuh (2004). Performance of jatropha oil blends in a diesel engine. Renewable Energy, 29: 1135-1145.

Murugesan A, Umarani C, Subramanian R, Nedunchezhian N (2009). Biodiesel as an alternative fuel for diesel engines – A review. Renewable Sustain. Energy Rev., 13: 653-662.

Nanthagopal K, Rayapati S (2009). Experimental investigation and performance evaluation of DI diesel engine fueled by waste oil- diesel mixture in emulsion with water. Therm. Sci., 13(3): 83-89.

Ramadhas AS, Jayaraj S, Muraleedharan C (2004). Use of vegetable oils as I.C engine fuels – A review. Renewable Energy, 29: 727-742.

Reed TB, Graboski MS, Gaur S (1992). Development and commercialization of oxygenated diesel fuels from waste cooking oils. Biomass Bioenergy, 3(2): 111-115.

Rehemann H, Phadatare AG (2004). Diesel engine emissions and performance from blends of karanja methyl ester and diesel. Biomass Bioenergy, 27: 393-397.

Sahoo PK, Das LM (2009). Combustion analysis of jatropha, karanja and polanga based biodiesel as fuel in a diesel engine Fuel 88: 994-999.

3

Pig-fat (Lard) derivatives as alternative diesel fuel in compression ignition engines

Ejikeme, P. M.[1]*, Anyaogu, I. D.[2], Egbuonu, C. A. C.[3] and Eze, V. C.[1]

[1]Industrial and Biomass Research Laboratory, Department of Pure and Industrial Chemistry, University of Nigeria, Nsukka, Enugu State, Nigeria.
[2]Department of Science Laboratory Technology, Federal Polytechnic Nasarawa, Nasarawa State, Nigeria.
[3]Nutritional and Toxicological Biochemistry Unit, Department of Biochemistry, University of Nigeria, Nsukka, Enugu State, Nigeria.

Prime steam lard obtained from pig fat by rendering was characterized and refined. The lard was transesterified with methanol and sodium hydroxide (as catalyst) at 40, 50, 60, and 70°C. Fuel properties of the fatty acid methyl esters (FAME) or biodiesel were determined alongside that of petroleum derived diesel obtained commercially. Results obtained showed that the saponification and iodine values of the lard were 198 and 64.08, respectively, while the flash point and viscosity were 203°C and 47.87 mm^{2}s^{-1}, respectively. While the refractive index of the methyl esters decreased with increase in temperature from 1.4480 to 1.4430 (1.4459 ± 0.0022), the specific gravity increased from 0.860 to 0.875. The flash point and the pH of the biodiesel samples were 134.25°C ± 1.71°C and 7.05 ± 0.03, respectively, and the viscosity was 4.23 ± 0.1 mm^{2}s^{-1}. The flash point, iodine value, viscosity, cetane index, and the specific gravity of the petrodiesel were 79°C, 84, 3.12, 46, and 0.84, respectively. The values obtained for the FAME are within the standard limits and compares well with that reported in the open literature. It can be used alone or in blends with petrodiesel to run compression ignition engines.

Key words: Pig fat, rendering, transesterification, fatty acid methyl esters, cetane index.

INTRODUCTION

Fatty acid methyl esters also known as biodiesel are mono-alkyl esters of long-chain fatty acids derived from lipids such as vegetable oils and animal fats, restaurant greases or used frying oils (Van Gerpen, 2005; Tapasvi et al., 2005; Dunn and Knothe, 2001). Large harvests of traditional crops, low farm prices, dependence on foreign energy sources and environmental problems due to combustion of fossil fuels have increased interests in renewable energy sources such as biodiesel (Tapasvi et al., 2005; Dunn and Knothe, 2001; Knothe and Steidly, 2005).

Vegetable oils were technically found to be suitable as diesel fuel because of their chemical structure (presence of long, saturated, unbranched hydrocarbon chains of fatty acids). The direct use of vegetable oils in compression ignition engines led to several problems such as fuel injector deposits, incomplete combustion and toxic substances formation as a result of their high viscosity amongst other factors (Ejikeme, 2008). Four methods, viz; dilution, thermal cracking, microemulsification and transesterification have been adopted to overcome the above problems and approximate the properties and performances of biodiesel to petro-diesel. Transesterification is the most popular method and it is used in both large and small scale productions.

Several environmental advantages of biodiesel as opposed to petro-diesel include its zero sulphur content and the consequent reduction in the SO$_2$ emissions, aromatic hydrocarbons and particulate emissions as well as reducing life-cycle of CO$_2$ emissions by over 78% as

*Corresponding author. E-mail: ejikemepaul13@yahoo.com; paul.ejikeme@unn.edu.ng.

opposed to petrodiesel, (Sinha and Agarwal, 2005; Pinto et al., 2005; Van Gerpen et al., 2007; Graboski and McCormick, 1998). Chhetri et al. (2008) reported that total change in the environmental (Cn_t), economic (Ce_t), and social (Cs_t) capitals; $d(Cn_t)/dt \geq 0$, $d(Ce_t)/dt \geq 0$, and $d(Cs_t)/dt \geq 0$, respectively, were higher after the introduction of biodiesel use in set studies.

The use of animal fats in biodiesel production has been reported by several authors. Chung et al. (2009) reported 81.3% FAME content in a transesterification reaction involving duck tallow with methanol using NaOH catalyst. In a paper, Raine et al. (2008) reviewed the resource, production and engine performance of New Zealand tallow (beef/mutton fat) sourced biodiesel and concluded amongst other things, that unmodified engine performance was equal to that of the petro-diesel-fuelled engine when using blends up to 100% biodiesel. Biodiesel made from tallow using the very similar processes to plant oils have been shown to have higher cetane number, implying cleaner and more efficient burning in diesel engines than plant oils (Miller-Klein Associates, 2006) though exhibiting higher cloud points because of their higher level of saturates. Chicken fat amongst other fats/oils were used to produce biodiesel that met standard specifications (Mattingly et al., 2004; Mattingly, 2005). Standard biodiesel have been produced from varying grades of beef tallow and chicken fat (Babcock et al., 2007), and the kinetics of biodiesel production from Chinese tallow tree oil was studied by Crymble et al. (2006).

Meat, fat and bone meals which were fed to cattle and other domestic animals have been banned in developed countries as they have been implicated as the main route for the spread of bovine spongiform encephalopathy, BSE (mad cow diseases), which also did not spare human beings. More recently, bird and swine flu have also been reported to be ravaging birds and pigs in many countries (WhipNet Technologies (2009). Rendered fats from beef (tallow) and pigs (lard) as well as birds could be channeled to the production of fatty acid methyl esters which is used as either a substitute or is blend with the conventional petro-diesel. For instance, the amount of rendered fats produced in Austria in 2004 alone was 25,000 tons and life cycle analysis for biodiesel from rendered fats is better than for rapeseed oil methyl ester (Apel, 2006).

A major aim of this work was therefore to use pig fat (lard) obtained from Nsukka, Enugu State-Nigeria to produce FAME's and characterize same to ascertain its suitability or otherwise as diesel fuel substitutes.

MATERIALS AND METHODS

Methanol 99.7% purity was a product of Merck Darmstadt, Germany, while NaOH was a product of Loba Chemie GmbH, Switzerland. All other reagents are of analytical grade unless otherwise stated. Mixed pig fat was collected from the Slaughter house at Ogige Market, Nsukka, Enugu State, Nigeria as

fat trimmings from meat cuts. It was rendered according to the modified method of Schneider (2006). The fat was chopped into small pieces of approximately 10 mm^2, put in a 5 L beaker with water of about a quarter the volume of the fat. The set up was heated gradually to boil with continuous stirring until the water evaporated completely. Further cooking led to the melting of the solid fat which was sieved through a sieve cloth, allowed to cool to about 50°C and water equivalent to twice the volume added and allowed to stand overnight to clarify some of the little protein bits. The thick slab of lard floating on the water was collected, dried, weighed and stored in a refrigerator for biodiesel production.

Biodiesel (FAME) production

The thick slab of lard obtained as described above was cut into small bits and melted over low and sustained heating procedure at 40°C. 180 cm^3 of the melted lard was put in the reaction vessel maintained at 40°C. 1.4 g NaOH dissolved in 138 cm^3 of methanol and heated to 40°C was added to the reaction vessel and stirred mechanically. The system was maintained at that temperature for 90 min and the reaction quenched by removing the vessel from the heating system and immersed in cold water. The methods of Ikwuagwu et al. (2000) as described elsewhere (Ejikeme, 2008) was used to separate the upper (biodiesel) layer from the lower (glycerol) layer. The methods of Journey to Forever as described by Alamu et al. (2007) was applied in the washing of the lower (biodiesel) layer for four times to completely remove the unreacted catalyst, etc.

Fatty acid methyl esters were also produced at temperatures of 50, 60, and 70°C. Some properties of the FAME's produced that were determined using standard methods include; flash point (ASTM D 93), heating value, iodine (ASTM D-1959-67) and pH values, refractive index (ASTM D 6751), viscosity (ASTM D 445) and specific gravity. The colour was determined by visual observation while the cetane index was calculated from the four variable equation of the ASTM (ASTM D 6751).

RESULTS AND DISCUSSION

Results of the various parameters measured at the various temperatures used in the transesterification process are given in Table 1. From the table, the yield of both the glycerol and the methyl esters did not vary significantly when students T test is applied to the values at 95% confidence interval or 0.05 level of significance. They are 24.00 ± 0.82 and 70.78 ± 1.71, respectively. The maximum yield of the methyl esters after 90 min. at 70°C was 73%. This value is 91% of the value (79%) reported by Xiu and Ting (2009) for fat melts of pig using 0.3 g SO_4^{2-}/TiO_2 heterogeneous catalyst and 1.5 g ethanol to 1 g of oil after 8 h. The yield obtained is though higher than 64.3 and 71% obtained with 210 ml/g and 250 ml/g, respectively of expanded graphite catalyst in the esterification of acetic acid with isoamyl alcohol (Xiu et al., 2008) by 13.2 and 2.8%, respectively. Chung et al. (2009) reported a percent conversion of 79.7, 62.3 and 79.3, respectively using KOH, NaOH and CH_3NaO catalysts in the transesterification of duck tallow with methanol. FAME content of their biodiesel at different temperatures did not vary significantly with temperature as observed in our work. At all the temperatures, the pH

Table 1. Yield of glycerol and biodiesel at different temperatures*.

T (°C)	Glycerol (%)	Ester (%)	Flash point (°C)	pH	Refractive index
40	25	69	132	7.03	1.4480
50	24	70	136	7.06	1.4470
60	24	71	135	7.03	1.4455
70	23	73	134	7.08	1.4430
Mean±SD	24.00±0.82	70.78±1.71	134.25±1.71	7.05±0.03	1.4459±0.0022

*SD = Standard deviation.

was almost constant with an average value of 7.05 ± 0.03. The reason for the near-neutral pH obtained at all temperatures is the number of washes that the methyl esters were subjected to. The refractive index varied by a value of 0.005 units only between the two extremes of temperatures studied.

The flash points recorded did not follow a specific trend with respect to the temperature. On the average, the flash point for methyl esters from lard as determined in this work was 134.25°C ± 1.71. Wörgetter et al. (2006) reported values as high as 169 and 182°C for animal fat and lard, respectively. The standard flash points in degrees Celsius for various countries are: >101 (Europe); >100 (Sweden, Italy, France and Austria); > 110 (Czech Republic and Germany); >120 (Australia) and >130 (USA-ASTM) (National Standards for Biodiesel 2009). The highest minimum standard value of >130 is comparable to the 134.25°C ± 1.71 obtained in our work and as such, it could be said that the flash point obtained from our work conformed to standard specifications. The value implies that less methanol was carried over from the production process.

Results of some of the fuel-related properties of the lard FAME, petrodiesel and ASTM standard values are given in Table 2. The specific gravity of 0.8668 obtained is greater than 0.859 reported for cotton seed oil methyl ester (Alhassan et al., 2005) and within the 0.86 to 0.90 ASTM standard specifications.

The viscosity of the product was 4.23 mm^{2}s^{-1}. Values of 4.14 and 4.54 mm^{2}s^{-1}, respectively at 40°C have been reported for animal fat and lard (Wörgetter et al., 2006), while Chung et al. (2009) reported kinematic viscosities of 5.5, 6.0 and 5.8 mm^{2}s^{-1} for duck tallow biodiesel using KOH, NaOH and CH$_3$NaO, respectively. The implication of the relatively low viscosity obtained in our work is that atomization of the biodiesel in the engine will be enhanced and the coking of the cylinders caused by highly viscous FAME may be averted by its use. The ASTM recommended viscosity range is 1.9 to 6.0 mm^{2}s^{-1} (National Standards for Biodiesel, 2009). The viscosity of the parent refined lard was reduced by about 91.2% in the lard methyl esters, thus obviating the problems associated with the use of underivatised and melted lard in powering diesel engines. Li and Li (2009) produced biodiesel from crude fish oil from the soapstock of marine fish and reported among other things greater kinematic viscosity than obtained for biodiesel from waste cooking oil. Mixtures of animal fats (beef tallow, choice white pork fat, poultry fat and yellow grease) with fuel oil exhibited rheological properties very near to those of pure petrodiesel than vegetable oil (Goodrum et al., 2003).

The calculated cetane index of the pig fat methyl ester was 54.8. This value is higher than 51.96 reported for soybean oil methyl esters (Ejikeme, 2008). Cetane number increases with increase in the number of carbon atoms for saturated carboxylic acids. This is exemplified by methyl caprylate (33.6); methyl caprate (47.7); methyl laurate (61.4); methyl palmitate (74.5) and methyl stearate (86.9) (Klopfenstein, 1985). The ASTM standard for cetane index is ≥ 40.

The presence of about 34% saturated fatty acids (20% palmitic and 16% stearic) in pig fat may have led to the relatively high value of cetane index obtained since the unsaturated components are mainly monounsaturated, oleic (~54%) and linoleic (~11%). Ikwuagwu et al. (2000) reported a cetane index value of 44.81 for rubber seed oil biodiesel. Rubber seed oil has a preponderance of unsaturated fatty acids relative to lard.

Generally, biodiesel from more saturated feedstocks have higher cetane numbers and better oxidation stability but poor cold flow properties (Van Gerpen et al., 2007). Either too high or too low a cetane number can cause operational problems (Knothe et al., 1997). In the former case, combustion can occur before the fuel and air are properly mixed resulting in incomplete combustion and smoke whereas in the latter case, engine roughness, misfiring, higher air temperatures, slower engine warm-up and incomplete combustion occurs.

The colours of vegetable and animal oils are usually transferred to the biodiesel made from them. This makes the use of synthetic dyes less important as opposed to petrodiesel which are colourless on fractionation from petroleum. The values of the fuel-related properties of the petrodiesel from Table 2 compares well with the pig fat methyl esters with respect to the benefits derivable from the use of biodiesel. The lower flash point of the commercial petrodiesel relative to the biodiesel produced is one of the advantages of the latter over the former in cases of accidents as stated earlier. The higher cetane index of the biodiesel means that its ignition temperature

Table 2. Fuel-related properties of pig fat biodiesel produced.

Properties	Refined lard	Biodiesel	Petro-diesel	ASTM values
Colour	Milkish yellow	Amber yellow liquid	Clear liquid	-
Sp. gravity	-	0.8668	0.8400	-
Viscosity (mm^2s^{-1}) (at 40°C)	47.87	4.23	3.12	1.9-6.0
Refractive index		1.4435	-	-
Flash point (°C) (D6751-07a)	206	134	79	≥93
Iodine value (mgI/g)	61.22	64.08	84	<125 (SS155436-96)
Cetane index		54.8	46	≥ 40
pH	6.75	7.05	-	-
Calorific value (kJ/g^{-1})	-	42.50	44.96	-
Saponification value (mgKOH/g)	198	-	-	-

is better than that of the commercial petrodiesel used for comparison. The higher calorific value of the petrodiesel implies that it has a little higher energy supply potential.

Also, the viscosities of both the petro- and biodiesel are not very different from each other. The viscosity of the latter is 1.356 times that of the former. This value is lower than the 1.684 to 1.712 times that of the petrodiesel reported for palm kernel oil biodiesel by Alamu in a PhD thesis submitted to the Mechanical Engineering Department of Ladoke Akintola University of Technology, Ogbomosho, Nigeria. The higher viscosity of biodiesel relative to petrodiesel has been reported to lead to decreased leakages of fuel in plunger pair as well as changing the parameters of fuel supply process (Lebedevas and Vaicekauskas, 2006). The reduction of overall cost of biodiesel was reported in a study using a mixture of 50% of both low grade animal fat and soybean oil (Canoira et al., 2008) and the biodiesel obtained were shown to have acceptable standards with a lower final cost.

Conclusions

i. The production of biodiesel from pig fat by transesterification with methanol and NaOH catalyst was assessed,
ii. The flash point, specific viscosity, cetane index and other fuel related properties of methyl esters of pig fat methyl esters are within the various limits provided for in the standards for biodiesel,
iii. The result of the present work therefore suggests that methyl esters which were produced using pig fat obtained from an abattoir in Nsukka could be used as biodiesel in running compression ignition engines,
iv. If the harvesting of this and other animal fats is properly organized in this part of the world, it could add to the energy mix and reduce the anticipated pressure in the use of edible vegetable oil in biodiesel production.

ACKNOWLEDGEMENT

The corresponding author gratefully acknowledges the preliminary work of Chibuzo Egu.

REFERENCES

Alamu OJ, Waheed MA, Jekayinfa SO (2007). Alkali-catalysed Laboratory Production and Testing of Biodiesel Fuel from Nigerian Palm Kernel Oil, Agricultural Engineering International: The CIGR E journal, Manuscript Number EE 07 009, IX, July.

Alhassan M, Isa AG, Garba MU (2005). Production and Characterisation of Biodiesel from Cotton Seed Oil, Proc. 6th Annual Engineering Conference, FUT Minna. June 2005. pp 36-45

Apel A (2006). Diploma Thesis submitted to the Department of Technology and Sustainable Product Management, ITNP, Vienna University of Economics and Business Administration, WU Wien.

Babcock RE, Clausen EC, Popp M, Mattingly B (2007). Biodiesel Production from Varying Grades of Beef Tallow and Chicken Fat. Final Report, Award No. MBTC-2058. Mark Blackwell Transportation Centre pp. 1-12. Available at: http://hdl.handle.net/123456789/6980.

Canoira L, Rodriguez-Gamero M, Querol E, Alacantara R, Lapuerta M, Olivia F (2008). Biodiesel from Low-Cost Animal Fat: Production Process Assessment and Biodiesel Properties Characterization, Ind. Eng. Chem. Res. 47(21):7997-8004.

Chhetri AB, Tango MS, Budge SM, Watts KC, Islam MR (2008). Non-Edible Plant Oils as New Sources for Biodiesel Production. Intl. J. Mol. Sci. 9:169-180.

Chung KH, Kim J, Lee KY (2009). Biodiesel Production by Transesterification of Duck Tallow with Methanol on Alkali Catalysts. Biomass Bioenergy 33(1):155-158.

Crymble SD, Hernandez R, French T, Zappi ME, Baldwin BS, Thomas D (2006). Kinetic Study of Biodiesel Production from Chinese Tallow Tree Oil, Poster Session: Catalysis and Reaction Engineering Division, AIChE Annual Meeting, San Francisco, CA, Nov. 15. No. 416.

Dunn RO, Knothe G (2001). Alternative Diesel Fuels from Vegetable Oils and Animal Fats. J. Oleo. Sci. 50(5):415-426.

Ejikeme PM (2008). Fuel Properties of the Derivatives of Soybean Oil. J. Chem. Soc. Niger. 33(1):145-149.

Goodrum JW, Geller DP, Adams TT (2003). Rhelogical Characterization of Animal Fats and Their Mixtures with #2 Fuel Oil. Bio. Bioenergy 24(3):249-256.

Graboski MS, McCormick RL (1998). Combustion of Fat and Vegetable Oil Derived Fuels in Diesel Engines. Progress Energy Combust. Sci. 24(2):131-132.

WhipNet Technologies (2009). Biodiesel from Animal Fat: Converting Chicken Fat to Biodiesel. Available at: http://e85.whipnet.net/alt.fuel/animal.fat.html, as at 27/05/09.

Ikwuagwu OE, Ononogbu IC, Njoku OU (2000). Production of Biodiesel using Rubber Seed Oil. Ind. Crops Prod. 12:57-62.

Klopfenstein WE (1985). Effect of Molecular Weights of Fatty Acid Esters on Cetane Numbers as Diesel Fuels. JAOCS 62(6):1029-1031.

Knothe G, Dunn RO, Bagby MO (1997). Biodiesel: the use of vegetable oils and their derivatives as alternative diesel fuels. In: ACS symposium series No. 666: Fuels and chemicals from biomass. Chap. 10, Eds. Saha, B.C. and Woodward, J., Washington, DC, USA pp. 172–208.

Knothe G, Steidley KR (2005). Kinematic Viscosity of Biodiesel Fuel Components and Related Compounds. Influence of Compound Structure and Comparison to Petrodiesel Fuel Components. Fuel 81:1059-1065.

Lebedevas S, Vaicekauskas A (2006). Research into the Application of Biodiesel in the Transport Sector of Lithuania. Transport 21(2):80-87.

Li CY, Li RJ (2009). Fuel Properties of Biodiesel Produced from the Crude Fish Oil from the Soapstock of Marine Fish. Fuel Process. Technol. 90(1):130-136.

Mattingly B (2005). Production of Biodiesel from Chicken Fat Containing Free Fatty Acids. Unpublished Masters Thesis, University of Arkansas, Department of Chemical Engineering. In Alptekin, et al. (2011). World Ren. Energy Congress, 8-13 May, Sweden pp. 319-326.

Mattingly B, Manning P, Voon J, Himstedt H, Clausen E, Popp M, Babcock R (2004). Comparative Esterification of Agricultural Oils for Biodiesel Blending. Final Report, Award No. MBTC-2052. Mark Blackwell Transportation Centre. Available at: http://www.markblackwell.org/research/finals/arc2052/Mark%20Black well%20Final%20Report.htm as at 6th Jan. 2007.

Miller-Klein Associates (2006). Use of Tallow in Biodiesel. Available at http://www.hgca.com/publications/document/use_of_tallow_in_biodie sel.pdf as at 26th Feb., 2009.

National Standards for Biodiesel (2009). A Sunday Energy Inc. Biodiesel, Minneapolis, Resource Document. Available at: http://bdresource.com/index.php?option=com_content&task=view&id =177&Itemid=30. As at July 14.

Pinto AC, Guarieiro LLN, Rezende MJC, Ribeiro NM, Torres EA, Lopes WA, Pereira PAP, de Andrade P (2005). Biodiesel: An Overview. J. Brazilian Chem. Soc. 16(6B):1313-1330.

Raine RR, Johnson TR, Blackett B, Farid MM, Behzadi S, Elder ST (2008). New Zealand Tallow Sourced Biodiesel – A Review of the Resource, Production and Engine Performance, U21 International Conference on Energy Technologies and Policy, Auckland, 8th-10th Sept. pp. 1-15.

Schneider D (2006). Rendering Lard 2.0. Available at: http://www.obsessionwithfood.com/2006-01-01-blog-archive.htm1#113709378997673043. As at 22/06/08.

Sinha S, Agarwal AK (2005). Performance Evaluation of a Biodiesel (Rice Bran Oil Methyl Ester) Fuelled Transport Diesel Engine. SAE Int. 1:1720.

Tapasvi D, Wiesenborn D, Gustafon C (2005). Process Model for Biodiesel Production from Various Feedstocks. Trans. ASAE. 48(6):2215-2221.

Van Gerpen JH (2005). Biodiesel Processing and Production. Fuel Process. Technol. 86(10):1097-1107.

Van Gerpen JH, Peterson CL, Goering CE (2007). Biodiesel: An Alternative Fuel for Compression Ignition Engines, Presented at the 2007 Agricultural Equipment Technology Conference, Louisville, Kentucky, USA, 11-14th Feb., ASAE Publication Number 913C0107.

Wörgetter M, Prank IH, Rathbauer J, Bacovsky D (2006). Local and Innovative Biodiesel, Final Report. Francisca Josephinum-Biomass Logistics and Technology. Wieselburg, pp. 32-33.

Xiu YP, Pu L, Yi SY, Hai LR, Fei G (2008). Esterification of Acetic Acid with Isoamyl Alcohol over Expandable Graphite Catalyst. E-J. Chem. 5(1):149-154.

Xiu YP, Ting TY (2009). Preparation of Biodiesel through Transesterification of Animal Oil and Alcohol Under the Catalysis of SO_4^{-2}/TiO_2. E-J. Chem. 6(1):189-195.

4

Cosolvent transesterification of *Jatropha curcas* seed oil

I. A. Mohammed-Dabo[1]*, M. S. Ahmad[2], A. Hamza[1], K. Muazu[3] and A. Aliyu[1]

[1]Chemical Engineering Department, Ahmadu Bello University Zaria, Nigeria.
[2]Center for Renewable Energy Research, Umaru Musa Yarádua University, Katsina, Nigeria.
[3]National Research Institute for Chemical Technology Zaria, Nigeria.

This is paper aimed at characterizing Nigerian *Jatropha curcas* seed oil in terms of % free fatty acid (FFA), viscosity, calorific, acid, iodine and saponification values and cetane number which were found to conform to those in literature. The GC-MS analysis of the raw oil indicated that oil consisted principally of palmitic, oleic, linoleic and stearic acids. The established operating conditions for the efficient cosolvent transesterification of Jatropha oil, using tetrahydrofuran as the cosolvent, were found to be 40°C, 200 rpm, 4:1 methanol-to-oil molar ratio, 1:1 cosolvent-to-methanol volume ratio, 0.5% w/w catalyst concentration and a time duration of 10 min. An optimum yield of 98% was obtained at these conditions and the result of the GC-MS analysis confirms the formation of methyl ester at these conditions. Properties of the biodiesel obtained at these optimum transestrification conditions were compared favourably with the ASTM D 6751-02 Standard B100.

Key words: *Jatropha curcas*, transesterification, cosolvent, biodiesel.

INTRODUCTION

The three dimensional energy threat of climate change, affordability and energy security has lead the world to discover the finite nature of fossil fuels, even though they will continue to dominate the global primary energy mix for several decades to come (World Energy Outlook, 2009; Meda et al., 2009). Growth in world population has resulted into surge in energy demand, and therefore the need for secured energy source that are renewable, environmentally friendly, affordable and above all sustainable has arise. Various feedstocks have been proposed for the production of biodiesesl, one of the contending candidates is the *Jatropha curcas*. This tree is selected due to its numerous advantages over others. *J. curcas* is non-edible as it contains compounds that are highly toxic. It is resistant to drought and pests. Seed yields under cultivation can range from 1,500 to 2,000 kg ha^{-1}, corresponding to extractable oil yields of 540 to 680 L ha^{-1} (58 to 73 US gallons per acre) and they have the potential to get as much as 1,600 gallons of diesel fuel per acre in a year. *J. curcas* tree can also be intercropped with other cash crops such as coffee, sugar, fruits and vegetables.

Biodiesel production through transesterification process involves reaction of the oil with an alcohol (mostly methanol) in the presence of a catalyst resulting into the formation of a diesel equivalent biofuel (mono alkyl ester). This transesterification process is associated with many problems, in which the principal of these problems is that the reactants (oils and alcohols) are not readily miscible because of their chemical structures. Oil disperses in the methanol medium, so the rate of collision of the glyceride and the methoxide (the mixture of methanol and the alkaline catalyst – KOH or NaOH) molecules becomes slower. This lowers the rate of collisions of the molecules and also the rate of reaction causing longer reaction times, higher operating expenses and labour, higher fixed capital investments and consequently higher product costs (Caglar, 2007).

To overcome this difficulty of mixing of the reactants, a single phase reaction is proposed (Boocock et al., 1998). The proposed model involves a cyclic solvent introduced into the reaction mixture which makes both the oil and

---

*Corresponding author. E-mail: iroali@mail.ru.

methanol miscible by reduction of mass transfer resistance. This solvent has numerous numbers of solvents with the boiling point up to 100°C (Caglar, 2007). Tetrahydrofuran (THF) (65.8°C) is preferred because of its close boiling point to that of methanol (64.4°C) so that after reaction, both methanol and THF is recycled in a single step to be used again. The addition of THF to create a single phase greatly accelerates the reaction so that 99.89 wt%, whose conversion is almost complete, is achieved in a very short reaction time, such as 10 min (Boocock et al., 1998). The primary concerns with cosolvent transesterification are the additional complexity of recovering and recycling the cosolvent, although this can be simplified by choosing a cosolvent with a boiling point near that of the alcohol being used (Caglar, 2007). Additional concerns have been raised about the hazard level associated with the cosolvents. Several cosolvent are considered to replace THF among which are ethyl acetate, diethyl ether, and 1,4-dioxane (Boocock et al.,1998) each offering a unique set of properties and advantages.

This transesterification process is associated with many principal problems of which the reactants (oils and alcohols) are not readily miscible because of their chemical structures. Oil disperses in the methanol medium, so the probability and the rate of collision of the glyceride and the methoxide (the mixture of methanol and the alkaline catalyst – KOH or NaOH) molecules becomes slower. This lowers the rate of collisions of molecules and also the rate of reaction causing longer reaction times, higher operating expenses and labour, higher fixed capital investments and consequently higher product costs (Caglar, 2007; Behzadi and Mohammed, 2007).

Many researchers have investigated the cosolvent transesterication of various edible vegetable oils, establishing various optimum operating conditions for those oils (Antoline et al., 2002). Out of the available researches, none has neither critically used *J. curcas* oil (which non-edible) for the cosolvent transesterification nor establish the process optimum parameters. This work therefore seeks to cosolvently transesterify Nigerian *J. curcas* seed oil, employing tetrahydrofuran as the cosolvent. Establishing the optimum transesterification conditions was equally an objective this research work was set to achieve.

MATERIALS AND METHODS

Materials

Dried *J. curcas* seeds were obtained from Oil Seed Research unit of the Institute for Agricultural Research (IAR), Ahmadu Bello University Zaria. Tetrahydrofuran (THF), anhydrous methanol, sodium hydroxide, tetraoxosulphate (IV) acid, hydrochloric acid, carbon tetrachloride, potassium iodide, potassium dichromate and Wijs solution were procured from Chemical Stores in Zaria, Nigeria. All the procured chemicals were of analytical grades.

Methods

Oil extraction

The Jatropha seed was weighed, grinded and the chaff separated from the kernel and both were weighed. Then the kernel was grinded and steamed in an oven to reduce the viscosity of the oil after which a mechanical extractor was used to extract the oil from the cake. Respective weights of the cake and the extracted oil were measured.

Oil characterization

The oil was characterized for physical and chemical properties such as viscosity, FFA content, acid, saponification, iodine and calorific values. Others are density and moisture content.

Viscosity determination

The viscosity of the raw oil was measured using Brookfield rotary digital viscometer NDJ-8S at 24.3°C. 200 ml of the oil was poured into a beaker and spindle, No 2 was attached to the viscometer, this was then lowered into the beaker and allowed to attain same temperature with the sample and viscometer was set at a speed of 60 rpm. While the viscometer was on, the reading at 25% shear rate was taken.

Free fatty acid determination

This is the percentage by weight of specified fatty acid in the oil. The method applied for this analysis, is the American Oil Chemists' Society (AOCS) method 5a-40. 1 g of sample was measured in a conical flask and dissolve with 25 ml of isopropyl alcohol. 3 drops of phenolphthalein indicator was added to the solution. The mixture was then titrated with 0.1 N sodium hydroxide solution shaken constantly until a pink colour persisted for 30 s. The percentage FFA was then calculated using the formula:

$$\%FFA = \frac{Titre\ value * Normality\ of\ NaOH * 28.2}{weight\ of\ sample} * 100 \tag{1}$$

Acid value determination

This is the number of milligram of KOH required to neutralize the free fatty acid in 1 g of the sample. 1 g of sample was weighed into a conical flask, 25 ml of the sample was poured into the conical flask and 3 drops of phenolphthalein was added and titrated against 0.1 N solution of potassium hydroxide. The acid value was calculated according to the following formula:

$$Acid\ number = \frac{Titre\ value * Normality\ of\ KOH * 56.1}{weight\ of\ sample} x100 \tag{2}$$

Saponification value determination

This is the milligram of KOH required to saponify 1 g of fat or oil. Saponification value is the measure of the molecular weight of the fatty acid. The AOCS method Cd 3-25 was employed. 2 g of the sample was weighed into a 250 ml conical flask and 50 cm^3 of 0.5 N ethanolic KOH (that has stayed overnight) was added to the sample. The mixture was then heated to saponify the oil. The unreacted KOH was then back titrated with 0.5 N hydrochloric acid

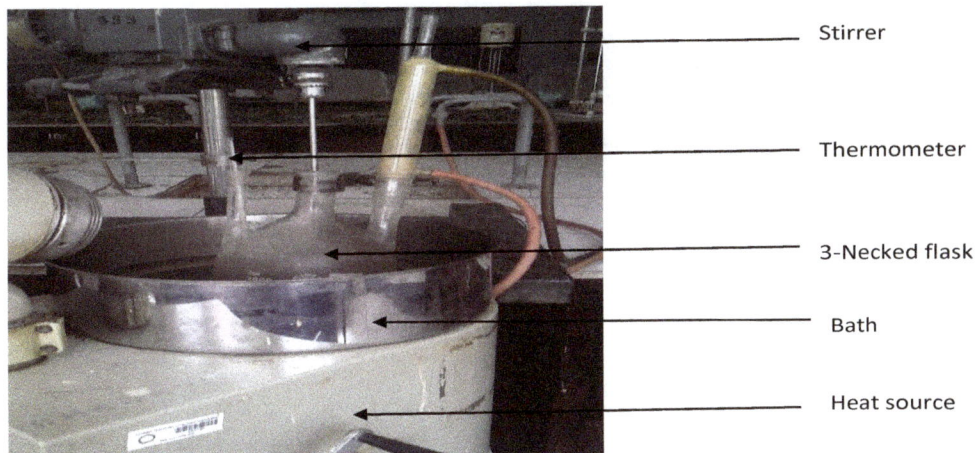

Figure 1. Esterification/Transeterification experimental set up.

using 2 to 3 drops of phenolphthalein indicator. The saponification value was calculated using the formula:

$$Saponification\ value = \frac{Titre\ value * Normality\ of\ NaOH * 56.1}{weight\ of\ sample} \quad (3)$$

Iodine value determination

This is the measure of the degree of unsaturation in relation to the amount of fat or oil. Iodine value is defined as the gram of iodine absorbed per 100 g sample. When unsaturated oil is heated, polymerization of the triglyceride occurs which leads to gum formation. Also, unsaturated compounds are susceptible to oxidation when exposed to air, thereby degrading the oil quality. Hence, the higher the iodine value, the greater the degree of unsaturation. The method employed was AOCS method 1b-87. 2 g of the sample weighed into a 250 ml conical flask, 10 cm^3 of CCl$_4$ was then added to dissolve the sample by swirling, 20 cm^3 of Wij's solution was added and stoppered immediately, then swirled and the mixture was allowed to stand in the dark at ambient temperature for 30 min. 15 cm^3 of 10% KI solution was then added with 100 cm^3 of distilled water to rinse the flask and the solution was then titrated with a solution of sodium thiosulphate using starch indicator. A blank sample was prepared and back titrated accordingly. The iodine value was calculated using the formula:

$$Iodine\ value = \frac{(Sample\ titre - blank\ titre) \cdot Normality \cdot 12.69}{weight\ of\ sample} \quad (4)$$

Specific gravity determination

A clean dried density bottle of 25 ml capacity was weighed (Wo), it was then filled with the oil, stopper was inserted and reweighed (W1), the oil was then substituted with water and weighed. The specific gravity was calculated using the following formula:

$$Specific\ gravity = \frac{mass\ of\ substance}{mass\ of\ equal\ amount\ of\ water} = \frac{W1 - W0}{W2 - W0} \quad (5)$$

Moisture content determination

The moisture content of the raw oil was also determined using oven

drying method. The oil was heated at 130°C for 1 h and filtered using muslin cloth to drive off the water and remove the solid particles respectively.

Oil pre-treatment

The oil extracted contain water and some solid particle, it was heated at a temperature of 130°C for 30 min to boil out the water (Singh, 2009), it was then allowed to cool and settle over night, and further, filtered with fine filter mesh to remove the solid particles. The removal of water from the oil is necessary because water content of above 0.5% facilitate saponification.

Oil esterification

The oil had an FFA content of 14.8%, and therefore, there is need for reduction below 0.5%, otherwise there will be high saponification (Van Gerpen, 2005). This reduction was achieved via esterification of the oil with methanol and using tetraoxosulphate (VI) acid as the catalyst. 50 g of the oil was measured and place into three neck round bottom flask and placed in a water bath and heated to a temperature of 60°C. 5% w/w acid was mixed with 20% w/w methanol and also heated to a temperature of 60°C and mixed with the oil. A suspended mechanical stirrer was inserted into the oil via one of the necks, the flask was equipped with a reflux condenser and a thermometer was dipped into the mixture via the remaining neck of the flask. The mechanical stirrer was set at 700 rpm and timing was started at this point. A picking pippeted was used to withdraw samples at one hour interval (1, 2, 3, 4 and 5 h) and was titrated against 0.1 N solution of KOH to determine the %FFA level. The experimental setup is shown in Figure 1.

Oil transesterification

The esterified oil was transesterified in 50 g batches to establish the optimum conditions for the reaction by varying the time at 10, 20, 30, 40, 50 and 60 min, and the temperature at 30, 40, 50 and 60°C. The methanol to oil molar ratio and the Methanol-to-Cosolvent (THF) volume ratio were varied at 3:1, 4:1, 5:1, 6:1 and 1:1, 1:2, 1:3 respectively. The effect of stirring speed was also investigated by varying it at 100, 200, 300, 400, 500, 600 and 700 rpm. The transestrification was carried out in the same set-up used for esterification (Figure 1).

Figure 2. Separation set up.

Table 1. Fatty acid composition of the raw *J. curcas* oil.

S/No.	Component	Composition (Wt %)	Molecular weight
1	Palmitic acid (16:0)	15.86	256.42
2	Linoleic acid (18:2)	37.24	280.45
3	Oleic acid (18:1)	37.13	282.46
4	Stearic acid (18:0)	9.76	284.48
Total		99.99	
Average molecular weight (g/mol)			1103.81

Table 2. Properties of the raw *J. curcas* oil.

S/No.	Property	Unit	Value
1	FFA	%	14.8
2	Saponification value	mg KOH/g oil	202.34
3	Iodine value	g of Iodine/100g of sample	100.56
4	Cetane number	-	50.65
5	Density	Kg/m^3	874
6	Calorific value	MJ/Kg	81.90
7	Viscosity at 25°C	mPa.s	151.4
8	Acid value	%	29.6
9	Moisture content	%	0.4

Ester purification

In each case the mixture was neutralized by addition of 1 drop of 0.6 N H_2SO_4 cooled in an ice bath and poured into a separating funnel to separate the fatty acid methyl ester (FAME) from the glycerol by gravity. The experimental setup is shown in Figure 2. The glycerol, which settled at the bottom, was drained off while the methyl ester was washed with hot distilled water and dried.

Ester characterization

Upon establishing the optimum conditions, the oil produced at the optimum conditions was characterized for fatty acid methyl ester using GC-MS and other properties were also measured in accordance with the standard methods to see if the biodiesel

produced had comparable properties with ASTM D 6751-02 standard B100.

RESULTS AND DISCUSSION

Characterization of the raw *J. curcas* oil

Tables 1 and 2 respectively present the compositional analysis and properties of the raw *J. curcas* oil. It can be observed from Table 1 that the oil consisted principally of 15.86, 37.24, 37.13 and 9.76% w/w palmitic, linoleic, oleic and stearic acids respectively. The fatty acid

Table 3. Results of esterification.

S/No.	Time (min)	FFA (%)
1	0	14.8
2	60	8.23
3	120	6.55
4	180	5.43
5	240	2.55
6	300	0.4

composition of the crude *J. curcas* seed oil used in this work is comparable to those reported by Makkar et al. (1997). This indicates that the oil contained more unsaturated fatty acids (linoleic acid and oleic acid) than the saturated fatty acids (palmitic acid and stearic acid). The oxidative stability of biodiesel is a function of the fatty acid composition of the parent oil. The higher the amount of saturated fatty acid in the parent oil the greater the oxidative stability of the oil and vice versa.

Properties of the raw *J. curcas* oil presented in Table 2 were found to be within the literature values (Alptekin and Canakci, 2008). These include; saponification value (202.34 mgKOH/g oil), iodine value (100.56 g of iodine/100g of oil), cetane number (50.6484), density (874Kg/m^3), viscosity (151.4 mPa.s at 25°C) and moisture content (0.4%). The calorific value of the raw *J. curcas* oil was found to be 37.23 MJ/Kg. However, the free fatty acid content of the raw Jatropha oil was found to be 14.8% which was not within the ASTM specified limit of ≤0.5% for biodiesel production. Therefore the raw *J. curcas* oil needs to be neutralized, through esterification.

Results of estrification

As stated previously, the free fatty acid content of the raw *J. curcas* oil was not within the specified ASTM standard for biodiesel production; therefore, there was the need to neutralize the FFA through esterification by reacting the oil with the methanol in the presence of tetraoxosulphate (IV) acid. Table 3 shows the obtained esterification results at various reaction times ranging from 0 to 300 min. This was done with the view of reducing the oil's FFA content to enable us carry out the transesterification on the oil. It can be observed from the Table 3 that at zero minute, the FFA content of the oil was 14.8%. This value decreased to 8.23% after the first 60 min of the reaction, 6.55% after 120 min of the reaction and 5.43% at 180 min of the reaction, this further dropped down to 2.55% at 240 min before it finally came down to 0.4% after 300 min of reaction, which is within the specified limit of ≤0.5% for transesterification process (Singh, 2009).

Optimizing the transesterification process

Having reduced the FFA of the raw oil, transesterifaction process was carried out on the oil with a view to establishing the optimum operating conditions for the biodiesel production. The variables optimized included reaction time, temperature and agitation. Others are the methanol-to-oil ratio, methanol-to-cosolvent ration and the catalyst (NaOH) concentration. It should be noted that in this research work THF was chosen as the cosolvent.

Effect of reaction time

Duration of reaction has been established as one of the critical parameters for biodiesel production. However, it has been found out that incorporation of a cosolvent to the transesterfication alcohol leads to the reduction of the reaction time. To investigate the effect of time on the process, batches of 50 g of the esterified oil were transesterified with a methanol/oil ratio of 3:1, catalyst (NaOH) concentration of 0.5% w/w of oil, 30°C reaction temperature and methanol-to-cosolvent ratio of 1:1% v/v at stirrer speed of 100 rpm. The reaction time was varied at 10, 20, 30, 40, 50 and 60 min. At these conditions, the highest yield of 73.5% was obtained at the 50 min reaction time. However, as can be observed from Figure 3, at the first 10 min, about 73% yield of biodiesel was obtained. Ideally, when optimizing a process, a point with the highest yield is usually considered as the optimum but here, considering the fact that the difference between the yield at 10 and 50 min was just 0.5%, 10 min was chosen as the optimum reaction time. This is in order to save cost of heating that would be required to heat the vessel for 50 min just to attain an extra 0.5% yield of the fatty acid methyl ester (FAME).

Effect of reaction temperature

Temperature is one of the significant factors affecting biodiesel yield. Reaction temperature must be lower than the boiling point of the alcohol in order to ensure that the alcohol will not leak out through vaporization. Having

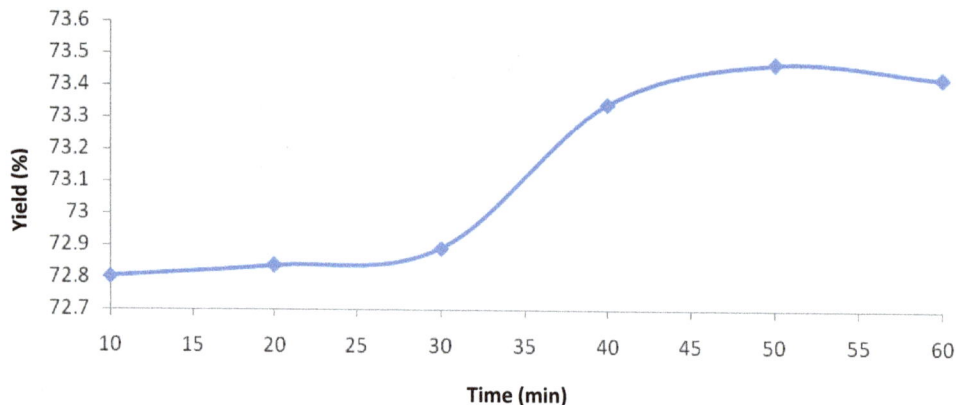

Figure 3. Effect of reaction time on the biodiesel yield.

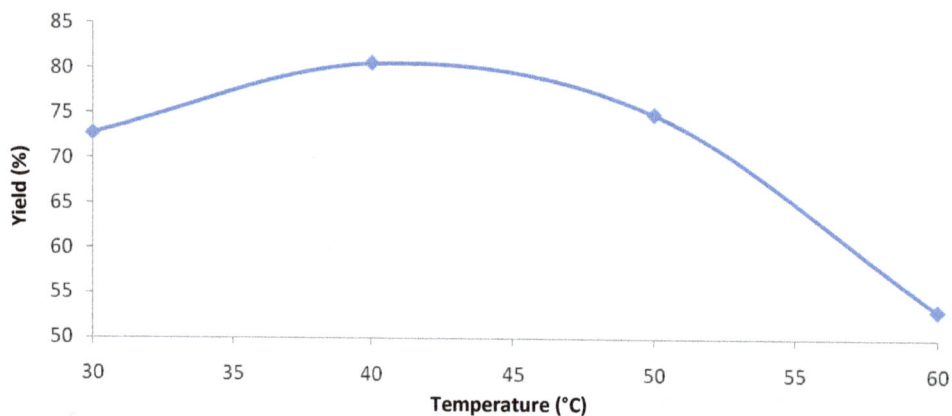

Figure 4. Effect of reaction tempearture on the biodiesel yield.

established 10 min as the reaction time, the oil was transesterified at the earlier stated condition for 10 min under various temperatures of 30, 40, 50 and 60°C. Figure 4 illustrates the effect of reaction tempearture on the biodiesel yield.

As can be observed from figure, the highest yield of 81% was obtained at a reaction temperature of 40°C. Beyond this temperature, the yield decreased drastically. The decrease in yield at temperature above 40°C could be as result of saponification reaction noticed at tempe-rature above 40°C. Therefore, the optimum temperature is 40°C as oppose to the 60°C reported by Demirbas (2009) for conventional transestrification (without a cosolvent)

Effect of methanol-to-oil ratio

Even though theoretically (from the transesterification reaction stoichiometry), methanol-to-oil ratio of 3:1 is required for transestrification reaction. However, researches have indicated that methanol-to-oil ratio

significantly affect the yield of biodiesel. Figure 5 shows the effect of Methanol-to-oil ratio on the fame yield. It can vividly be noticed that biodiesel yield is increased when methanol to-oil-ratio is raised beyond 3:1 and reaches a maximum. Increasing the alcohol-to-oil ratio beyond the optimum did not increase the yield; instead it will increase the cost of recovery of the methanol and causes difficulty in glycerol separation consequently resulting in lower yield. When the Methanol-to-oil ratio was varied at 3:1, 4:1, 5:1 and 6:1 the highest biodiesel yield 98% was obtained at 4:1 ratio. Further increase in the ratio resulted into decrease in the yield. Therefore, the optimum methanol-to-oil ratio was chosen as 4:1. It should be noted that, this transesterification stage was conducted at 40°C, methanol-to-cosolvent ratio of 1:1% v/v, stirrer speed of 100 rpm, and catalyst concentration of 0.5% w/w of oil for 10 min.

Effect of catalyst concentration

Investigating the effect of catalyst concentration (NaOH)

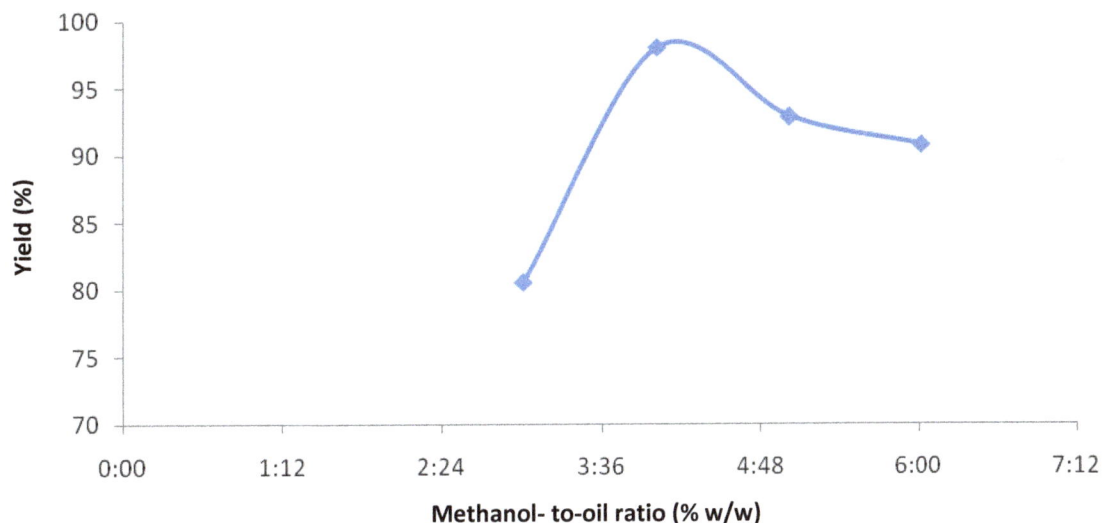

Figure 5. Effect of methanol-to-oil ratio on the biodiesel yield.

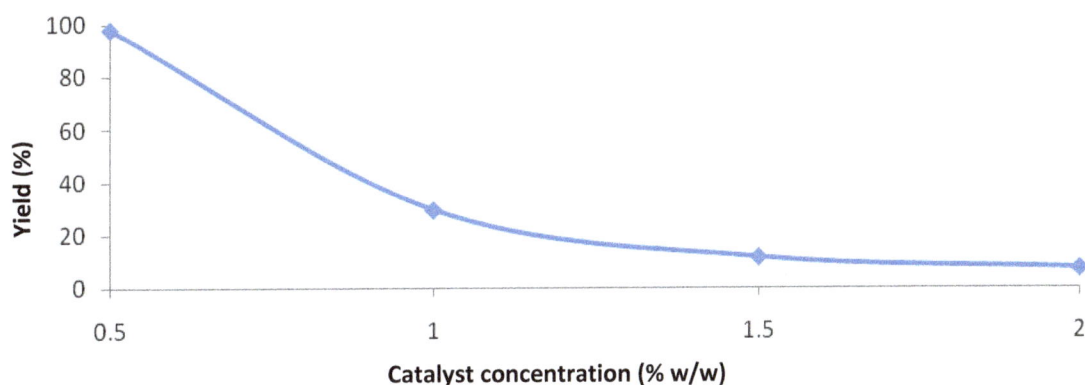

Figure 6. Effect of catalyst (NaOH) concentration on the biodiesel yield.

on the yield of biodiesel was carried out by varying the concentration of NaOH at 0.5, 1.0, 1.5 and 2.0% w/w of oil and keeping the other variables stated above constant. The result shown in Figure 6 indicates that the optimum catalyst concentration was 0.5% w/w. Further increase in concentration of catalyst beyond this facilitated saponification reaction and caused difficulty in the biodiesel separation from glycerol (Shay, 1993). This resulted in lowering the yield of the biodiesel at higher catalyst concentration. This finding corroborates with the findings of Encinar et al. (2010) that the introduction of a cosolvent to the transesterification alcohol reduces the amount of catalyst required.

Effect of methanol/cosolvent volume ratio

As opposed to conventional transesterification cosolvent transestrification offers significant reduction in mass transfer resistance (Encinar et al., 2010). At this stage of the work, methanol-to-cosolvent (Tetrahydofuran) ratio was varied at 1:1, 1:2 and 1:3% v/v while keeping other reaction variables constant. Figure 7 presents the influence of methano-to-cosolvent ratio on the biodiesel yield. The figure indicates that for this investigation, the optimum methanol-to-cosolvent volume ratio is 1:1 which gives a yield of 98%. Further increase in the ratio resulted in the decrease of the biodiesel yield. This could probably be as a result of the dilution effect on the reagents.

Effect of agitation

In conventional transestrification reaction, agitation is a significant factor affecting yield of biodiesel. Several works have been conducted to investigate this effect, Vincent (2005) and Antoline et al. (2002) reported 600 rev/min as the optimum impeller speed. The effect of

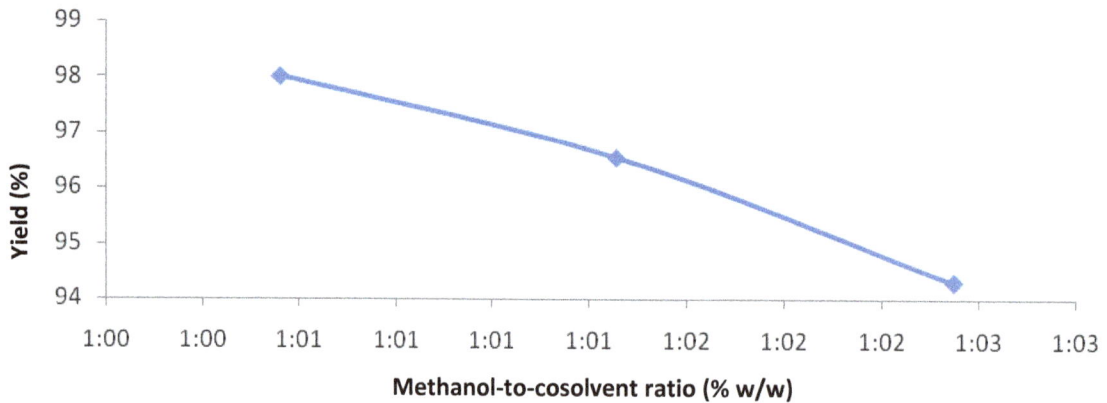

Figure 7. Effect of methano-to-cosolvent ratio on the biodiesel yield.

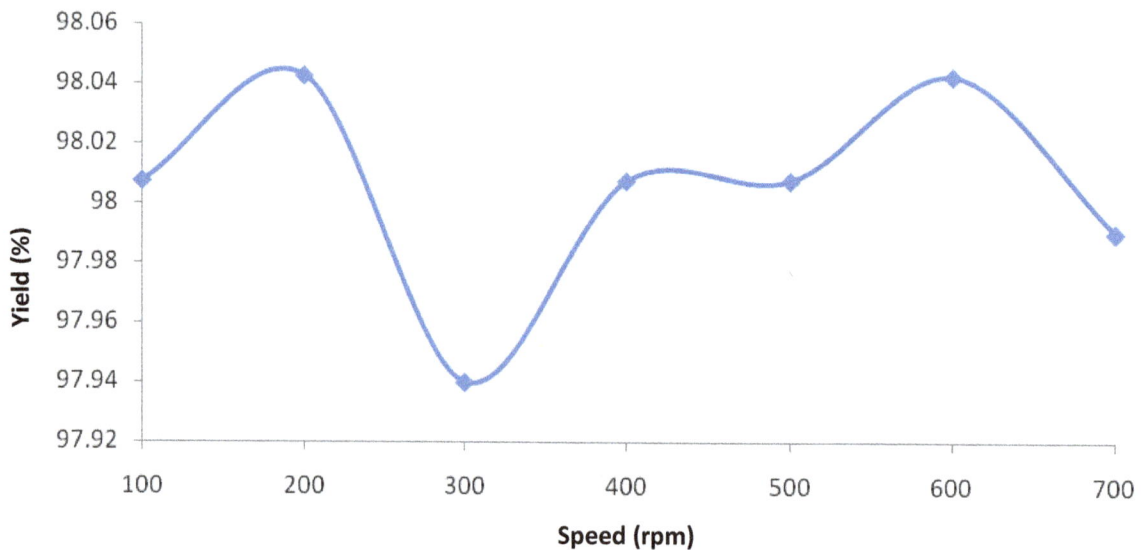

Figure 8. Effect of agitation on the biodiesel yield.

stirrer speed on the yield of biodiesel was investigated by varying the speed between 100 and 700 rev/min. Figure 8 shows effect of agitation on the *J. curcas* biodiesel yield. It can be observed from the figure that there is no definite trend. However, the highest yield value of 98.04% was obtained both at 200 and 600 rpm. Since actually mixing is only needed to establish one phase, once this is established, there is no need for agitation. In cosolvent transestrification, cosolvent contributes to overcome region of slow rate and therefore agitation becomes insignificant. In this case however, the optimum agitation may be considered as 200 rev/min based on the highest FAME yields obtained at this condition. However, it should be noted that the difference in biodiesel yield between 100 rpm (98.0076%) and 200 rpm (98.0426%) was just 0.035%. So based on the results of this investigation, agitation of the reaction mixture for about

100 to 200 rpm especially for the first few minutes of the reaction could be sufficient.

Characterization of the produced *J. curcas* biodiesel

Sample of the biodiesel produced at the optimum conditions established previously was characterized to ascertain its composition and relevant properties. Table 4 presents the composition of the FAME produced from the *J. curcas* oil as extracted from the obtained various peaks of the GC-MS used. It can be noticed from the table that the produced biodiesel consisted principally of the fatty acid methyl esters of palmitic, linoleic, oleic and stearic acids, hence confirming a high quality product. It can equally be observed that traces of other methyl esters were also observed in the composition of the produced

Table 4. Peak report of FAME produced at the optimum conditions.

Peak number	%area	Component
1	0.17	Myristic acid, methyl ester
2	2.83	Palmitoleic acid, methyl ester
3	20.01	Palmitic acid, methyl ester
4	0.18	Cyclopropaneoctanoic acid, methyl ester
5	0.35	Margaric acid, methyl ester
6	44.95	Linoleic acid, methyl ester
7	17.86	Oleic acid, methyl ester
8	11.51	Stearic acid, methyl ester
9	0.72	Ricinoleic, methyl ester
10	0.79	Arachidic, methyl ester
11	0.24	Bet-monolinolein
12	0.08	Stearic acid diglycerin ester
13	0.12	Docosanoic acid, methyl ester
14	0.18	Lignoceric acid, methyl ester
Total	100.00	FAME

Table 5. Properties of biodiesel produced at the optimum conditions.

S/No.	Property	FAME	ASTM D6751-02	Petrodiesel
1	Cetane number	51.43	47 min	46
2	acid number, mg KOH/g	0.75	0.80 max	0.35
3	Total glycerol, % w/w	0.0	0.24	NIL
4	Water content, % v/v	0.0	0.05	0.02
5	Density, Kg/m^3	869	875-900	850
6	Calorific value, MJ/Kg	38.53	-	42
7	Viscosity at 40°C, mm^2/sec	4.6	1.9-6.0	2.6
8	Saponification value, mg KOH/g oil	200.28	-	NIL
9	Iodine number, g /100g oil	103.2	-	-

biodiesesl. These traces could not affect the quality properties of the fuel as indicated in Table 5.

The produced biodiesel at the optimum conditions was analysed to ascertain its quality with respect to the standard biodiesel and petrodiesel properties. Table 5 shows the comparative properties of the produced *J. curcas* oil biodiesel with respect to the ASTM biodiesl and petrodiesel. It can be observed from the table that the produced biodiesel (FAME) virtually conforms to the properties specified by the American standard (ASTM D6751-02) and most of the properties tally with those of the conventional diesel oil (petrodiesel). The produced oil can therefore conveniently be used in a diesel engine.

CONCLUSIONS

The following conclusions were drawn from this research work:
1. The raw Jatropha seed oil was found to contain basically four major fatty acids namely: palmitic acid, oleic acid, linoleic acid and stearic acid. The unsaturated fatty acids were in higher quantity than the saturated acid and therefore the oil is good for biodiesel production.
2. Properties of the crude *J. curcus* seed oil such as viscosity, saponification value, Iodine value, cetane number, calorific value, flash point etc., were found to have comparable values with those obtained in the literature with the exception of the % FFA
3. The transesterification optimum conditions were found to be 40°C, 200 rpm, 0.5% w/w catalyst concentration, 4:1 methanol-to-oil molar ratio, 1:1 methanol-to-cosolvent volume ratio and time of 10 min.
4. Transesterification proceeds to completion at an even lower impeller stirring speed of 200 rpm as compared to 600 rpm reported for conventional transesterification.
5. Overall, the cosolvent transesterification offers an avenue for reduction in cost of material (in terms of catalyst concentration) and energy (in terms of stirring speed) in biodiesel production as well as significant reduction in time of reaction for a possible continues process of biodiesel production.

6. The biodiesel produced was found to have properties that are within the ASTM standard for biodiesel and have small of the cosolvent used.

REFERENCES

Alptekin E, Canakci M (2008). Characterization of the key fuel properties of methyl ester-diesels. Fuel, 88(1): 75-80.

Antoline G, Tinaut FV, Briceno Y, Castano V, Perez C. and Ramirez AI (2002). Optimization of biodiesel production by sunflower oil estrification. Bioresour. Technol., 83(2): 111-114.

ASTM D 6751-02 (2008). Biodiesel standard. United State of America Biodiesel National Board, November 2008.

Behzadi S, Mohammed MF (2007). Production of Biodiesel Using a Continuous Gas-Liquid Reactor, Department of Chemical and Materials Engineering University of Auckland, Auckland New Zealand. Available at: http://docs.business.auckland.ac.nz/Doc/Paper-Production-of-biodiesel-using-a-continuous-gas-liquid-reactor.pdf

Boocock DGB, Samir KK, VM, Lee C, Buligan S (1998). Fast Formation of High Purity Methyl Esters from Vegetable Oils. J. Am. Oil Chem. Soc. (JAOCS), 75(9): 1167- 1172.

Caglar E (2007). Biodiesel production using co-solvent. European Congress of Chemical Engineering. Copenhagen: Izmir Institute of Technology.

Demirbas A (2009). Progress and recent trends in biodiesel fuels. Elsevier J. Energy Conversion Manage., 50 (2009) 14-34.

Encinar JM, Gonzalez JF, Pardal H, Martinez GM (2010). Transestrification of rapeseed oil with methanol in the presence of various co-solvents. Proceedings of the Third International Symposium on Energy from Biomass and Waste. Venice, Italy, Nov., 8-11.

Makkar HPS, Aderibigbe AO, Becker K (1997). Comparative evaluation of non-toxic varieties of *Jatropha curcas*. J. Food Chem., 62(2): 202 – 218.

Meda CS, Venkata RM, Mallikarjunand MV, Vijaya KR (2009). Production of Biodiesel from Neem Oil. Int. J. Eng. Studies, 1(4): 295–302.

Shay EG (1993). Diesel fuel from vegetable oils: Status and Opportunities. Biomass Bioenergy, 4:227-42.

Van GJ, Gerhard K, Jurgen K (2005). Handbook of Biodiesel, AOCS Press Champaign Illinois, 2005

Vicent G, Martinez M, Aracil J, Esteban A (2005). Kinetics of sunflower oil methanolysis. Ind. Engine. Chem. Res., 44(15): 5447-5454.

World Energy Outlook (2009). Reference Scenario, IEA/OECD.

5

Experimental and theoretical evaluation of the performance parameters and emission characteristics of bio diesel using C.I engine for various injection pressure

S. Sundarapandian

Department of Mechanical Engineering, Sethu Institute of Technology, Kariapatti, Tamil Nadu State, India.

Cost and limited reserves of conventional fossil fuels have intensified the search for alternative fuels for use in internal combustion engines. A possible alternative engine fuel is vegetable oil because it is a clean burning, renewable, non-toxic, biodegradable and environmentally friendly transportation fuel. It can be used in a neat form without any modification of the engine. Vegetable oils are produced from crops such as soybean, peanut, sunflower, cotton, jatropha, mahua, neem, coconut, linseed, mustard, *Millettia pinnata*, rapeseed and castor oil plant. A theoretical model was developed to evaluate the performance characteristics and combustion parameters of vegetable oil esters like Jatropha, Mahua and Neem for various injection pressures and compared to diesel fuel. From the investigation it was concluded that the performance of vegetable oil esters such as Jatropha, Mahua, and Neem were much better. Thus the developed model was highly capable for simulation work with bio-diesel as a suitable alternative fuel for diesel engines.

Key word: IC engines, injection pressure, tranesterification, emission, performance, brake power.

INTRODUCTION

An experimental investigation was carried out to assess the performance parameters and emission levels of the three different vegetable oil esters in a single cylinder with four strokes constant speed computerized diesel engine test rig. The performance characteristics and emission levels such as NO_x, CO, HC, and smoke were measured using eddy current dynamometer, crypton computerized emission analyzer instruments (Model-EN2-390) and Bosch Smoke meter. From the experimental results, it is concluded that in terms of performance characteristics and emission levels, vegetable oil esters can be regarded as the best alternative fuel instead of diesel fuel.

THEORETICAL CONSIDERATION

Description of the four zone model

The present four-zone model is developed without deviating much from the basic concepts of the two zone model like the jet penetration, volume of spray, preparation rate, reaction rate for the purpose of heat release, the effect of impingement of the spray on the cylinder walls etc. In essence the burning zone of the two-zone model is further subdivided to provide a total of four distinct zones, namely:

(1) Fuel zone
(2) Stoichiometric burning zone
(3) Product plus air zone
(4) Air zone- unburnt zone, Figure 1.

The main advantage of this model is that it can truly represent the temporal and spatial variations of the fuel-air ratio and temperature (Venkatraman and Devaradjane, 2010).

Where $\dfrac{dM_{fi}}{d\theta}$ =Fuel injection rate

$\dfrac{dM_{fr}}{d\theta}$ = Fuel reaction rate

$\dfrac{dM_{ae}}{d\theta}$ = Air Entrainment Rate

$\dfrac{dM_{ac}}{d\theta}$ = Air consumption rate for stoichio burning

$\dfrac{dM_{sp}}{d\theta}$ = stoichiometric product movement rate

Theory of spray formation and combustion heat transfer

The theory used to simulate the combustion process in the combustion chamber of a diesel engine mainly involves the jet formation, calculation of jet penetration and spray volume, estimation of fuel burning rate and finally the heat transfer between the cylinder contents and the surroundings (Heywood,1989).

Fuel jet penetration

The development of the fuel spray or jet in the diesel combustion chamber is based upon the theory of a steady flow semi-infinite free jet, and finally modified by the use of empirical factors of transient (real) jets. Assuming that fuel jet penetration model is developed by modifying the transient (real) jet equation, the fuel jet penetration is X_{max}.

$$XU_{max} = 7.414\sqrt{K} \tag{1}$$

Where x = jet penetration. (m)
U_{max} = centre line velocity (m/sec)

K = kinematics momentum flux (m^4/sec^2)

$$XdX = 7.414\sqrt{K}\,dt \tag{2}$$

$$X_{max} = 0.685 \times 2.420 \left(\left[\frac{\Delta p}{\rho_a} \right]^{0.5} dn.t \right)^{0.5} \tag{3}$$

Volume of fuel spray

The volume of fuel before impingement consists of conical part of the half cone angle θ and the bell shaped part. The volume of the conical part of the spray is calculated analytically and for the bell shaped part of the jet, numerical integration is used. The volume flow rate at any section along the axis of the spray can be computed by integrating the product of velocity and the area.

$$\text{Volume flow rate} = 2\pi r_o u_{max} \int_0^{r_a} \frac{r}{r_o} \frac{u}{u_{max}} dr \tag{4}$$

$$\frac{r}{r_0} = \varepsilon \tag{5}$$

$$r_0 = X \tan\theta \tag{6}$$

$$U_{max} = 7.414 \frac{\sqrt{k}}{x} \tag{7}$$

From the free jet theory.

$$\text{Volume flow rate} = 5.9902 \, \tan^2\theta\sqrt{K}\,X.KF \tag{8}$$

Combustion and heat release

In this model the combustion period is assumed to consist of two periods. They are pre-mixed period and diffusion period. The combustion of the reacted fuel is assumed to be ideal. It produces only H_2O and CO_2 for the purpose of calculating the cylinder pressure and the energy level of the spray as a whole. Whitehouse model incorporating the rate of preparation of the fuel, surface area of fuel droplets and partial pressure of oxygen in the cylinder was used for this work. The preparation rate equation is as follows:

$$P = K'M_i^{1-x}M_u^x PO_2^L \tag{9}$$

$$P = K'M_i^{1/3}M_u^{2/3}PO_2^{0.4} \quad (Kg/^\circ CA) \tag{10}$$

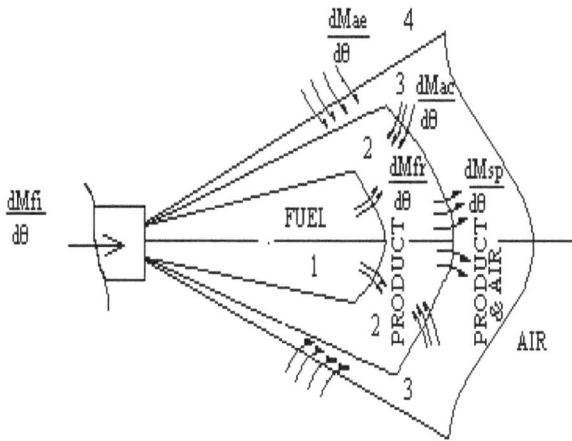

Figure 1. Schematic representation of combustion zones.

Heat transfer

$$M_i = n \times \rho \frac{1}{6} \pi D_0^3 \qquad (11)$$

$$M_u = n \times \rho \frac{1}{6} \pi D^3 \qquad (12)$$

Total surface area at any instant $= n \times \pi \times D^2$

Arrhenius type of equation is used for reaction rate of the prepared but unburnt fuel, which is as follows:

$$R = K'' \frac{PO_2}{N\sqrt{t}} e^{-ACT/T} \int (P - R) dx \quad (Kg/°CA) \qquad (13)$$

The heat transfer rate is calculated by using Annand's formula (1963). This formula seems more fundamental than the available alternatives. The equation considers net heat transfer as the summation of both radiative and convective heat transfer.

$$\frac{dQ_{loss}}{dt} = \left(S_c . H . (T - TW) + S_r . C . \left(T_4 - TW^4 \right) \right) \qquad (14)$$

where $dt = \frac{d\theta}{6N}$, $H = a\frac{K}{D} R_e^b$, $R_e = \frac{\rho \times U \times D}{\mu}$, $K = \frac{C_p \mu}{0.7}$ (15)

Method of estimating the final cylinder pressure

The final cylinder pressure equation can be written as

$$P_2 = \left\{ P_1 + P_1 \left(\frac{T - t_{b1}}{t_{b1}} \right) \left(\frac{V_{b1}}{V_{u1} + V_{b1}} \right) \right\} \left(\frac{V_1}{V_2} \right)^{\frac{C_p}{C_v}} \qquad (16)$$

P_2= First estimated value of the final pressure
P_1= Pressure after entrainment
$T - t_{b1}$ =Temperature rise due to constant volume heat addition
V_1, V_2= Cylinder volume at beginning and end of the step.

Energy equations

The energy equation can be written as:

$$\frac{dE_{cyl}}{d\theta} = DQF - DW - DQC - DQR \qquad (18)$$

$DQC - DQR$ = Heat transfer to the system
DQF = Energy associated with the external flow to the system
DW = External work done by the system

EXPERIMENTAL WORK

The performance tests were carried out on a single cylinder, four stroke naturally aspirated and water cooled kirloskar computerized diesel engine test rig. Diesel engine was directly coupled to an eddy current dynamometer. The engine and dynamometer were interfaced with a control panel which was connected to a computer. This computerized test rig was used for recording the test parameters such as fuel flow rate, air flow rate, temperature and load for calculating the engine performance, brake thermal efficiency and emissions like CO, HC, NOx and smoke (Yaman et al., 2001), shown in Figure 2 and Table 1.

RESULTS AND DISCUSSION

The performance and emission characteristics curves were carried out experimentally like predicting various performance characteristics such as thermal efficiency, specific fuel consumption for different vegetable oil esters such as jatropha, mahua and neem. The experimental results of the vegetable oil esters were compared with diesel fuel. From the results it was concluded that the performances of the vegetable oil esters are more or less equal to diesel. But pollutants like HC, CO, Nox and Smoke are reduced nearly 18% when compared to diesel. The results are shown in Figures 3 to 8.

Experimental brake thermal efficiency for various injection pressures

The brake thermal efficiency is also predicted with respect to various injection pressures for different

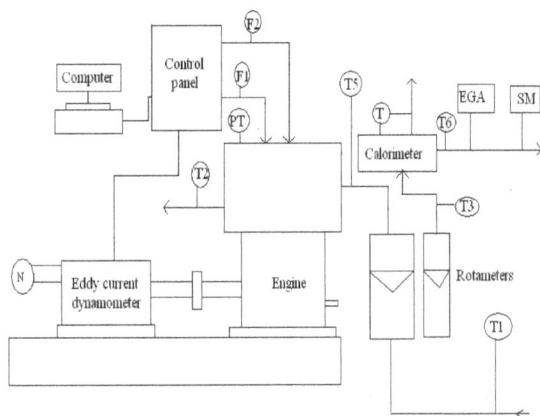

Figure 2. Schematic of experimental set-up.

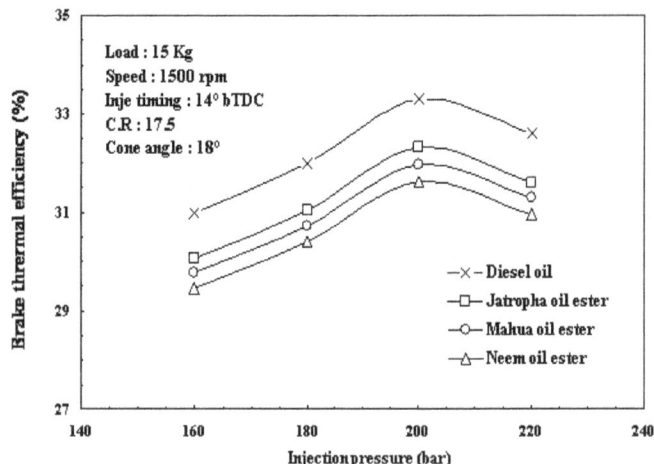

Figure 3. Comparision of brake thermal efficiency of three different biodiesel with respect to injection pressure.

Table 1. Engine Specifications.

Engine parameters	Specifications
Engine type	Kirloskar, Four stroke
No of cylinder	Single
Bore	87.5 mm
Stroke	110 mm
Cubic capacity	661 cc
Compression ratio	17.5
Rated speed	1500 rpm
Dynamometer	Eddycurrent,Water cooling

vegetable oil esters and compared with diesel fuel. Injection pressure is one of the very important operating variables, which affects the brake thermal efficiency (Nagarhalli and Nandedkar, 2011). The brake thermal efficiency is predicted for various injection pressures, and the injection pressure tried was 140 bars to 240 bars. When the injection pressure rose to 200 bar, it was found that the brake thermal efficiency has reached its maximum.

Fuel particles were uniformly mixed with air particles at an injection pressure of 200bar. When the injection pressure was increased above 200bar, the wall-wetting problem was created. This problem leads to a decrease in the performance of the engine. When the cone angle is decreased below 200 bar the fuel penetration and velocity were is also decreased and it generates more unburnt hydrocarbon. So injection pressure of 200 bar is found as the optimum value for maximum efficiency in the engine.

Experimental brake specific fuel consumption for various injection pressures

The brake specific fuel consumption found for diesel is

0.257 kg/kw-h; whereas for jatropha oil ester it is 0.277 kg/kwhr; for mahua oil ester, 0.286 kg/kw-h; and for neem oil ester, 0.291 kg/kw-h for optimum engine conditions such as 75% of load, injection pressure of 200bar.

Because at optimum engine condition fuels burn completely. Misfiring does not occur at this time. This increases the temperature and the number of moles of the burned gases in the cylinder. This effect increases the pressure to give increased thermal efficiency and decreases the specific fuel consumption. The specific fuel consumption is predicted with respect to injection pressure for various vegetable oil esters and compared with diesel fuel.

Experimental carbon monoxide for various injection pressures

Figure 5 shows the comparison of predicted emission result of carbon monoxide emission with various injection pressures for different vegetable oil esters and compared with diesel fuel. The carbon monoxide for vegetable oil esters are nearly 18 % reduced than diesel. The main difference in ester-based fuel compared to diesel is the oxygen content and cetane number. As the ester based fuel contains some oxygen which acts as a combustion promoter inside the cylinder, results in better combustion than diesel fuel. Hence carbon monoxide, which is present in the exhaust due to incomplete combustion, reduces drastically. The reduction of carbon monoxide in case of ester is lowered when compared to diesel.

Experimental hydrocarbon for various injection pressures

Figure 6 shows the comparison of predicted result of

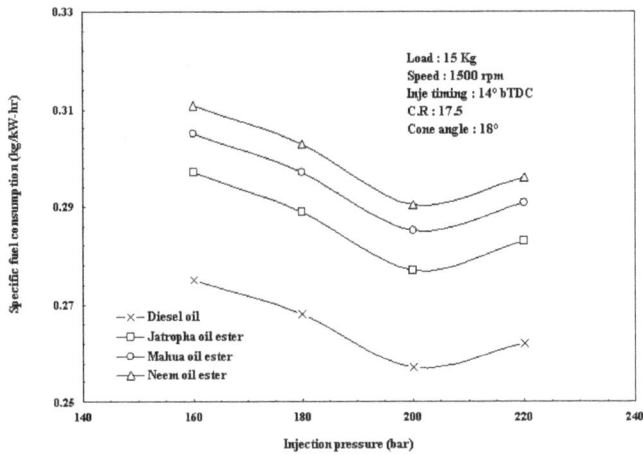

Figure 4. Comparision of specific consumption of three different biodiesel with respect to injection pressure.

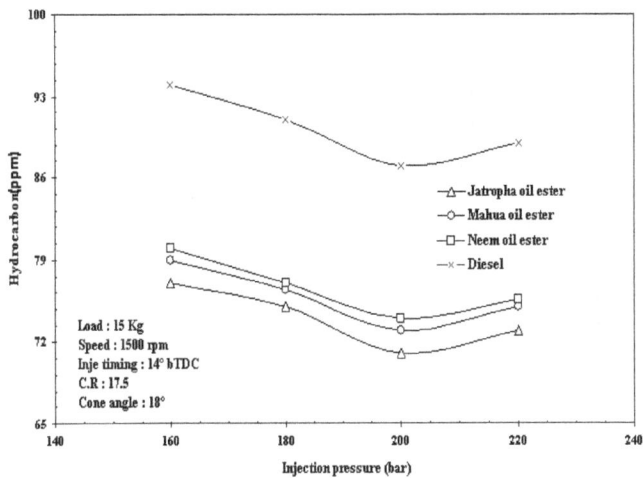

Figure 6. Comparison of carbonmonoxide for three different biodiesel with respect to injection pressure.

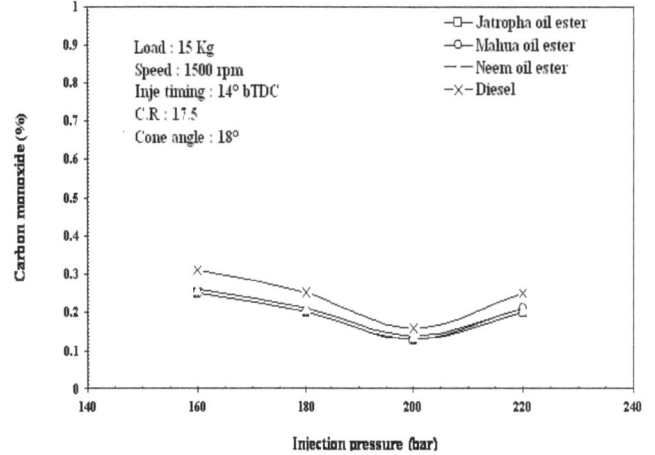

Figure 5. Comparision of hydrocarbon for three different biodiesel with respect to injection pressure.

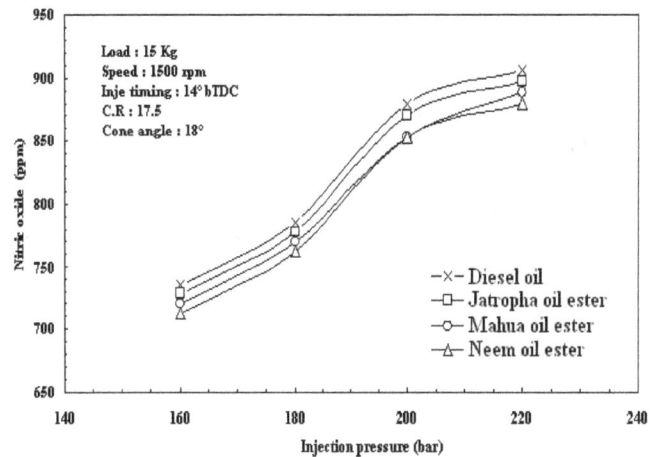

Figure 7. Comparision of predicted nitric oxide of three different biodiesel with respect to injection pressure.

hydrocarbon emission with various injection pressures for different vegetable oil esters and compared with diesel fuel. The hydrocarbon emission for diesel is about 119 ppm; while for jatropha oil ester, mahua oil ester and for neem oil ester it is 98, 100, and 101 ppm respectively. Cetane number of the fuel plays a vital role in ignition process.

As cetane number of ester-based fuel is higher than diesel, it exhibits a shorter delay period and the fuel undergoes better combustion. Here, the oxygen content of the fuel comes into picture as it enhances the combustion process. Therefore overall result of oxygen content and cetane number of the fuel leads to low CO and HC emission. Thus it is very clear from the graph that esters emit lower Hydrocarbon emission than that of diesel.

Experimental Nitric oxide for various injection pressures

Figure 7 shows the comparison of predicted result of NO_x formation with various injection pressures for different vegetable oil esters and compared with diesel fuel. The NO_x for diesel is about 985 ppm; while for jatropha oil ester, mahua oil ester and for neem oil ester it is 967, 960 and 955 ppm respectively.

In a direct injection naturally aspirated four-stroke diesel engine NO_x emission is sensitive to oxygen content, adiabatic flame temperature and spray properties. A change in any of those properties may change the NO_x production. Further, more fuel chemistry effects in the flame region could account for a change in NO_x production. NO_x formation is increased with increase in temperature (Sathiyagnanam and Saravanan, 2011).

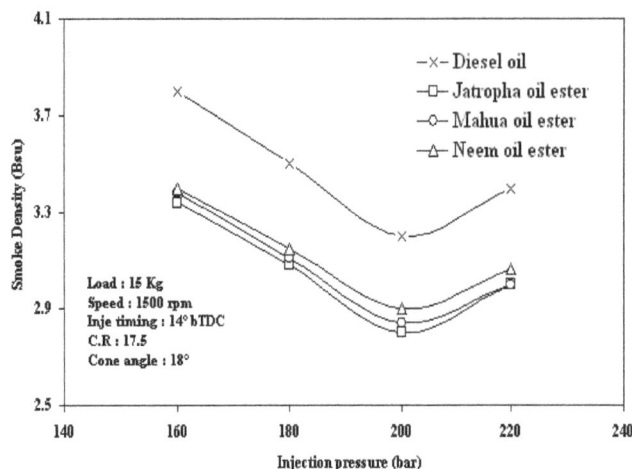

Figure 8. Comparision of predicted smoke density of three different biodiesel with respect to injection pressure.

EXPERIMENTAL SMOKE DENSITY FOR VARIOUS INJECTION PRESSURES

Figure 8 shows the comparison of experimental results of smoke with various engine injection pressure for different vegetable oil esters and compared with diesel fuel. The smoke for diesel is 4 bsu; for jatropha oil ester, 3.5bsu; for mahua oil ester, 3.6bsu; and for neem oil ester, 3.6bsu. The smoke formed due to incomplete combustion was much lower for esters compared to diesel. This is because of better combustion of esters. The main difference in ester-based fuel compared to diesel is the oxygen content and cetane number. As the ester based fuel contains some oxygen which acts as a combustion promoter inside the cylinder, thus resulting in better combustion than diesel fuel.

Conclusion

1. The brake thermal efficiency is reduced to about 3% for Jatropha, 4% for Mahua and 5% for Neem oil ester when compared to diesel. It is concluded that the brake thermal efficiency for vegetable oil ester slightly decreased when compared to diesel.
2. The specific fuel consumption for vegetable oil esters increased to about 8, 11 and 13% respectively for jatropha, mahua and neem oil ester when compared to diesel. It is concluded that the specific fuel consumption for vegetable oil ester is slightly increased than diesel.
3. The Carbon monoxide reduced by 18% for jatropha, 17 % for mahua, and 16 % for neem oil ester when compared to that of diesel. It is concluded that the carbon monoxide for vegetable oil ester is less when compared to diesel fuel.

4. The concentration of Hydro carbon decreased by 18 % for jatropha oil ester, 16 % for mahua oil ester and 15 % for neem oil ester when compared to diesel fuel.
5. The formation of Nitric oxides decreased by 1.8% for jatropha oil ester, 2.5% for mahua oil ester, and 3 % for neem oil ester when compared to diesel fuel.
6. The Smoke level decreased by 12 % for jatropha oil ester, 11 % for mahua oil ester, and 10 % for neem oil ester when compared to diesel fuel. Hence, it is concluded that in terms of performance characteristics, vegetable oil esters serve as a potential substitute for diesel fuel, (Babu and Devaradjane, 2003).

REFERENCES

Babu AK, Devaradjane G (2003). "Vegetable oils and their derivatives as Fuel for CI Engines" An overview. Anna University, SAE 2003-01-0767.
Heywood JB (1989). "Internal Combustion Engine Fundamentals". McGraw Hill Book Co. pp 540-565, 625-670.
Sathiyagnanam AP, Saravanan CG (2011). Experimental studies on the combustion characteristics and performance of a direct injection engine fueled with bio diesel blends with SCR – WCE. pp. 2231-2236.
Venkatraman M, Devaradjane G (2010). Effects of Compression ratio, Injection timing and Injection pressure on a DI diesel engine for better performance and emission fueled with diesel and biodiesel blends. IJAER, ISSN- 0976-4259. 1(3): pp. 3-7.
Yaman K, Ueta A, Shimamoto Y (2001). Influence of physical and chemical properties of biodiesel fuels on injection combustion and experimental emission characteristics in direct injection compression ignition engine" IJER. 2(4): 249-281.
Nagarhalli MV, Nandedkar VM (2011). Effect of Injection pressure on emission and performance characteristics of karanja biodiesel and its blends in C.I engine" Int. J. Appl. Eng. Res. Ddindigul, India. 1:4.

6

Epoxidation of *Ximenia americana* seed oil

M. H. Shagal, J. T. Barminas, B. A. Aliyu and S. A. Osemeahon

Department of Chemistry, Modibbo Adama University of Technology, P. M. B. 2076, Yola, Adamawa State, Nigeria.

Epoxidized *Ximenia americana* seed oil may be well suited in the partial replacement of petroleum products in synthesis of polyol. However, a process is needed to obtain this material from *X. americana* seed oil at sufficient conversion and scale to assess products. Therefore, *X. americana* seed oil was extracted using hot-water process and finally epoxidized in a solvent-free process in a three neck flask with the use of acetic acid as oxygen carrier, sulphuric acid as a catalyst and hydrogen peroxide. The studied parameters were iodine value conversion and oxirane oxygen content at varying time from 0 to 6 h and different temperatures 40 to 80°C. Epoxidized *X. americana* seed oil with up to 95.95 conversions was produced at a time of 6 h and temperature of 80°C. The result shows that the conversion of iodine value increased with reaction time and temperature. The highest amount of oxirane oxygen content of 3.60% was achieved at 4 h reaction time and temperature of 60°C. Interestingly, the oxirane oxygen content increased with reaction time and temperature then decreased after having achieved the optimal point.

Key words: Oxirane, temperature, epoxidation, oil, *Ximenia americana*, seed.

INTRODUCTION

Polyurethanes were first discovered by Professor Otto Bayer in 1973 and are used in a wide variety of applications (Chian and Gan, 1998). They have been exploited as coating, thermoset and thermoplastic materials, adhesives and rigid or non-rigid foams (Guo et al., 2006; Lligadas et al., 2006; Wang et al., 2009). Hence, the worldwide demand for polyols is projected to increase each year. Polyurethanes are generally produced from the reaction between polyol and diisocyanate (Pechar et al., 2006). Polyols are normally derived from petroleum feedstocks and are known as petroleum-based polyols (Tu et al., 2007). As the demand for polyols is increasing whilst the amount of petroleum is declining, alternative raw material for the production of polyols are needed.

In recent years, the price of crude oil has escalated raising many concerns over the stability and the sustainability of petroleum resources. The rising cost of crude oil also impacts the cost of polyurethane products because majority of the raw materials, such as polyols and isocyanates used in foams are petroleum derivatives (Flora, 2011). Finding an alternative feedstock for polyurethane has become highly desirable for both economic and environmental reasons. Natural oils have been shown to be a potential bio-renewable feedstock for polyurethane (Barret et al., 1993; Heidbreder et al., 1999; Petrovic et al., 2005).

The hope is that using renewable resources as feedstock for chemical processes will reduce the environmental footprint by reducing the demand on non-renewable fossil fuels currently used in the chemical industry and reduces the overall production of carbon dioxide, the most notable greenhouse gas. There is resurgence of interest in the use of annually renewable feedstock like vegetable oil for fuel and industrial applications. The use of agricultural feedstock allows us to manage our own carbon cycle more efficiently and reduce carbon (IV) oxide (CO_2) emissions (Cheng and Rayson, 1999).

Most of bio-based polyols for polyurethanes are synthesized from vegetable oils (Desroches et al., 2012). A vegetable oil polyols can be used as a replacement for conventional polyols, reacting with isocyanates to produce flexible slabstock polyurethane foam (PUF), elastomer and coating (Narine et al., 2007a, b). Polyol from vegetable oils and lactic, glycolic, or acetic acids, provide bio-based building blocks for further polyurethanes syntheses by reaction with diisocyanates. The obtained polyurethanes are partially bio-based and may be applied as binders and coatings (Boutevin et al., 2012). The use of petrochemical polyols is disadvantageous in terms of production, energy and costs. From both economic and environmental point of view, it is desirable to replace petroleum polyols with renewable resources (Veenendaal, 2007; Petrovic, 2008).

The potential for polyols derived from vegetable oils to replace petrochemical-based polyols began garnering attention beginning around 2004, partly due to the rising costs of petrochemical feedstock and partially due to an enhanced public desire for environmentally friendly green products (Niemeyer et al., 2006). Over the last few decades, polyol derived from petroleum oil as feedstock has been used to produce polyurethane to fulfill the world's needs for this polymer. Recently, a number of studies have been conducted to investigate the use of vegetable oils as a renewable and sustainable feedstock to replace the use of petroleum oil. Many researchers have concluded that vegetable oil can be exploited as an alternative raw material to substitute for petroleum oil to produce polyol. Examples include soybean oil, safflower oil, olive oil, canola oil, cottonseed oil, palm oil and rapeseed oil (Petrovic et al., 2003). It has been reported that that High-functionality polyols for application in polyurethanes were prepared by epoxide ring-opening reactions from epoxidized sucrose esters of soybean oil in which secondary hydroxyl groups were generated from epoxides on fatty acid chains (Pan and Webster, 2012). Vegetable oil-based polyol is a sustainable material which can be efficiently produced. *Ximenia americana* seed oil is a potential feedstock source.

X. americana "Wild plum or Plum" in English or locally called "Tsada" in Hausa and "Chabbuli "in Fulani is a medicinal plant that is bushy and spiny shrub, 4 to 5 m high with an open crown. The fruits are green but turn golden yellow or red when ripe. The fruit when eaten is refreshing and has an almond acid taste. The plant is found from Senegal to Cameroon including Northern parts of Nigeria (Arbonnier, 2004). *X. americana* is widely spread in tropical Africa and tropical America. It is a medium size deciduous tree which is widely cultivated for its yellow, pleasant acid fruits and as a live fence. The tree has large pencils with small white flowers and plum like fruits, which is up to 3.5 cm long. The tree has leaves with common stalk in the range of 30 to 60 cm long. The leaves are 5.8 cm long and are almost paired up in opposite direction. The flowers are green and usually found between March to April and July to August. The fruit are ellipsoid and are usually 2.5 to 3.5 cm long. They are normally found between April to May and July to August and December (Keay, 1989).

The aim of this research work is to assess the potential of *X. americana* seed oil as a chemical feedstock in the polyurethane industry by chemically modifying the *X. americana* seeds oil through epoxidation process.

MATERIALS AND METHODS

Sample collection and preparation

The fruits of *X. americana* were collected locally from the forest around Yola metropolis. The voucher specimen was identified and authenticated by Dr. D. A. Jauro of Forestry Department, School of Agriculture and Agricultural Technology, Modibbo Adama University of Technology, P.M.B. 2076, Yola. The fruits were then washed, air-dried at room temperature and cracked to remove the hard shell in other to obtain the seed. The seeds were washed with distilled water and air-dried at room temperature, after which they were crushed/pounded using mortar and pestle to paste. The pounded materials were stored in a closed container pending use in the extraction processes.

Extraction of *Ximenia americana* seed oil

The crushed seeds were manually/traditionally treated with hot water and continuously stirred in other to get the oil. The seed oil obtained was stored in closed container prior to epoxidation.

Experimental procedure

The epoxidation process was adapted from Goud et al. (2007). Iodine value was determined by applying the Wijs method (Ketaren, 2005; Siggia, 1963; Sudarmaji et al., 1997). The oxirane content was determined by the method adapted from Siggia (1963).

The calculation of chemicals required for the epoxidation reaction using acetic acid as the oxygen carrier is summarized below. Based on the literature, a typical fatty acid composition profile (Eromosele and Eromosele, 2002) for *X. americana* seed oil is presented in Table 1.

Total mole of *X. americana* seed oil (XASO) is expressed as concentration of double bonds (DB) in the oil ⟶ (n_t):

Volume of XASO (V) = 200 ml.

Density of XASO (ρ) = 0.867 g/ml (Eromosele et al., 1993).

Mass of XASO (M) = ρ.V = (0.867) (200) = 173.4 g.

Table 1. Fatty acid composition and their molecular weights present in *X. americana* seed oil.

Fatty acids	Molecular formula	Composition (wt %)	Molecular weight (g/mol)
Linoleic	$(C_{18}H_{32}O_2)$	1.34	280.45
Linolenic	$(C_{18}H_{30}O_2)$	10.31	278.43
Arachidonic	$(C_{20}H_{32}O_2)$	0.60	304.47
Eicosatrienoic	$(C_{20}H_{34}O_2)$	3.39	306.48
Erucic	$(C_{22}H_{42}O_2)$	3.46	338.57
Nervonic	$(C_{24}H_{46}O_2)$	1.23	366.62
Oleic	$(C_{18}H_{334}O_2)$	72.09	282.46

$$\text{n Linoleic acid} = \frac{(0.0134) \times (173.4)}{280.45} = 0.008285$$

$$\text{n Linolenic acid} = \frac{(0.1031) \times (173.4)}{278.43} = 0.064208$$

$$\text{n Arachidonic acid} = \frac{(0.006) \times (173.4)}{304.47} = 0.003417$$

$$\text{n Eicosatrienoic acid} = \frac{(0.0339) \times (173.4)}{306.46} = 0.019180$$

$$\text{n Erucic acid} = \frac{(0.0346) \times (173.4)}{338.57} = 0.017721$$

$$\text{n Nervonic acid} = \frac{(0.0123) \times (173.4)}{366.62} = 0.005818$$

$$\text{n Oleic acid} = \frac{(0.7209) \times (173.4)}{282.46} = 0.442555$$

$$n_t = 0.008285 + 0.064208 + 0.003417 + 0.019180 + 0.017721 + 0.005818 + 0.442555 = 0.56$$

∴ Total mole of XASO = 0.561 mol.

Acetic acid

Mole ratio of acetic acid to DB = 0.5:1.

Glacial acetic acid (99.5 wt%), molecular weight = 60.05, density = 1.05 g/ml.

Mole of acetic acid $(0.5) \, n_t = (0.5)(0.561) = 0.281$ mol.

Mass of acetic acid = $(0.281)(60.05) = 16.87$ gram.

Mass of glacial acetic acid $\left(\frac{100}{99.5}\right)(16.87) = 16.96 \, gram$

∴ Volume of glacial acetic acid required = $\frac{16.96}{1.05} = 16.15$ ml

Hydrogen peroxide

Mole of hydrogen peroxide to DB = 1.5:1

Hydrogen peroxide (6 wt%), molecular weight = 34.01, density = 1.10 g/ml.

Mole of hydrogen peroxide = $(1.5)(0.561) = 0.842$ mol.

Mass of hydrogen peroxide = $(0.842)(34.01) = 28.64$ gram.

Mass of hydrogen peroxide solution = $\left(\frac{100}{6}\right)(28.64) = 477.33 \, g.$

∴ Volume of hydrogen peroxide needed = $\frac{477.33}{1.10} = 434$ ml.

Sulfuric acid catalyst

Mass of sulfuric acid in the mixture is 3% of total mass of hydrogen peroxide and acetic acid.

Sulfuric acid (98 wt. %), molecular weight = 98.08, density = 1.84 g/ml.

Mass of sulfuric acid = $\left(\frac{3}{100}\right)(H_2O_2 + CH_2COO_S)$

$= \left(\frac{3}{100}\right)(28.64 + 16.87 == 1.37 \, g.$

Mass of sulfuric acid solution = $\frac{1.37}{1.84} = 0.74$ mol.

Determination of Iodine value and conversion

In order to determine conversion of iodine value, the iodine value of *X. americana* seed oil was calculated using the following equation:

$$\text{Iodine value} = \frac{(B - S) \times M \times 12.69}{W}$$

Where: S = Volume of $Na_2S_2O_3$ solution required for titration of the sample (ml), B = Volume $Na_2S_2O_3$ solution required for titration of the blank (ml), W = Weight of sample used (g), M = Molarity of the $Na_2S_2O_3$ (0.1M).

Initial value of *X. americana* seed oil (IV_0) expressed as iodine value at t = 0 (h)

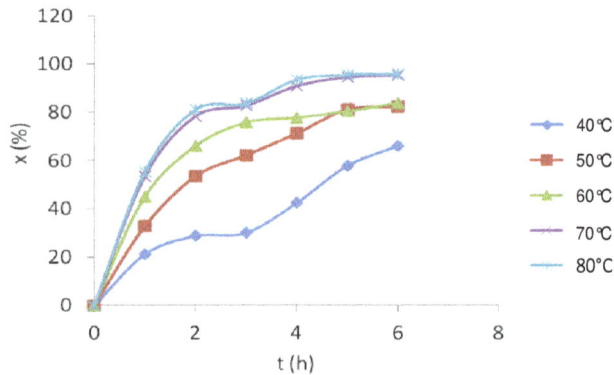

Figure 1. Effects of reaction time (t) on reaction conversion (X) at different temperatures.

M = 0.10 M, W = 0.20 g, B_1 = 19.65 ml, B_2 = 19.50 ml, B_3 = 19.65 ml, ∴ B_{AV} = 19.60 ml.

S_1 = 4.30 ml, S_2 = 4.35 ml, S_3 = 4.25 ml, ∴ S_{AV} = 4.30 ml.

$$IV_0 = \frac{(19.60-4.30)(0.1)(12.69)}{0.20} = 97.08 \text{ g I/100 goil.}$$

Conversion of iodine value (% x):

% X = (IV_0 - IV / IV_0)(100%)

Where: IV_0 = Initial Iodine Value, IV = Iodine value at certain condition.

1) Calculation at T = 40 ℃.

a) Reaction time (t) = 1 h

W = 0.2 g, S_1= 7.7 ml, S_2= 7.5 ml, S_3= 7.6 ml. S_{AV}= 7.6 ml

$$IV = \frac{(19.60-7.6)(0.1)(12.69)}{0.20} = 76.14 \text{ g } I_2/100 \text{ g oil}$$

$$\therefore \% X == \frac{(97.08-76.14).100\%}{97.08} = 21.57\%$$

b) Reaction time (t) = 2 h

W = 0.20 g, S_1 = 8.7 ml, S_2 = 8.7 ml, S_3 = 8.7 ml, S_{AV} 8.7 ml.

$$IV = \frac{(19.60-8.7)(0.1)(12.69)}{0.20} = 69.16 \text{ gram.}^{I_2}/_{100} \text{ g oil.}$$

$$\% X = \left(\frac{97.08-69.16}{97.08}\right)100\% = 28.76\%.$$

c) Reaction time (t) 3 h

W = 0.20 g, S_1 = 8.90 ml, S_2 = 8.90 ml, S_3 = 8.90 ml, S_{AV} 8.90 ml.

$$IV = \frac{(19.60-8.90)(0.1)(12.69)}{0.20} = 67.89 \text{ g}^{I_2}/_{100} \text{ g oil}$$

$$\therefore \% X = \left(\frac{97.08-67.89}{97.08}\right)100\% = 30.07\%.$$

d) Reaction time (t) = 4 h

W = 0.20 g, S_1 = 10.8 ml, S_2 = 10.9 ml, S_3 = 10.7 ml, S_{AV} 10.80 ml.

$$IV = \frac{(19.60-10.8)(0.1)(12.69)}{0.20} = 55.84 \text{ g}^{I_2}/_{100} \text{ g oil}$$

$$\therefore \% X = \left(\frac{97.08-55.84}{97.08}\right)100\% = 42.49\%.$$

e) Reaction time (t) = 5 h

W = 0.20 g, S_1 = 13.20 ml, S_2 = 13.10 ml, S_3 = 13.20 ml, S_{AV} 13.16 ml.

$$IV = \frac{(19.60-13.16)(0.1)(12.69)}{0.20} = 40.86 \text{ g}^{I_2}/_{100} \text{ g oil}$$

$$\therefore \% X = \left(\frac{97.08-40.86}{97.08}\right)100\% = 57.91\%.$$

f) Reaction time (t) = 6 h.

W = 0.20 g, S_1 = 14.50 ml, S_2 = 14.55 ml, S_3 = 14.50 ml, S_{AV} 14.52 ml.

$$IV = \frac{(19.60-14.43)(0.1)(12.69)}{0.20} = 32.80 \text{ g}^{I_2}/_{100} \text{ g oil}$$

$$\therefore \% X = \left(\frac{97.08-32.80}{97.08}\right)100\% = 66.21\%.$$

The reaction conversion at 40℃ and other reaction temperatures (50, 60, 70 and 80℃) are summarized in Figure 1.

Determination of oxirane oxygen content

The number of oxirane groups indicated by the percentage of oxirane oxygen content was calculated using the equation below:

$$\text{Oxygen content} = \frac{(B-S) \times M \times 16 \times 100}{W \times 1000}$$

Where: S = Volume of Na_2OH used for sample (ml), B = Volume Na_2OH used for blank (ml), M = Molarity of the NaOH = 0.1M, W = Weight of sample used (g), Volume NaOH used for blank (ml).

B_1 = 18.7 ml, B_2 = 19.00 ml, B_3 = 18.8 ml, ∴ B_{AV} = 18.83 ml.

1) Calculation of T = 40 ℃.

a) Reaction time (t) = 1 h.

W = 0.2 g, S_1 = 17.8 ml, S_2 = 17.6 ml, S_3 = 17.8 ml, ∴ S_{AV} = 17.73 ml.

$$\% \text{ Oxirane} = \frac{(18.83-17.73)(0.1)(16)(100)}{0.20 \times 1000} = 0.88\%.$$

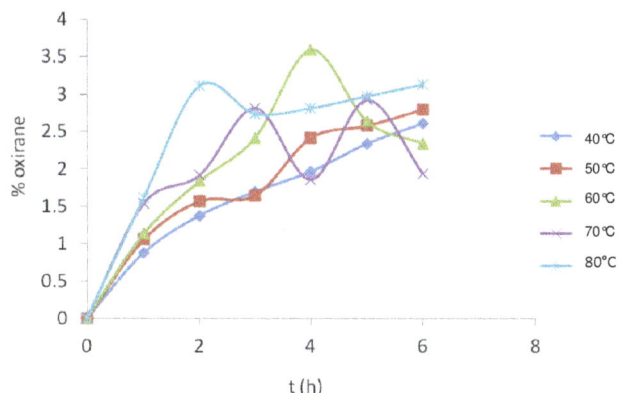

Figure 2. Effects of reaction time (t) on oxirane content (%) at different temperatures.

b) Reaction time (t) = 2 h

$W = 0.2$ g, $S_1 = 17.1$ ml, $S_2 = 17.2$ ml, $S_3 = 17.0$ ml, $\therefore S_{AV} = 17.1$ ml.

$$\% \text{ Oxirane} = \frac{(18.83 - 17.1)(0.1)(16)(100)}{0.20 \, X \, 1000} = 1.38\%.$$

c) Reaction time (t) = 3 h

$W = 0.2$ g, $S_1 = 16.7$ ml, $S_2 = 16.7$ ml, $S_3 = 16.7$ ml, $\therefore S_{AV} = 16.7$ ml.

$$\% \text{ Oxirane} = \frac{(18.83 - 16.7)(0.1)(16)(100)}{0.20 \, X \, 1000} = 1.70\%.$$

d) Reaction time (t) = 4 h

$W = 0.2$ g, $S_1 = 16.4$ ml, $S_2 = 16.3$ ml, $S_3 = 16.4$ ml, $\therefore S_{AV} = 16.37$ ml.

$$\% \text{ Oxirane} = \frac{(18.83 - 16.37)(0.1)(16)(100)}{0.20 \, X \, 1000} = 1.97\%.$$

e) Reaction time (t) = 5 h

$W = 0.2$ g, $S_1 = 15.90$ ml, $S_2 = 16.0$ ml, $S_3 = 15.8$ ml, $\therefore S_{AV} = 15.9$ ml.

$$\% \text{ Oxirane} = \frac{(18.83 - 15.9)(0.1)(16)(100)}{0.20 \, X \, 1000} = 2.34\%.$$

f) Reaction time (t) = 6 h

$W = 0.2$ g, $S_1 = 15.70$ ml, $S_2 = 15.50$ ml, $S_3 = 15.65$ ml, $\therefore S_{AV} = 15.62$ ml.

$$\% \text{ Oxirane} = \frac{(18.83 - 15.62)(0.1)(16)(100)}{0.20 \, X \, 1000} = 2.57\%.$$

The amount of oxirane oxygen content at 40°C and other reaction temperatures (50, 60, 70 and 80°C) are summarized in Figure 2.

RESULTS AND DISCUSSION

The iodine value and the oxirane oxygen content are the important properties in the characterization of epoxidized vegetable oils. The iodine value indicates the remaining unsaturation after epoxidation reaction, while the oxirane oxygen content indicates the epoxy groups present in the products. In the preparation of polymer, epoxy resins with a lower iodine value and higher oxirane oxygen content are desired. The iodine values of 6 treatments were reduced. The reductions in iodine values indicated the consumption of the unsaturation during epoxidation, but they did not represent solely conversion to epoxy groups because epoxy rings degradation generates side products.

The effects of reaction time and temperature on iodine value and reaction conversion are shown in Figure 1. The result indicated that the conversion of iodine value in the XASO increases linearly with the rise of reaction time and temperature and at a faster rate with temperature than with reaction time. Unsaturated double bonds present in the oil are converted to oxirane rings through the epoxidation reaction as indicated by the decrease in the iodine value. This value in the XASO represents the concentration of double bonds and it decreases with reaction time. Therefore, the reaction conversion increases with reaction time and temperature.

The effect of reaction time and temperature on oxirane content is shown in Figure 2. The result shows that an oxirane content increase with reaction time and temperature and then that value reaches the maximal level. Following this, the oxirane content decreases with reaction time and temperature. This finding supports previous research with the utilization of cottonseed oil as a raw material to produce epoxidized oil (Dinda et al., 2008). The result can be explained by the fact that through epoxidation reaction, double bonds in the oil were converted to epoxidized oil. Maximum oxirane content of 3.60% was achieved at 4 h reaction time and temperature of 60°C. Hence, the results of this study indicate that the optimal operating condition for epoxidation reaction was achieved at a reaction time of 4 h and a temperature of 60°C.

Another important finding was that at higher reaction times and temperature than the optimal conditions, it will result lower oxirane content. A possible explanation for this observation might be that higher reaction times and temperature favour a high rate of oxirane ring opening thereby producing epoxidized oil with lower oxirane content (Purwanto et al., 2006). Therefore, side reaction products may be formed as the oxirane ring may be decomposed due to reaction mixture contain materials that are likely to react with the oxirane rings such as sulfuric acid, acetic acid and water (Milchert and Smagowicz, 2009).

Reaction temperature higher than 60°C result in a lower lower oxirane content indicated by a reduced amount of oxirane content with reaction time. This result may be explained by the fact that epoxidation reaction using peroxy acid in this case peroxyacetic acid is highly exothermic (Millchert and Smagowicz, 2009). Hence, higher temperature during the epoxidation may cause the decomposition rate of epoxy groups to be higher than the formation rate. As a result, lower epoxy groups will be produced.

Conclusion

This research report has considered the possibility to use *X. americana* seed oil as a potential feedstock for polyol production to increase its economic value. The optimal condition for epoxidation reaction using peroxyacetic acid was achieved at a reaction time of 4 h and temperature of 60 °C. That means, if we epoxidized *X. americana* seed oil following these conditions then subjected the epoxidized oil to hydroxylation reaction, definitely the polyol to be obtained shall be the one with high hydroxyl value (most important requirement for good polyols).

RECOMMENDATION

It is recommended that further research should be undertaken to perform hydroxylation reaction of the epoxidized oil to produce polyol, to perform the epoxidation reaction using a metal catalyst, and to monitor the unsaturation and epoxy groups in the epoxidized oil by fourier transform infrared (FTIR) spectroscopy.

REFERENCES

Arbonnier M (2004). Trees, shrubs and lianas of West African dry zones. Cirad Margraf Publishers, The Netherlands. pp. 189-426.

Boutevin B, Caillol S, Desroches M, Boutevin G, Loubat C, Auvergne R (2012). Synthesis of new polyester polyols from epoxidized vegetable oils and biobased acids. Eur. J. Lipid Sci. Technol. 114(12):1447-1459.

Barret LW, Sperling LH, Murphy CJ (1993). Naturally functionalized triglyceride oils in interpenetrating polymer networks. J. Am. Oil Chem. Soc. 70(5):523 534.

Cheng OT, Rayson Y (1999). Degredation kinetic of biofluids. Int. J. Lubri. 5:45-50.

Chian KS, Gan LH (1998). Development of rigid polyurethane foam from palm oil. J. Appl. Polym. Sci. 68(3):509-515.

Desroches M, Escouvois M, Auvergne R, Caillol S, Boutevin B (2012). From vegetable oils to polyurethanes: synthetic routes to polyols and main industrial products. Polym. Rev. 52:38-79.

Dinda S, Patwardhan AV, Goud VV, Pradhan NC (2008). Epoxidation of cottonseed oil by aqueous hydrogen peroxide catalysed by liquid inorganic acids. Bioresour. Technol. 99(9):3737-3744.

Eromosele CO, Eromosele IC (2002). Fatty acid compositions of seed oils of *Haemotostaphisbarteri* and *Ximenia americana*. Bioresour. Technol. 82(3):303-304.

Eromosele IC, Eromosele CO, Innazo P, Njerim P (1993). Studies on some seeds and seed oils. Bioresour. Technol. 64:245-247.

Flora EF (2011). Property of polyurethane: from Soy-derived Phosphate Ester. World Academy of science. Eng. Technol. 76:235-238.

Goud VV, Patwardhan AV, Dinda S, Pradhan N (2007). Kinetics of epoxidation of Jatropha oil with peroxyacetic acid and peroxyformic acid catalysed by acidic ion exchange resin. Chem. Eng. Sci. 62(15):4065-4076.

Guo A, Zhang W, Petrovic ZS (2006). Structure-property relationships in polyurethanes derived from soybean oil. J. Mater. Sci. 41(15):4914-4920.

Heidbreder A, Hofer R, Grutzmacher R, Westfechtel A, Blewtt CW (1999). Oleochemical products as building blocks for polymers. Fett/Lipid 101(11):418-424.

Keay RWJ (1989). Trees of Nigeria. Claredon Press, Oxford. P. 367.

Ketaren S (2005). Edible oils and fats, UI-Press, Jakarta. P. 145.

Lligadas G, Ronda IC, Galia M, Cadiz V (2006). Novel silicon-containing polyurethanes from vegetable oils as renewable resources.Synthesis and properties. Biomacromolecules 7(8):2420-2426.

Milchert E, Smagowicz A (2009). The influence of reaction parameters on the epoxidation of rapeseed oil with paraceticacid. J. Am. oil Chem. Soc. 86(12):1227-1233.

Narine SS, Kong X, Bouzidi L, Sporns P (2007a). Physical properties of polyurethanes produced from polyols from seed oils. I. Elastomers. J. Am. Oil Chem. Soc. 84:55-63.

Narine SS, Kong X, Bouzidi L, Sporns P (2007b). Physical properties of polyurethanes produced from polyols from seed oils: II. Foam. J. Am. Oil Chem. Soc. 84:65-72.

Niemeyer T, Patel M, Geiger E (2006). "A Further Examination of Soy-Based Polyols in Polyurethane Systems". Salt Lake City, UT: Alliance for the Polyurethane Industry Technical Conference. P. 2-4.

Pan X, Webster DC (2012). New Biobased High Functionality Polyols and Their Use in Polyurethane Coatings. Chemsuschem. 5(2):419-429.

Pechar TW, Sohn S, Wilkes GL, Ghosh S, Frazier CE, Fornof A, Long TE (2006). Characterization and comparison of polyurethane networks prepared using soybean-based polyols with varying hydroxyl content and their blends with petroleum-based polyols. J. Appl. Polym. Sci. 101(3):1432-1443.

Petrovic Z, Guo A, Javni I (2003). Process for the preparation of vegetable oil based polvols and electroinsulating casting compounds created from vegetable oil-based polyols, United States Patent, 6:354-573.

Petrovic ZS (2008). Polyurethanes from vegetable oils. Polym. Rev. 48(1):109-155.

Petrovic ZS, Zhang W, Javni I (2005). Structure and properties of polyurethane prepared from triglyceride polyols by ozonolysis. Biomacromolecules, 6(2):713-719.

Purwanto E, Fatmawati A, Setyopratomo P, Junedi, Rosmiati M (2006). Influence of epoxidation reaction period and temperature on the quality of polyol synthesized from soybean oil', in Proceedings of the 13th Regional Symposium on Chemical Engineering 2006 — Advanced in Chemical Engineering and Biomocular Engineering. Nanyang Technological University, Singapore. pp. 277-279.

Siggia S (1963). Quantitative organic analysis, 3rd edn, John Wiley & Sons, hic, New York. P. 103.

Sudarmaji S, Haryono B, Suhardi D (1997). Analysis procedure for food and agricultural materials, 4th edn, Liberty, Yogyakarta. P. 234.

Tu Y, Kiatsimkul P, Suppes G, Hsieh F (2007). Physical properties of water- blown rigid polyurethane foams from vegetable oil-based polyols. J. Appl. Polym. Sci. 114(5):2577-2583.

Veenendaal B (2007). Renewable content in the manufacture of polyurethane polyols-An opportunity for natural oils. Polyurethanes Mag. Int. 4(6):352-359.

Wang C, Yang L, Ni B, Wang L (2009). Thermal and mechanical properties of cast polyurethane resin based on soybean oil. J. Appl. Polym. Sci. 112(3):1122-1127.

7

Influence of wood humidity of three species (*Prosopis africana*, *Anogeissus leiocarpa* and *Tectona grandis*) on the production of charcoal by the traditional wheel

Akossou A. Y. J.[1], Gbozo E.[1], Darboux A. E.[1] and Kokou K.[2]

[1]Département d'Aménagement et de Gestion des Ressources Naturelles, Faculté d'Agronomie, Université de Parakou, BP 123, Parakou Bénin.
[2]Laboratoire de Botanique et d'Ecologie, Faculté des Sciences, Université de Lomé, BP 1515 Lomé Togo

Charcoal production is an increasing practice which however, remains traditional in Benin despite the introduction of new techniques of carbonization. The objective of this study is to evaluate the effect of species humidity on the production process, in the perspective of improving the yield in relation to farming practices. To achieve this, three wood species appearing in four humidity conditions were used in randomized complete block design with two replications. The humidity of wood negatively affects the process; while an increased density had a positive effect on the process. The mixture of species and mixture of wood at different state of humidity gave good results compared to the use of a single species in the dry state. The yield of dry wood was estimated to an average of 31.38% for *Prosopis africana*, 21.5% for *Anogeissus leiocarpa*, 15.25% in wet or mixture of wood of different humidity and 21.5% in semi-wet or dry for *Tectona grandis* and 22% for the mixture of species.

Key words: Carbonization, charcoal, wood humidity, mass yield, Benin.

INTRODUCTION

Biomass which is the fourth largest energy source in the world, provides about 13% of world's energy consumption (Hall et al., 2000). In general, wood represents about 86% of household energy consumption (Broadhead et al., 2001) and the demand is expected to rise to 45% over the next 30 years due to the population growth and the increasing needs in sub-Saharan Africa (De Montalembert and Clement, 1983). For many urban areas in developing countries, charcoal provides a reliable, convenient and accessible source of energy in meeting household requirements as well as a variety of other needs (Goldstein, 1981; Demirbas, 2001). Charcoal also has unique cooking properties that make households

go for it even when other fuels are also available (Seidel, 2008).

Wood constitutes the main fuel for Beninese households and provides 89% of domestic energy. Ninety three percent (93%) of the Beninese population use wood energy in rural areas, against 80% in urban areas (Dossou, 1992, 1996). Daily consumption of charcoal is estimated at 0.35 kg per person in urban areas as against 0.15 kg per person in rural areas (Agbo and Mama, 2001). About 3 million tons of fuelwood are consumed each year in Benin; and due to demographic growth, the need for fuel wood is expected to increase. Faced with growing demand for fuelwood and its lack, solutions differ from one author to another. According to Malty (1999), wood transportation is the main factor in the increasing cost of wood during periods of scarcity, as consumers have to travel long distances in search for them; logically its price increases. Wood as energy source is no longer competitive with other sources of energy such as oil, gas, etc. The pressure on forest resources is naturally regulated by the law of supply and demand, without any particular action. But the author recognizes that the law of supply and demand cannot effectively regulate the pressure on forest resources, so producers should be able to influence it by imposing the prices. However, this is often not the case.

De Gier (1989) has proposed three alternatives to fuel wood crisis such as: the identification of other sources of energy; increasing the supply of wood and adoption of energy conservation methods. To prove the ineffectiveness of the first solution in developing countries, Anderson and Firhwick (1984), reported that as long as wooded areas and forests have not disappeared, the costs of cooking with wood fire or coal are lower than those of other commercial fuels as oil and gas. Compared to the second solution, efforts to promote community and private plantations for the production of wood as source of energy (coal, firewood) marketed in Benin, comes exclusively from the natural fallow or wastelands (Ogouvidé et al., 2006).

In addition, the requirements in quality (heavy and slow burning coal) are actualized by species characterized by great density, slow growing and therefore particularly vulnerable to exploitation, this being a serious constraint for producers (Girard, 2002). The most used species in Benin are among others *Anogeissus leiocarpa*, *Pterocarpus erinaceus*, *Prosopis africana*, *Vitellaria paradoxa* and *Pseudocedrela kotschyi* (Idjigbérou, 2007). The same species are considered the most exploited in the Sudanian savanna and dry forest areas of Togo (Kokou et al., 2009). Adam (1990) and Schenkel et al. (1997) reported that "The Wheel" is still the most prevalent traditional method for charcoal production in developing countries despite the introduction of new techniques of carbonization through development projects.

In Africa, a limited number of people consider charcoal production to be their main economic activity, while a majority engage only occasionally as a means to generate income, particularly in times of financial stress, such as when making large payments for medical bills; funeral expenses; food supplies in the event of poor harvests; marriage ceremonies or school fees (BTG, 2010). Other reasons for non-adoption of new production techniques are the cost of investment, difficulties of installation and availability of equipments. Therefore, practices and current production of charcoal are characterized by a low yield estimated at 15 and 20% (150 to 200 kg of charcoal per ton of wood) (Girard, 2002). However, the yields obtained during the

carbonization of wood are the best possible, as the producer must work with rigor and professionalism.

According to FAO (1983), it is possible to save nearly 2 million cubic meters of wood, thereby increasing efficiency in the production of charcoal with the use of improved wheel. With improvements to the traditional wheel on the number of vents reduced to 4 instead of 7, the stacking of wood, the choice of species with high calorific value and the tightness of the wheel, brought the coal mass yield from 11 to 20% (Mama, 2006). By entering this logic to improve yields of charcoal produced by the traditional wheel to allow producers to generate more income and reduce pressure on forest resources, this paper aims to assess the effect of the wood humidity content of three species based on the endogenous production practices.

MATERIALS AND METHODS

Experimental site

The experiment was carried out on a site located in the district of Savè in central Benin. The district of Savè is a transition zone between the Guinean climate with four seasons and the Sudanian climate with two seasons. It enjoys a climate with contrasting seasons characterized by noticeable fluctuations in temperature and an average annual rainfall of 1200 mm. The average annual temperature is 27°C; the average relative humidity is 31% for the minimum and 98% for the maximum. The soil type is tropical ferruginous concretions of crystalline basement or hydromorphic along the river.

Experimental design

The experimental design used was a randomized complete block design with two replications. Controlled factors are:

(i) The species of wood with four levels, these species are among the species most used in the carbonization process in Benin (A. leiocarpa, P. africana, T. grandis and their mixture in proportions of 1/3);
(ii) Four moisture conditions, wet wood (humidity approaching 45%), semi-dry wood (humidity around 30%), dry wood (humidity approaching 15%) and a mixture of wood at different humidity levels.

The combination of the levels of two factors (humidity and species) gives a total of 4 × 4 = 16 objects. Each object was repeated twice. Thus, the implementation of this experimental design required 32 outbreaks of carbonization. For each outbreak of carbonization, the load volume is a stere of wood. In this context, the wood was cut in pieces of one meter long. Characteristics of wood used in the experiment are presented in Table 1. The carbonization was carried out according to the method of the traditional wheel.

Data collected

The data collected include the duration of carbonization, the total weight of the charcoal obtained, the weight of unburnt wood remaining; the mass yield calculated from the mass of a stere of green wood and mass of charcoal obtained on one hand; the mass of the stere of anhydrous wood and mass of charcoal obtained on

Table 1. Mean values and standard errors (in parentheses) of humidity, diameter of the logs, and mass (kg) of a stere of wet wood and anhydrous wood for different levels considered.

| Species | Humidity (%) | | Diameter (cm) | Total mass load / stere (kg) | Mass of dry wood / stere (kg) |
	State	Value			
A. leiocarpa	Wet	40.26 (0. 52)	15.98 (0.16)	757.73 (73.02)	453.10 (47.54)
	Semi-humid	24.00 (0.06)	17.76 (0. 59)	759.13 (35.85)	576.97 (26.86)
	Dry	12.63 (0.32)	15.86 (0.30)	575.36 (9.85)	502.77 (10.42)
	Mixed	27.61 (0.57)	16.02 (0.01)	708.94 (9.28)	513.32 (10.72)
P. africana	Wet	38.59 (0.16)	15.98 (0.30)	913.40 (5.06)	560.96 (1.72)
	Semi-humid	24.52 (0.68)	16.02 (0.00)	864.02 (35.77)	652.43 (32.84)
	Dry	13.07 (0.21)	16.46 (0.10)	633.41 (11.26)	550.61 (8.47)
	Mixed	24.70 (0.11)	15.82 (0.22)	784.55 (9.55)	590.79 (6.34)
Tectona grandis	Wet	41.14 (0.80)	14.86 (0.78)	625.08 (5.77)	367.91 (1.58)
	Semi-humid	24.12 (0.26)	14.32 (0.78)	547.22 (14.24)	415.31 (12.20)
	Dry	12.55 (0.32)	15.68 (0.53)	478.09 (0.36)	418.11 (1.83)
	Mixed	26.08 (0.96)	15.80 (0.15)	531.63 (21.70)	392.79 (10.96)
A. leiocarpa + T. grandis + P. africana	Wet	37.99 (1.82)	15.38 (0.01)	836.39 (15.09)	518.98 (24.55)
	Semi-humid	23.69 (0.26)	15.50 (0.36)	698.68 (1.13)	533.16 (0.98)
	Dry	12.53 (0.17)	13.93 (0.30)	682.16 (190.60)	596.36 (165.56)
	Mixed	25.60 (0.70)	14.42 (0.20)	719.02 (20.60)	534.95 (9.80)

the other hand. Yields are given by the following formulas:

$$RM_{bh} = M_{ca} / M_{bh} \text{ and } RM_{ba} = M_{ca} / M_{ba}$$

with

RM_{bh} = Wet mass yield

RM_{ba} = Dry mass yield

M_{ca} = Mass of dry coal

M_{bh} = Mass of wet wood

M_{ba} = Mass of dry wood

The mass yield also known as yield or yield weight on dry wood is only valid for a rigorous comparison (Girard, 1992), because from a stere of green wood of the same weight and different humidity, the quantities of charcoal they produce will be different, only dry wood being transformed into charcoal (Mundhenk et al., 2010).

Statistical analysis

Data analyses were performed using SAS software. Factors (species and moisture) were subjected to analysis of variance with two factors of classification, as response variables were the duration of carbonization, the weight of charcoal obtained, unburnt weight and yield. Duncan's tests comparing mean was used to

detect possible differences between the levels of a factor.

RESULTS

Quantity of charcoal

Quantities of charcoal obtained per a stere of wood vary from one humidity state to another according to the species. The highest values of quantities of charcoal were observed with wood in the semi-wet of species *A. leiocarpa*, *P. africana*, *T. grandis*, and with wet wood for the mixture of species. The lowest values were obtained with dry wood for the species *A. leiocarpa* and wet wood for species *P. africana*, *T. grandis* and mixture of species. However, comparing the values revealed no difference in the degree of wood humidity regardless of the species, except *T. grandis* which had a higher quantity of charcoal at semi-wet state of 98.15 kg/stere.

Whatever the state of humidity, the quantity of charcoal obtained by stere can be estimated to 109.11 kg/stere for *A. leiocarpa*, to 182.9 kg/stere for *P. africana* and 116.89 kg/stere for the mixture of the three species. After carbonization of a stere of wood, *P. africana* presented the highest value in the quantity of charcoal. The mixture of species provided an intermediate value. *T. grandis* had the lowest values (Table 2). For a given species or a given state of humidity, the values associated with the same letter were not statistically different from each other

Table 2. Mean values and standard deviations (in parentheses) of mass (in kg) of charcoal produced per stere of wood with different considered level of humidity.

Species	Humidity			
	Wet	Semi - wet	Dry	Mixture of humidity
A. leiocarpa	108.70^{aA} (8.9)	118.00^{aA} (9.9)	104.25^{aAB} (2.47)	105.50^{aA} (3.50)
P. africana	165.75^{aB} (6.01)	191.85^{Ab} (13.22)	190.20^{aC} (9.48)	183.80^{aB} (1.13)
T. grandis	52.55^{aC} (3.75)	98.75^{bA} (3.89)	80.00^{aB} (5.66)	62.50^{aC} (3.54)
A. leiocarpa + T. grandis + P. africana	108.05^{aA} (8.13)	114.00^{aA} (4.24)	125.00^{aA} (8.49)	120.50^{aA} (0)

Table 3. Mean values and standard deviations (in parentheses) of mass (in kg) of unburnt wood per stere of wood for different levels considered.

Species	Humidity			
	Wet	Semi - wet	Dry	Mixture of humidity
A. leiocarpa	11.00^{aA} (2.12)	6.90^{aA} (1.27)	1.25^{aA} (1.77)	6.75^{aA} (3.18)
P. africana	3.50^{aA} (4.24)	10.00^{aA} (4.24)	10.50^{aA} (3.54)	15.25^{bA} (3.89)
T. grandis	10.35^{aA} (1.67)	4.55^{aA} (2.05)	6.00^{aA} (2.83)	13.75^{aA} (3.89)
A. leiocarpa + T. grandis + P. africana	12.50^{aA} (2.12)	6.50^{aA} (0.71)	6.75^{aA} (1.48)	8.50^{aA} (1.43)

at the level α = 5%. Lowercase letters denote the comparisons between humidity state for a given species, and the capital letters indicate the comparisons between species of a given humidity level.

Quantity of unburnt wood

For the same humidity, the quantities of unburnt wood did not differ from one specie to another. Similarly, for the same species, the humidity of wood had no influence on the mass of unburnt wood species except *P. africana*, whose blend of wood with varying degrees of humidity of wood had a value higher than the other degrees of humidity (15.25 kg/stere) (Table 3).

For a given species or a given state of humidity, the values associated with the same letter were not statistically different from each other at the level α = 5%. Lowercase letters denote the comparisons between humidity state for a given species, and the capital letters indicate the comparisons between species of a given humidity level.

Duration of carbonization

The species of wood had basically no effect on the duration of carbonization. On the contrary, humidity influenced this duration. It was estimated at about four days for wet wood, semi wet and mixture of humidity and

three days for dry wood shown in Figure 1. In the dry state, the species *P. africana* had duration of carbonization relatively higher than other species (about four days).

Yield of charcoal

The analysis of results showed that there was no interaction between the factors of humidity and species regardless of the yield considered (p = 0.09, p = 0.07, p = 0.08 respectively of yields of wood put in the oven, dry wood and dry wood minus wood unfired). The evaluation of the influence of the species reveals that *P. africana* presented highest yield of dry wood. It was estimated at 31.38%.

The yield of dry wood of *A. leiocarpa* was estimated at 21.50%; for *T. grandis*, it was estimated at 18.50% and for the mixture of the three species, it was estimated at 21.86%. The assessment of the effect of humidity in each species showed that the efficiency increased when the wood humidity decreased. The highest yield of hydra wood was obtained in the dry state (Table 4).

DISCUSSION

The efficiency of a wheel is determined by a number of factors, including wood humidity, the dimensions of pieces of wood, the size of the wheel, the species of

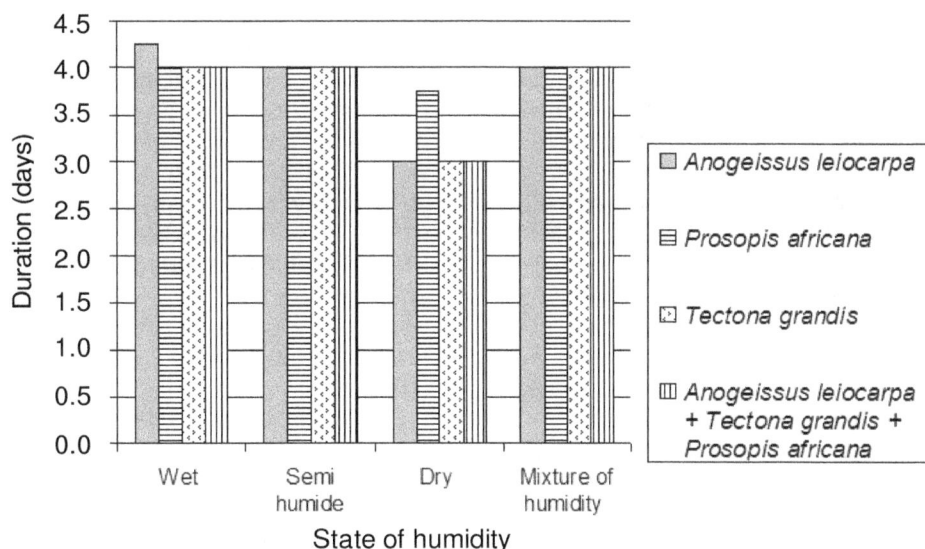

Figure 1. Duration of carbonization of each species according to the state of humidity.

Table 4. Mean values of yields of wet wood, dry wood and dry wood minus unburnt wood for different combinations levels considered.

Species	Humidity		Yield (%)		
	State	Value (%)	Wood put in the oven	Dry wood	Dry wood minus unburnt wood
A. leiocarpa	Wet	40.26 (0.52)	14.50[b]	24.00[a]	24.50[a]
	Semi-wet	24.00 (0.06)	15.50[b]	20.50[a]	20.50[a]
	Dry	12.63 (0.32)	18.50[a]	21.00[a]	21.00[a]
	Mixture of humidity	27.61 (0.57)	15.00[b]	20.50[a]	21.00[a]
P. africana	Wet	38.59 (0.16)	18.50[b]	30.00[a]	30.00[a]
	Semi-wet	24.52 (0.68)	22.00[b]	29.50[a]	30.00[a]
	Dry	13.07 (0.21)	30.00[a]	34.50[a]	35.50[a]
	Mixture of humidity	24.70 (0.11)	23.50[a]	31.50[a]	32.00[a]
T. grandis	Wet	41.14 (0.80)	8.50[c]	14.50[b]	14.50[b]
	Semi-wet	24.12 (0.26)	18.00[a]	24.00[a]	24.00[a]
	Dry	12.55 (0.32)	17.00[a]	19.00[a]	19.00[a]
	Mixture of humidity	26.08 (0.96)	12.00[b]	16.00[b]	16.50[ab]
A. leiocarpa + T. grandis + P. africana	Wet	37.99 (1.82)	13.00[a]	21.00[a]	21.50[a]
	Semi-wet	23.69 (0.26)	16.50[a]	21.50[a]	21.50[a]
	Dry	12.53 (0.17)	19.50[a]	22.50[a]	22.50[a]
	Mixture of humidity	25.60 (0.70)	17.00[a]	23.00[a]	23.00[a]

For a given species, the yield values associated with the same letter were not statistically different from each other at the level α = 5%.

wood used in the wheel, climatic conditions as humidity of the air, wind, etc. (Schenkel et al., 1997). In this study, the wood humidity and species were considered. Climatic conditions were considered using an average climatic condition of the subject area, because the study site is located in the transition zone of Benin, between the Sudano-Guinean and Sudanian zone.

The influence of physicochemical characteristics of wood on the process of charcoal production has been reported by several authors including Larzilliere (1978),

FAO (1983) and Bugnicourt and Mhlanga (1985). The quantity of charcoal obtained depends on physico-chemical characteristics of wood used as raw material. Thus, the quantity of charcoal obtained with P. africana, the heavier wood between the species investigated in this study, was different from other species. P. africana presented the quantities of charcoal ranging from 165 to 191 kg/stere. A. leiocarpa (heavyweight wood) presented an intermediate situation with quantities of charcoal per stere ranging from 104.25 to 118 kg/stere. T. grandis, the lighter species had the lowest quantities of charcoal per stere with values ranging from 52.55 to 98.75 kg/stere.

As a result of the lack of availability of raw material, producers in their practice often mixed species in the outbreaks of carbonization. The mixture of species of different densities can increase to some extent the quantity of charcoal produced. The quantities of charcoal obtained per stere (108 to 125 kg) are greater than those obtained with A. leiocarpa and T. grandis. Similar results were obtained by Mundhenk et al. (2010) in community forest Sambande, Kaolack in Senegal (102 kg/stere with 45% of humidity).

Similarly, a stere of wood in Madagascar gave 90 kg of charcoal (Bugnicourt and Mhlanga, 1985). The results obtained are relatively higher than those found by Ogouvidé (2010), in the same study area for the species Senna siamea, and A. leiocarpa which produced 62 and 72.73 kg/stere with traditional wheel respectively. Although the yield with the mass of wood put into the oven is to be considered for calculation, the mass yield of dry wood is preferable for comparisons, because it eliminates the effect of humidity factor especially since this factor strongly influences the performance.

Thus, differences in yields between species would be even greater if this factor is taken into account. The quantities of dry wood per stere for each species appear on average in a proportion of 75%. P. africana, showed the highest yield of dry wood which is an average of 31.38%. A. leiocarpa and the mixture of species gave intermediate values of yields of anhydrous wood with average of 21.5 and 22%, respectively. T. grandis considered as a light wood among the species studied appears with the lowest yield which is 15.25% in wet and mixed states and 21.5% in the semi-wet and dry states. The yield of T. grandis also known as Teak, is relatively high compared to the yield found in the literature on teak wood carbonization.

Coulibaly and Lessard (2006) evoked in an experimental study of charcoal production of commercial timber from thinnings of teak plantations in Tene forest in Côte d'Ivoire, a yield on dry wood of 15%. Ogouvidé and Mama (2007) by testing the adaptability of the circular type of Casamance wheel on the wood species A. leiocarpa in the charcoal production areas of Central Benin, found a yield of 25 and 19% for the Casamance and traditional wheel, respectively. In the same area and to the same species, PBF2 (2009) referred to a yield of

22%. Mundhenk et al. (2010), indicate in a general case, a yield of 15-17% for the traditional wheel. However, there may be exceptions where yields are up to 45% (Robinson, 1988). In Senegal CILSS (2008), PERACOD (2007), and World Bank (2008) reported yields of dry wood from 30 to 35%. In Chad, Hughes (2001) gave a yield of dry wood from 25 to 30% for the Casamance wheel and 18 to 23% for traditional wheel. CILSS-PREDAS (2004) reported a yield of dry wood from 16 to 20% for the traditional wheel and 20 to 40% for improved wheel. The same trends of yields are noted by other authors including FAO (1983), Leclerc (2002) and Dejonc (2003).

Strong disparities between the yields can be explained by several factors. On the whole, the dry wood with humidity levels below 15% gives the best yields while the wet wood with humidity above 20% have average yields. The differences between species were also mentioned by other authors. Rameau (2009) reported values as 28, 26 and 22% yield respectively for charcoal of beech, oak and fir. These differences may also come from the expertise of coal makers whose practices can help offset some of the technical constraints of the density or the humidity of wood. For example, the mastery of the difficulties of firing charges of wet or semi-wet wood especially for dense species reduces the consumption of a significant portion of wood. This may explain some of the results including equality between the quantities of charcoal obtained for Teck and P. africana in semi-wet state. Indeed, the consumption of wet and semi-wet wood of dense species leads to a significant loss of wood during carbonization process. However, in the dry state, while hardwoods are difficult to ignite, they burn slowly and are therefore more suited to heating. This difficulty of firing wet or semi-wet wood results in an increase in the duration of carbonization.

Wood carbonization is a three stages process. The first step is the drying of wood at 100 °C, during this step, the complete evaporation of the water contained in the timber to make it dry is performed. During the second stage, the temperature of the dry wood is raised to about 280 °C. In the last stage, is exothermic decomposition of the timber which is stopped at about 400 °C when there is no external energy (FAO, 1983). The first step will be much longer as there is more quantity of water to be evaporated. If the wood is already dry, this step is greatly reduced and the duration of carbonization is consequently shorter.

The duration of carbonization of dry wood is estimated at three days and four days per stere of wet wood. There is a wide disparity in the duration of carbonization reported in the literature. Mundhenk et al. (2010) noted in their study on the traditional wheel, two to three days duration of carbonization respectively of 30 and 50% of humidity in community forest Sambandé in Senegal. Briane and Haberman (1984) gave in France the duration of two days for a weel of 7 stere with a degree of wood humidity ranging from 29.9 to 42.9%. Hibajene (1994) mentioned in Zambia the duration of twenty-seven to thirty-one days to 6 stere and wood degrees of humidity ranging from 28.2 to 34.4%.

Schenkel (1991), mentioned 43 days for 15 stere of wood with the wood humidity ranging from 70.2 to 118.5%. This disparity is due to the fact that experiments are carried out in highly variable conditions (experimental sites, operator qualification, precision measurements). There is also a lack of information on these tests carried out in situ including characteristics such as humidity, size of logs and species.

The influence of humidity on the duration of carbonization has been mentioned by other authors including Larzilliere (1978), FAO (1983), Leclerc (2002), Jeffrey and Stedford (1983), CRA-Wallonie (2009) and Girard (1992). Mildred and Wilfrid (2003) made nuance and pointed out that wood humidity affects much more than the density of the species. For these authors, the humidity above 20% is already considered high and remarkably influences the yield of charcoal obtained. Comte (1975), and Roos and Roos (1979) pointed out, however, that the dry wood can have up to 25% of humidity. This may explain the differences in performance observed in the four states of humidity. CRA-Wallonie (2009) showed the strong influence of humidity on the yield; and the authors qualify as the secondary effects, other physical parameters as density, size, shape, etc. They also mentioned that humidity, as the heating rate increases, have a negative impact on the flow of mass and energy. This emphasizes the important aspect of the calorific value of charcoal. Indeed, consumers are looking out more for a heavy and hard charcoal with a slow consummation rate. This leads producers to make a selection of species to be cut for charcoal production.

Selective cutting means that certain trees providing good quality charcoals are selected and cut for charcoal production. Preference and suitability of trees used for charcoal production varies with size, availability and accessibility of the tree species. Large tree species (>20 cm diameter) with high caloric values are the most preferred, due to the large quantity of dense and hard charcoal they produce (Beukering et al., 2007).

Upon completion of the experiment, a survey carried out among producers in the study area revealed that charcoals are obtained from dense species such as *P. africana*, *P. erinaceus*, *V. paradoxa*, *Burkea africana*, *A. leiocarpa* and *Senna siamea*. Ogouvidé (2010) showed that the charcoal from most of these species has a calorific value higher than 90% in accordance with the standards of 52 to96% defined by Schenkel et al. (1997).

Conclusion

Production of charcoal in Benin is a rural activity, which is increasing due to the demand of charcoal which is increasing widely in major urban centers. Although, this activity has a great deal of income contribution of coalman, it is characterized by a low yield with a high pressure on forest resources. Looking for an improvement of the production process in relation to farming practices helped to highlight the influence of wood humidity and species used in the process. The humidity of wood negatively affects the process, while the density has a positive effect on the process. The mixture of species and mixture of wood at different state of humidity also yielded good results than the use of a single species in the dry state. The best yields are obtained in the dry state with about 13% humidity. These results are close to the yield provided by the new carbonization techniques such as Casamance wheel whose performance varies from 30 to 35% (PERACOD, 2007).

REFERENCES

Adam A (1990). Etude sur la filière du charbon de bois au Burundi. Ministère de l'Energie et des Mines; GTZ: Programme spécial Burundi. P. 150.

Agbo J, Mama VJ (2001). Synthèse et analyse des données sur le bois-énergie en République du Bénin. Rapport Projet GCP/INT/679/EC. P. 55.

Anderson D, Firhwick R (1984). Fuelwood consumption and deforestation in Africa countries. In World Bank Staff Working Paper No. 704.

Beukering PV, Kahyarara G, Massey E, di Prima S, Hess S, Makundi V, van der Leeuw K (2007). Optimization of the charcoal chain in Tanzania. Amsterdam, Institute for Environmental Studies, Netherlands. P. 46.

Briane D, Haberman A (1984). Essais comparatifs de six systèmes de carbonisation artisanale. Association Bois de Feu, Paris. P. 188.

Broadhead J, Bahdon J, Whiteman A (2001). Wood fuel consumption modelling and results. Annex 2 in Past trends and future prospects for the utilization of wood for energy. Working Paper No: GFPOS/WP/05, Global Forest Products Outlook Study. FAO, Rome. P. 5.

BTG Biomass Technology Group (2010). Making charcoal production in Sub Sahara Africa sustainable. Netherlands Programmes Sustainable Biomass. P. 59.

Bugnicourt J, Mhlanga L (1985). Energie populaire dans le tiers-monde. Cahiers d'étude du milieu et d'aménagement du territoire 20-21-22. Dakar : ENDA. P. 405.

CILSS (2008). Impacts des investissements dans la gestion des ressources au Sénégal : Synthèses des études de cas. Dakar: République du Sénégal, Institut Sénégalaise des Recherches Agricoles, Universiteit Amsterdam. P. 45.

CILSS-PREDAS (2004). Atelier régional sur la capitalisation de l'expérience sahélienne sur la carbonisation et l'agglo-briquetafe. Bamako 15-18 juin, actes de l'atelier : document de synthèse, 27 p.

Comte DE (1975). Charbon de bois A qui fait pour les entreprises peu importantes: Une Formation Illustrée Manuel. Geneva: Bureau International de Travail. P. 26.

Coulibaly B, Lessard J (2006). Expérimentations de production de charbon de bois commercial à partir des produits d'éclaircies de plantations de Teck dans la forêt de la TENE. Séminaire FORAFRI de Libreville. P. 7.

CRA-Wallonie (2009). Carbonisation des matières lignocellulosiques: modélisation des flux massiques et énergétiques. Projet de recherche. Available at. http://www.cra.wallonie.be/index.php?l=fr&page=19&id=41. As at26/08/2012

De Gier (1989). Woody biomass estimation natural wood and shrulands. In 3rd international course on the design of community forestry. Lecture notes block B2 Rural energy and community forestry. P. 120.

De Montalembert MR, Clément J (1983). Fuelwood supplies in the developing countries. FAO. Forestry Paper 42, Rome. Available at http://www.fao.org/docrep/X5329E/X5329E00.htm. As at 26/08/2012.

Dejonc E (2003). Précis illustré de mécanique - La mécanique pratique - Guide mécanicien, J. Rotschild éd. P. 576.

Demirbas A (2001). Carbonization ranking of selected biomass for charcoal, liquid and gaseous products. Energy Convers. Manag. 42: 1229-1238.

Dossou B (1992). Problématique et politique du bois-énergie au Bénin. Thèse de Doctorat Ph.D Université de Laval, Québec. P. 414.

Dossou B (1996). Les foyers économiques et les nouvelles sources d'énergie pour la lutte contre la désertification au Bénin. Communication présentée au Séminaire national sur le thème "Energie domestique et lutte contre la désertification au Bénin : Rôle de la femme". INFOSEC - Cotonou, du 15 au 17 Octobre 1996.

FAO (1983). Techniques simples de carbonisation. Rome. 41:152.

Girard P (1992). Analytical performance tests for charcoal making technics and equipement. Holt Roh-Werkst. 50:479-484.

Girard P (2002). Quel futur pour la production et l'utilisation du charbon d bois en Afrique? Unasylva (FAO), 211 (La dendroénergie), 30-34.

Goldstein IS (1981). Organic Chemical from Biomass. (CRC press, Boca Raton FL). P. 310.

Hall DO, House JI, Scrase I (2000). Overview of biomass energy, in Industrial Uses of Biomass Energy – The Example of Brazil, Rosillo-Calle, F., Bajay, S. and Rothman, H. (eds), Taylor & Francis, London. P. 26.

Hibajene SH (1994). Assessment of earth kiln charcoal production technology. Stockholm Environment Institute. P. 39.

Hughes D (2001). L'amélioration de la carbonisation dans les zones d'interventions de l'Agence Energie Domestique Environnement (AEDE). Rapport première mission. P. 32.

Idjigbérou ES (2007). Impact de la production de charbon de bois sur la diversité floristique des formations végétales du Centre et du Nord Bénin. Mémoire d'ingénieur de la Faculté d'Agronomie de l'Université de Parakou. P. 175.

Jeffrey LW, Stedford W (1983). Team Compares Charcoal Production Methods, VITA News. pp. 8-11.

Kokou K, Nuto Y, Atsri H (2009). Impact of charcoal production on woody plant species in West Africa: A case study in Togo. Sci. Res. Essay 4(9):881-893.

Larzilliere M (1978). Notice sur le débit des bois de feu, leur mode de vente et les procédés de carbonisation usités en France. France, Ministère de l'Agriculture et du commerce. P. 61.

Leclerc JR (2002). Optimisation d'un procédé de carbonisation de bois. Le design Experimental. P. 19.

Malty M (1999). Mort annoncée du bois-énergie à usage domestique. Article publié dans la revue Marchéage et gestion de l'environnement (MARGE). P. 8.

Mama VJ (2006). Amélioration des techniques endogènes de production du charbon de bois: Étude des paramètres de conversion de quelques essences calorifiques. Communication présentée à la deuxième édition de l'Atelier scientifique national, Centre Guy-Riobé de Parakou, du 11 au 14 janvier 2006.

Mildred DR, Wilfrid S-J (2008). Le charbon de bois: mythes et réalités. Bulletin trimestriel de la Care-Haïti et du Bureau des Mines et de l'Energie. No 8, édition Internet : Available at http://www.olade.org.ec/haiti/synergie/numero8/charbonjav.html As at 26/08/2012

Mundhenk P, Gomis O, Coumba MSY (2010). Comparaison des rendements de production de charbon de bois entre la meule traditionnelle et la meule Casamance dans la forêt communautaire de Sambandé. Programme pour la promotion des energies renouvelables, de l'électrification rurale et de l'approvisionnement durable en combustibles domestiques (PERACOD), P. 21.

Ogouvidé TF (2010). Mise au point d'une meule à cheminée de type casamançaise «Casa GV» adaptée aux zones de production du charbon de bois au Bénin. Rapport d'étude. P. 21.

Ogouvidé TF, Adégbola YP, Mama VJ (2006). Contraintes à la plantation du bois-énergie et à l'utilisation des énergies alternatives en milieu urbain au Bénin. Rapport d'étude, INRAB. P. 13.

Ogouvidé TF, Mama VJ (2007). Test d'adaptabilité du four casamançais dans les régions de production du charbon de bois du Centre-Bénin. Rapport d'étude. P. 11.

PBF 2 (2009). Réalisation d'action test sur le four casamançais dans les zones de production du charbon de bois au Bénin. Rapport de test du Projet Bois de Feu Phase 2. P. 20.

PERACOD (2007). Document de plan d'aménagement et de gestion de la forêt communautaire de Sambande. Dakar: Programme pour la promotion des énergies renouvelables, de l'électrification rurale et de l'approvisionnement durable en combustibles domestiques (PERACOD). P. 122.

Rameau A (2009). Fabrication du charbon de bois. La science illustrée no. 208. Extrait du Gloubik Sciences, http://www.gloubik.info/sciences/spip.php?article352.

Robinson A (1988). Techniques de carbonisation en Somalie dans la région de Bay. Unasylva 40(159):42-49.

Roos W, Roos U (1979). Survey de Four Simple Systems et Recommandations pour la sélection de Kilns. German À propos Technologie Échange Rapport. P. 49.

Schenkel Y (1991). Essais d'accompagnement des Bateke. Expérimentation de carbonisation par le secteur privé. Rapport, Centre de Recherche Agronomiques Zaïre Trading and Engineering. P. 10.

Schenkel Y, Bertaux P, Vanwijnsbergh S, Carré J (1997). Une évaluation de la technique de carbonisation en meule. Biotechnol. Agron. Soc. Environ. 1(2):113-124.

Seidel A (2008). Charcoal in Africa, importance, problems and possible solutions. Eschborn, GTZ, Household Energy Programme. P. 18.

World Bank (2008). Etude sur la filière charbon de bois au Sénégal. P. 62.

8

Performance and emission characteristics of compression ignition (CI) engine with dual fuel operation (diesel + compressed natural gas (CNG))

E. Ramjee, K. Vijaya Kumar Reddy* and J. Suresh Kumar

Department of Mechanical Engineering, JNTUH College of Engineering, Kukatpally, Hyderabad, Andhra Pradesh, India.

The main objective of this work is to study the performance and emission characteristics of compression ignition (CI) engine using compressed natural gas (CNG) for the following conditions. (i) At constant speed by varying injection pressure and load (ii) Dual fuel combustion phenomenon. The conventional fuels like petrol and diesel for internal combustion engines are getting exhausted at an alarming rate, due to tremendous increase in the vehicular population. Further, these fuels cause serious environmental problems as they release toxic gases into the atmosphere at high temperatures and concentrations. Some of the pollutants released by the engines are un-burnt hydro carbons (UBHC), CO, NOx, smoke and particulate matter. In view of this and many other related issues, these fuels will have to be replaced by alternative and eco-friendly fuels. A 4-stroke, single-cylinder, vertical, stationary diesel engine has been considered for the purpose of experimentation. It is modified suitably to run on the dual fuel mode. CNG gas-air mixer is incorporated on the intake side of the engine. The fuel injection system is also altered so that it injects only the pilot fuel. The engine performance is better on CNG compared to pure diesel up to engine loads of about 75.67%.

Key words: Dual-fuel, emissions, compressed natural gas (CNG), performance.

INTRODUCTION

There is an urgent need to save the conventional fuels like diesel and petrol, so as to reduce the oil import bills of an oil-dependent country and also to mitigate the menace of the environmental pollution. These fuels will have to be replaced by suitable alternative fuels like alcohols and various gaseous fuels (Ghazi, 1980). Among various gaseous fuels, compressed natural gas (CNG), liquid petroleum gas (LPG) and hydrogen are prominent. Dual-fuel operation is found to be one of the prominent methods of conserving diesel and petrol (Dong et al., 2001; Carlucci et al., 2008). Natural gas is found in Free State beneath the earth crest at high pressures and more commonly in association with the crude oil as the most volatile fraction (Mohamed, 2004). About 60% of the natural gas is produced along with the crude oil as an associated gas and rest as the free gas. It primarily consists of methane (about 80 to 90% by volume), and small quantities of other hydrocarbons like ethane, butane and paraffin's and other gases like CO_2 and N_2 (Anyogita et al., 2004). Natural gas has a good potential as substitute fuel for internal combustion engines. It may contain some impurities like hydrogen sulphide and moisture, in small proportions (Anyogita et al., 2004; Talal et al., 2010). It is normally stored in the gaseous form at a high pressure of about 200 bar for transportation. Natural gas is usually shipped in the liquefied form. The composition of natural gas varies from well to well and at the given production well it also varies from time to time (Papagiannakis et al., 2009). Typical composition of natural gas by volume is: Methane 87.3%, Ethane 7.1%, Propane 1.8%, Butane 0.7%, Nitrogen 2.2% and Carbon Dioxide 0.9%.

*Corresponding author. E-mail: kvijayakumarreddy@gmail.com.

Abbreviations: CNG, Compressed natural gas; CI, compression ignition; LPG, liquid petroleum gas.

Experimental set-up

A single-cylinder, 4-Stroke, water-cooled, vertical, stationary diesel

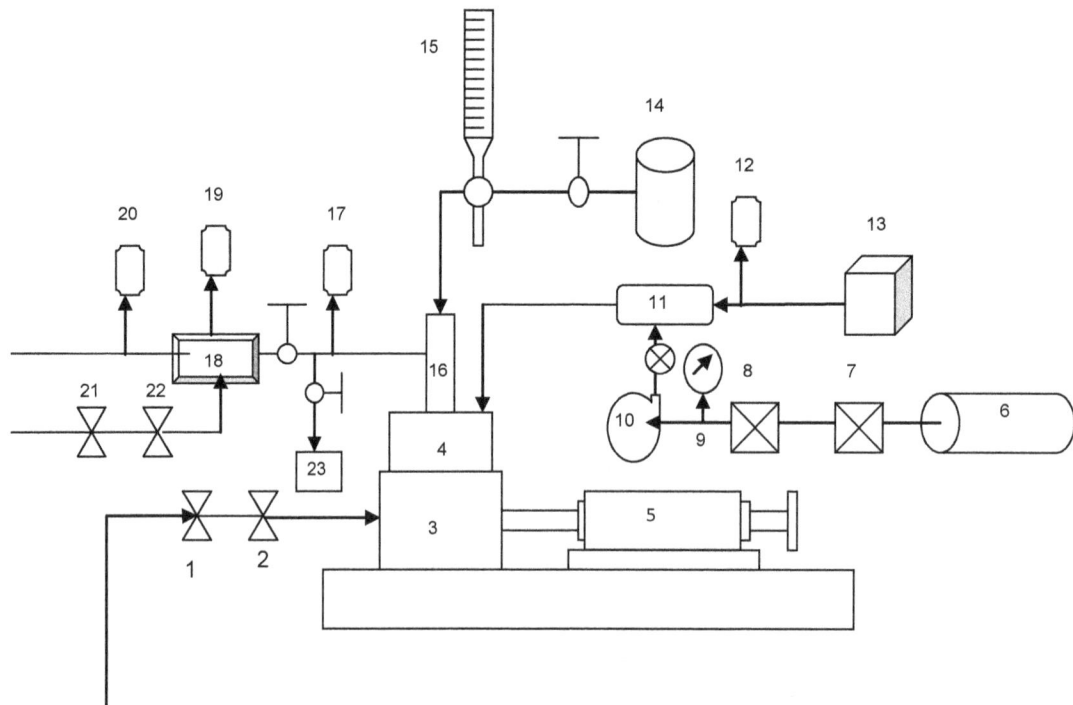

Figure 1. Layout of experimental set up, [1. Cooling Water flow meter 2. Inlet water temperature Sensor 3. Engine block 4. Cylinder head 5. Hydraulic Dynamo meter 6. CNG cylinder 7.On/off valve, 8. Solenoid valve 9. Pressure gauge 10. Regulator 11. Gas-air mixer 12. Manometer 13. inner box 14. Fuel tanks 15. Diesel flow measuring unit 16. Fuel injection system 17. Exhaust gas temperature sensors at inlet calorimeter 18. Calorimeter, 19. Calorimeter outlet water temperature sensors 20. Temperature at outlet calorimeter 21. Calorimeter water flow meter 22. Calorimeter inlet water temperature sensor 23. Emissions analyzer point].

engine of 3.7 kW rated power is considered for the purpose of experimentation. The experiments are conducted on diesel engine with diesel and dual fuel at different injection pressures of 180, 210 and 240 bars. A hydraulic dynamometer is used for loading the engine. The experiments intended to investigate the engine performance and emission levels with diesel and dual fuel (natural gas + diesel).

In the experimentation, load, pilot fuel quantity, fuel injection pressure and speed are the parameters selected for variation. Initially, the engine is operated on the diesel baseline mode at a constant speed of 1500 rpm at different loads. At each operating condition, dynamometer load, air flow rate, fuel flow rate, exhaust gas temperature and cooling water flow rate are recorded, tabulated and plotted. Then the engine is operated on the dual fuel mode at different loads for different injection pressures. The amount of diesel pilot fuel is kept constant while the engine speed is controlled by increasing the flow rate of natural gas to the engine until the specified engine speeds are obtained. All the experimental data for engine loads are recorded, tabulated and plotted. In the calculations of energy contributions of diesel and CNG, the lower calorific value of diesel is taken as 43626 kJ/kg while that of CNG is taken as 47132 KJ/kg. The layout of experimental set up is shown in Figure 1.

RESULTS AND DISCUSSION

The effect of fuel injection pressure on the engine performance and emission characteristics are

investigated at different engine loads. The summary of the outcome is dealt in the following sections: Figure 2 shows the variation of exhaust gas temperature with brake power. From the figure, it is observed that the exhaust gas temperature is increased with increase in brake power. This reveals that more amount of dual fuel being consumed per hour compared to diesel. From the Figure 3, it is observed that initially with increasing brake power, the brake specific fuel consumption of dual fuel and diesel are decreased and then increases with increase in brake power. The brake specific fuel consumption with dual fuel is less than that of diesel throughout the range of brake power at all injection pressures. This is mainly due to the combined effects of relative fuel density, viscosity and calorific value of the dual fuels. From Figure 4, it was observed that initially with increase in brake power, the brake thermal efficiency of the engine is increased with both dual fuel and diesel. The brake thermal efficiency obtained with dual fuel is higher than that of diesel at all injection pressures.

Figure 5 shows the variation of volumetric efficiency with diesel and dual fuel at the injection pressures 180 bar, 210 bars and 240 bar and conclusions drawn between them are given below. The volumetric efficiencies of dual fuel at injection pressures of 180 bars and 240 bars are lower up to 75% load compared to

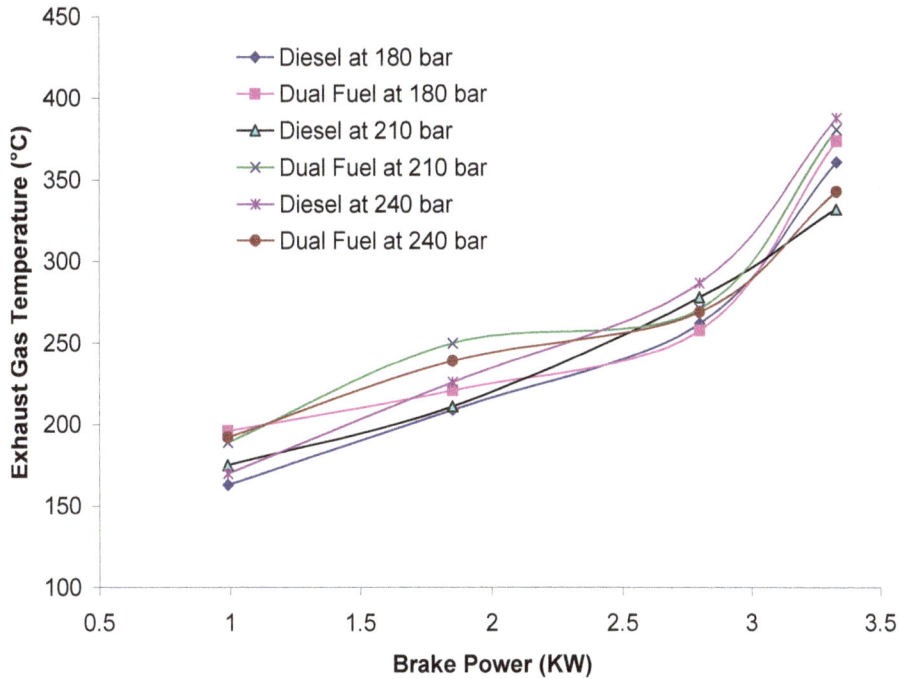

Figure 2. Brake power versus exhaust gas temperature with dual fuel operation.

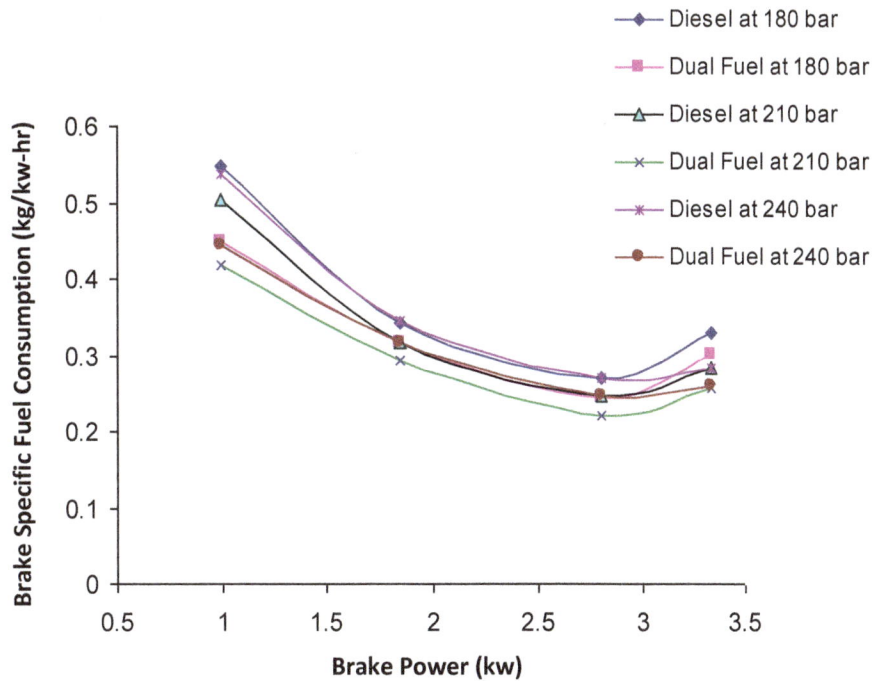

Figure 3. Brake power versus brake specific consumption with dual fuel operation.

diesel and increases tremendously at the engine load of 90% due to engine knocking.

At 210 bar injection pressure, the volumetric efficiency of the engine is increases as the load increases, this trend has continued up to 75% of load and then decreases up to 90% of the load, similarly the trend has been decreases up to 50% of the load and then increases up to 75%, further decreases up to 90% of the

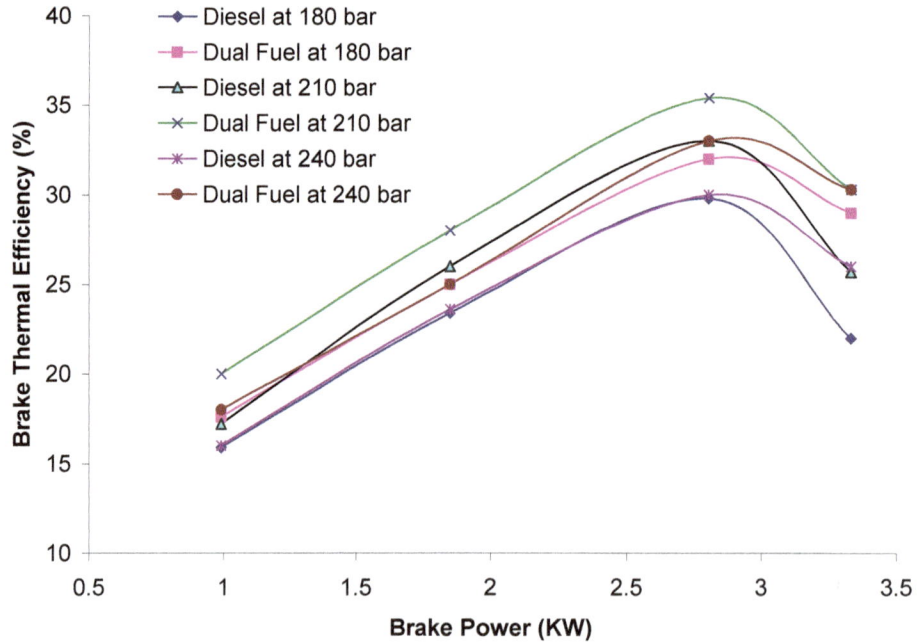

Figure 4. Brake power versus brake thermal efficiency with dual fuel operation.

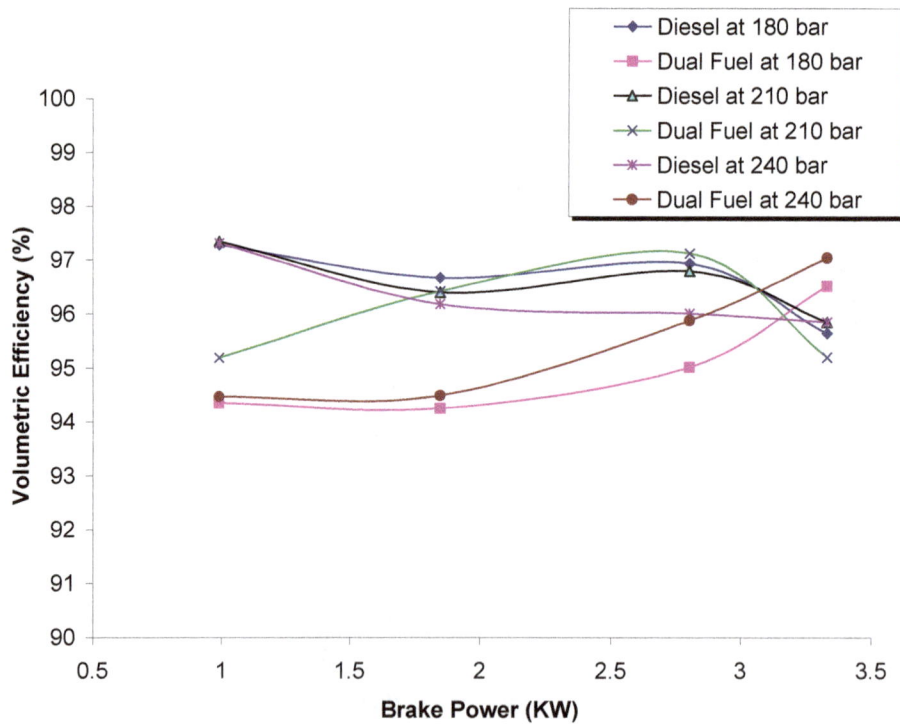

Figure 5. Brake power versus volumetric efficiency with dual fuel operation.

engine load. The decrease of volumetric efficiency in dual fuel is due to the larger volume of inlet air occupied by the CNG. The variation of NO_x emission with brake power is shown in Figure 6. From the figure, it is observed that

the amount of NO_x is increased with increase in brake power for both (diesel and dual) the fuels. The temperature of combustion chamber is also increases with increase in brake power and NO_x formation is

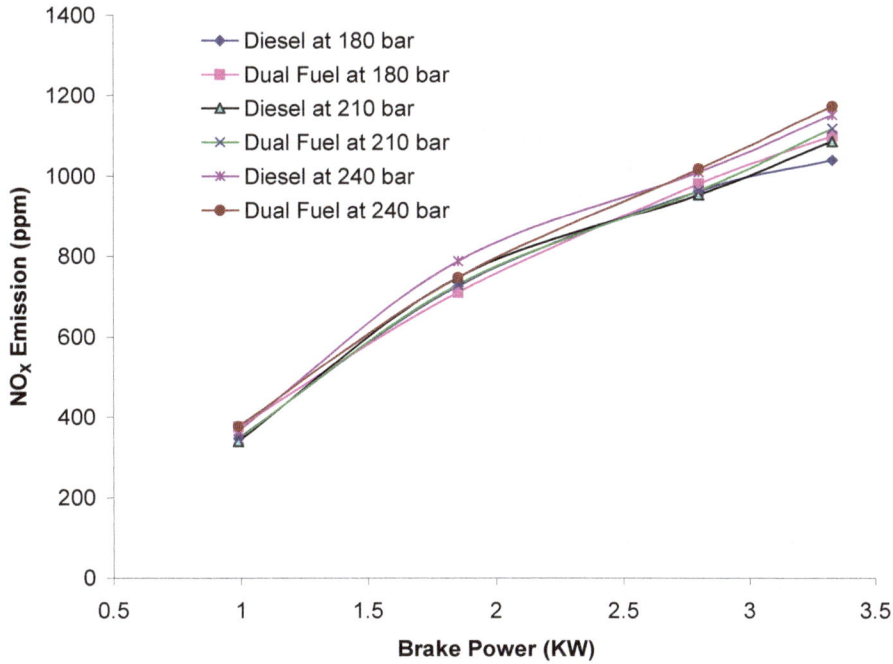

Figure 6. Brake power versus emission of NO_2 with dual fuel operation.

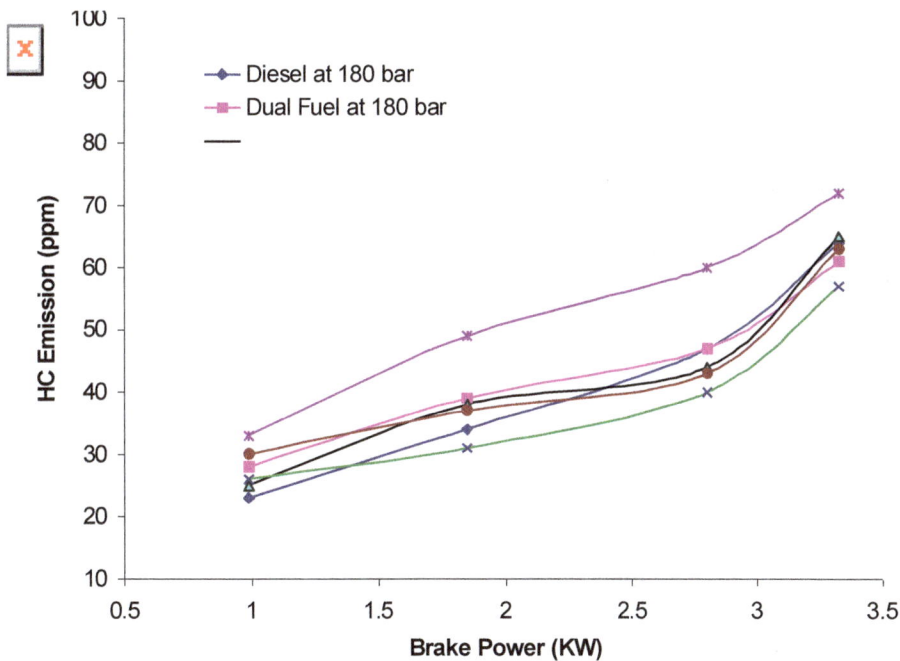

Figure 7. Brake power versus HC emission with dual fuel operation.

strongly temperature dependent Phenomenon. The NO_X emission for dual fuel is less when compared to diesel. Figure 7 shows the variation of UBHC with brake power. The UBHC is increased with increase in brake power for both (diesel + dual) fuels. It is observed that the emission of un-burnt hydro carbons for dual fuel is less than the diesel fuel. The lean mixture and exhaust gas temperature of CNG are responsible for less un-burnt hydro carbon emission when compared to diesel. The variation of smoke density with the brake power for different fuel injection pressures is illustrated in Figure 8. From this figure it was observed that, the smoke density is increased with an increase in the fuel injection pressure in both the cases that is, with diesel and dual

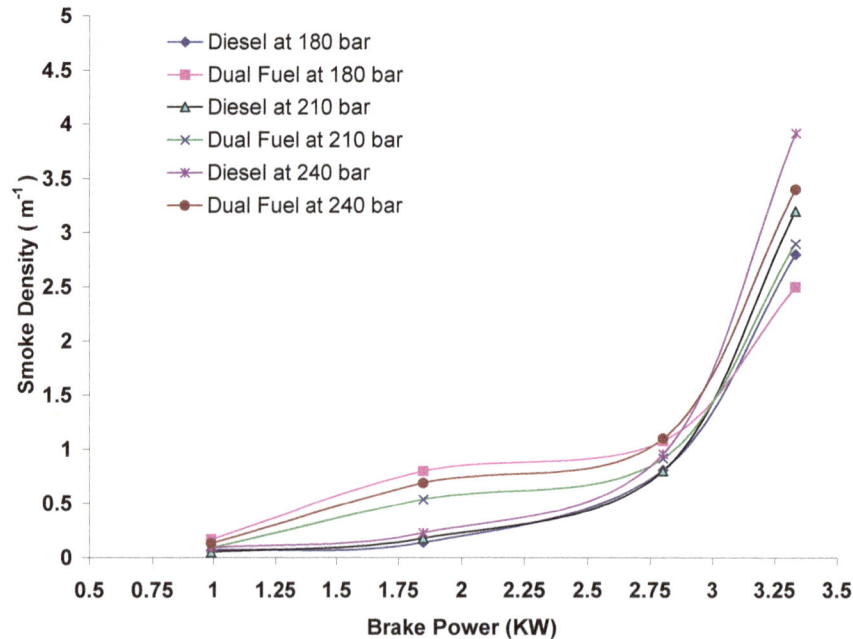

Figure 8. Brake power versus smoke density with dual fuel operation.

fuel. Further, the smoke density is decreased with increase in CNG content at all injection pressures.

Conclusions

Dual fuel operation is more convenient and economical for conserving the precious conventional liquid fuels. A diesel base line engine could be operated on the dual fuel mode with minor changes. The following conclusions have been made from the above work. The exhaust gas temperature for dual fuel operation is higher than diesel at the injection pressures of 180 and 210 bar. The exhaust gas temperature obtained for dual fuel and diesel are 374, 381 and 343°C and 361, 332 and 388°C, respectively. The brake specific fuel consumption with dual fuel is less than that of diesel throughout the range of brake power at all injection pressures. The brake specific fuel consumption is decreasing with increase in load up to 75.67% of rated load and then increasing beyond 75.67% of rated load for both fuel modes. Higher brake thermal efficiency is obtained at 210 bar injection pressure for diesel and dual fuel operations.

The Brake Thermal Efficiency is decreased with increase in Brake Power, when the engine is operated beyond 75.67% rated load. Prior to 75.67% load the Brake Thermal Efficiency is also increased with increase in Brake Power. The volumetric efficiency of dual fuel at 180 bar and 240 bar injection pressure is less up to 75% load compared to diesel and then increases tremendously due to engine knocking. At 210 bar injection pressure, the volumetric efficiency for both fuels is increases up to 75% load and then decreases. This

decrease in volumetric efficiency, in dual fuel, is due to the larger volume of inlet air occupied by the CNG. The emissions of NO_X is increased with increase in brake power for both (diesel and dual) the fuels at all injection pressures. The un-burnt hydro carbons are increased with increase in brake power for both (diesel + dual) fuels. The emission of un-burnt hydro carbons for dual fuel is less compared to diesel fuel. The smoke density is increases with increase in brake power at all defined fuel injection pressures for both the diesel and dual fuels.

REFERENCES

Anyogita S, Prakash A, Jain VK (2004). "A Study of Noise in CNG Driven Modes Transport in Delhi." Elsevier J. Appl. Accoustics 65:195-201.

Carlucci AP, De Risi A, Laforgia D, Naccaratoi F (2008). "Experimental Investigation and Combustion Analysis of a DI Dual-Fuel Diesel-Natural Gas Engine." Energy 31:256-263.

Dong J, Gao X, Li G, Zhang X (2001). Study on diesel-LPG dual-fuel engines. SAE Intl. pp. 36-79.

Ghazi AK (1980). A review of combustion processes in the dual-fuel engine-The gas diesel engine. In: Progressive Energy Combustion Science pp. 277-285.

Mohamed YES (2004). "Combustion Noise Measurements and Control from a Small Diesel and Dual-Fuel Engines." SAE International Number 2004-32-0072.

Papagiannakis RG, Kotsiopoulos PN, Zannis TC, fantis EAY, D.T. Hountalas DT, Rakopoulos CD (2009). "Theoretical Study of the Effect of Engine Parameters on Performance and Emissions of a Pilot Ignited Natural Gas Diesel Engine." Energy pp. 1-10.

Talal FY, Buttsworth DR, Khalid HS, Yusif BF (2010). "CNG-Diesel Engine Performance and Exhaust Emission Analysis with the aid of Artificial Neural Network." J. Appl. Energy 87(5):1661-1669.

9

Continuous ethanol production by *Kluyveromyces* sp. IIPE453 immobilized on bagasse chips in packed bed reactor

Sachin Kumar[1], Surendra P. Singh[3], Indra M. Mishra[4] and Dilip K. Adhikari[2]*

[1]Sardar Swaran Singh National Institute of Renewable Energy, Kapurthala-144 601, India.
[2]Biotechnology Area, Indian Institute of Petroleum, Dehradun- 248 005, India.
[3]Department of Paper Technology, Indian Institute of Technology, Roorkee, Saharanpur Campus- 247 001, India.
[4]Department of Chemical Engineering, Indian Institute of Technology, Roorkee- 247 667, India.

Continuous ethanol production eliminates much of the unproductive down-time associated with batch process and increases the productivity. Many studies showed that the cell immobilization leads to improve fermentation rates by the high cell concentrations, option of reusability and protection of cells from toxic effects of low pH, temperature, inhibitors, etc. The thermotolerant yeast *Kluyveromyces* sp. IIPE453 was immobilized on sugarcane bagasse chips and packed in a column. The maximum volumetric productivity 21.87 ± 0.75 g l^{-1} h^{-1} was achieved with ethanol concentration of 17.5 ± 0.6 g/l and sugar utilization of $76 \pm 2.4\%$ at dilution rate of 1.25 h^{-1} by feeding 50 g/l glucose concentration. The maximum 18.65 ± 0.75 g l^{-1} h^{-1} volumetric productivity was achieved with ethanol concentration of 37.3 ± 1.5 g/l and sugar utilization of $54\pm6.5\%$ at dilution rate of 0.5 h^{-1} by feeding 150 g/l glucose concentration.

Key words: *Kluyveromyces* sp., ethanol fermentation, continuous process, immobilization, sugarcane bagasse chips.

INTRODUCTION

The environment concern over the use and depletion of fossil fuels, the search for alternative fuel is desirable (Liang et al., 2008). Ethanol has attracted worldwide attention due to its potential use as a transportation fuel (Kumar et al., 2009a). Ethanol is traditionally produced in the batch fermentations by yeasts, mostly *Saccharomyces cerevisiae* and their interspecies hybrids, which provide the low productivity (Gunasekaran and Raj, 1999; Rebroš et al., 2005). High ethanol productivity from cheaper and renewable sources and minimum energy input are important aspects in the alcoholic fermentation research. Techniques such as continuous culture; cell immobilization and recycling of cells have been explored to achieve these objectives (Sheoran et al., 1998).

The continuous process can achieve substantial improvements in the efficiency of the process and product quality, subsequently higher productivities, lower operating costs, reduced product losses and environmental advantages (Bakoyianis et al., 1997; Verbelen et al., 2006). However, continuous process with free cells has disadvantages of higher cost of cell recycling, high contamination risk, susceptibility to environmental variations and the limitations of the dilution rate due to wash-out condition (de Vasconcelos et al., 2004). Cell immobilization facilitates the larger area of contact between cells and nutrient medium, potential for high fermentation rates offered by the high cell concentrations, option of reusability of cells, protection of cells from toxic effects of low pH, temperature, inhibitors, tolerance to high osmolalities etc. (Banat et al., 1998; Tata et al., 1999; Kocher et al., 2006). In many studies

---

*Corresponding author. E-mail: adhikari@iip.res.in.

Abbreviations: D, Dilution rate, h^{-1}; P_f, ethanol concentration in outlet, g/l; q_p, volumetric ethanol production rate; g l^{-1} h^{-1}; q_{sp}, specific ethanol production rate, g g^{-1} h^{-1}; q_s, specific sugar uptake rate, g g^{-1} h^{-1}; S_o, glucose concentration in feed, g/l; S_f, residual glucose concentration in outlet, g/l; $\bar{x}$, average biomass concentration, g/l; $Y_{P/S}$, ethanol yield on glucose consumed, g/g; η, sugar utilization, %.

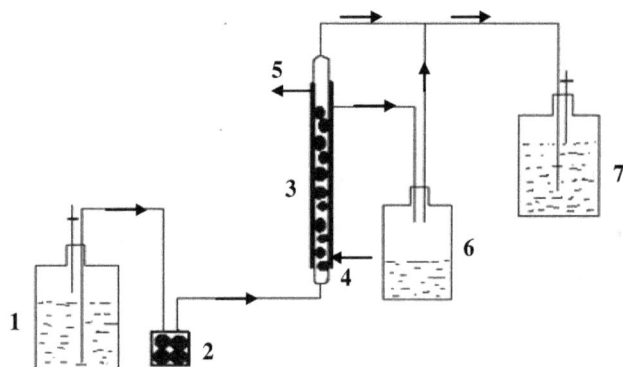

Figure 1. Schematic diagram of continuous ethanol fermentation process in a packed column; 1. Feed vessel; 2. Peristaltic pump; 3. Jacketed column; 4. Hot water in; 5. Hot water out; 6. Product vessel; 7. CO_2 absorber.

the immobilized microorganisms were found to be more stable than the free microorganisms, particularly when utilized over prolonged periods of time (Love et al., 1998; Göksungur and Zorlu, 2001).

The development of immobilized-cell processes, using low-cost support and low operational (immobilization) cost would be desirable for economical production process (de Vasconcelos et al., 2004). The selection of a suitable support for cell immobilization is difficult. A number of factors for example nature of support and its compatibility with the microorganisms, environmental conditions, etc. are known to influence the cell support matrix interactions (Bakoyianis et al., 1997). The four methods for immobilization are categorized as (Liang et al., 2008): (i) the methods involving binding of biocatalyst to a water insoluble support by using ionic or covalent bonding or adsorption; (ii) the methods involving multiple covalent bonding; (iii) the methods involving entrapment or encapsulation; and (iv) combinations of these methods.

A number of studies on ethanol production by immobilized microorganisms using *Saccharomyces cerevisiae*, *Zymomonas mobilis* and *Kluyveromyces marxianus* in organic or inorganic supports have been reported. The reported supports used in immobilization are calcium alginate (Bakoyianis et al., 1997; Sheoran et al., 1998; Love et al., 1998; Göksungur and Zorlu, 2001; Kocher et al., 2006; Valach et al., 2006), calcium pactate (Valach et al., 2006; Kesava and Panda, 1996), porous glass beads (Love et al., 1998; Tata et al., 1999), corn cobs (Kocher et al., 2006), sugarcane bagasse (de Vasconcelos et al., 1998,2004; Kocher et al., 2006), wood shavings (Kocher et al., 2006), kissiris (Bakoyianis et al., 1997; Nigam et al., 1997; Love et al., 1998), γ-alumina (Bakoyianis et al., 1997), delignified cellulosic material (Shindo et al., 2001), κ-carrageenan beads (Krishnan et al., 2000), sorghum bagasse (Yu et al., 2007) and zeolite (Kourkoutas et al., 2002). de Vasconcelos et al. (1998, 2004) and Liang et al. (2008)

satisfactory demonstrated the use of sugar-cane stalks as a support for yeast cells in alcoholic fermentation.

The technical barriers, such as substrate and product inhibition, CO_2 hold up in the gel beads are generally encountered during the ethanol fermentation in immobilized yeast reactor. Therefore, reduction in mass transfer rate, floatation of beads and their accumulation near the exit of the reactor results in a considerable low productivity (Sheoran et al., 1998). However, CO_2 hold up could be over come by increasing height to diameter ratio and making high porosity of bed by using sugarcane stalks (de Vasconcelos et al., 2004). In the present study we have reported the ethanol fermentation at 50°C by thermotolerant yeast *Kluyveromyces* sp. IIPE453. The yeast was immobilized on sugarcane bagasse chips and packed in a column. The effect of different dilution rates and sugar concentrations in feed on volumetric productivity and sugar utilization rate were observed and compared with the reported literature.

MATERIALS AND METHODS

Microorganisms and culture conditions

A thermotolerant yeast, *Kluyveromyces* sp. IIPE453 (Kumar et al., 2009b; Kumar et al., 2010), was grown in Bioflow-110 bioreactor (ca. 5 L) on medium SM containing (g/l), di-sodium hydrogen orthophosphate, 0.15; potassium di-hydrogen orthophosphate, 0.15; ammonium sulphate, 2.0; yeast extract, 1.0; glucose 20. The temperature and pH were controlled at 50°C and 5.0, respectively, during the process. The dissolved oxygen was controlled by agitation and 1 vvm aeration rate at 40% of saturation to obtain maximum cell mass.

Cell immobilization

Cells of *Kluyveromyces* sp. IIPE453 were immobilized by adsorption on sugarcane bagasse chips of size 4 to 6 mm. The bagasse chips were washed with distilled water twice and dried before using for immobilization. 10 g dry bagasse chips were suspended in 500 ml salt medium (SM) with glucose concentration 5 g/l and cell concentration 4 g/l. The suspension was incubated overnight at 50°C in shaker at 150 rpm in two flasks of 2 L. The bagasse chips were separated from cell suspension and washed twice to remove free cells with sterile distilled water. The bagasse chips with immobilized cells were packed in a jacketed column of 50 cm height and 2.5 cm i.d. (Figure 1). The immobilized bagasse occupied approximately 160±3 cm^3 volume with void volume 50±5 cm^3 and void fraction 0.31±0.04. The height of the bed was 32.5±0.5 cm.

Fermentation conditions

The medium for fermentation was same as used for growth except ammonium sulphate, 1.0 g/l. The medium was passed through the packed bed column. The liquid was fed upward from bottom in the column. A high accuracy positive displacement peristaltic pump was used to vary the liquid feed flow rate. The dilution rates were varied from 0.5 to 1.25 h^{-1} with an increment of 0.25 h^{-1} and glucose concentrations were varied from 50 to 150 g/l with an increment of 50 g/l. The product stream was collected from the upper part of the column. The temperature in the packed bed column was maintained at 50°C by passing the hot water through a jacket. Samples were taken at different time intervals and estimated for glucose and ethanol concentration in the broth.

Analytical methods

Glucose was analyzed by a HPLC using High Performance Carbohydrate Column (Waters) at 30°C and detected by a Waters 2414 refractive index detector. The acetonitrile and water mixture (75:25) as a mobile carrier at a flow rate of 1.4 ml min^{-1} was used. Ethanol was analyzed by gas chromatography using Ashco Neon II Gas Analyzer with a 2 m long x 1/8" dia Porapak-QS column with mesh range 80/100. The nitrogen gas was used as a carrier. The injector temperature, oven temperature and flame ionization detector temperature were kept at 220, 150 and 250°C, respectively.

The cell mass concentration in bagasse chips was analyzed on the basis of protein present in cell mass. 5 g dried immobilized bagasse chips were crushed in 50 ml 1 N sodium hydroxide and incubated at 80°C in water bath for 2 h. After incubation the suspension was centrifuged and protein concentration in supernatant was estimated by Folin-Lowery method. The same method was followed for estimation of protein in cell-free bagasse chips. The difference of protein concentration in bagasse chips immobilized with cells and cells free bagasse provides the loading of cells in bagasse. The protein concentration in the known quantity of yeast cells was also analyzed.

Mathematical modeling

The material balance on substrate concentration over a differential height of the column can be expressed as (Ozmihci and Kargi, 2008a):

$$-FdS = q_s \overline{X} dV = q_s \overline{X} A dz \tag{1}$$

Or,

$$-FdS = \frac{q_{sp}\overline{X}}{Y_{P/S}} dV = \frac{q_{sp}\overline{X}}{Y_{P/S}} A\, dz \tag{2}$$

where, F is the flow rate of the feed (l/h); dS is the difference between sugar concentration over the differential height (g/l); q_s is the specific rate of sugar uptake (g g^{-1} h^{-1}); q_{sp} is the specific rate of ethanol formation (g g^{-1} h^{-1}); $\overline{X}$ is the average biomass concentration (g/l); $Y_{P/S}$ is the product yield coefficient (g/g); dV is the differential volume (l); A is the cross-section area of the column (m^2); and dz is the differential column height (m). Assuming q_{sp}, $\overline{X}$ and $Y_{P/S}$ are constant. On integrating equation (Equation 2).

$$-F\int_{S_o}^{S} dS = \frac{q_{sp}\overline{X}}{Y_{P/S}} A \int_{0}^{H} dz \tag{3}$$

$$S_o - S = \frac{q_{sp}\overline{X}}{Y_{P/S}} \frac{A}{F} H \tag{4}$$

Where, S_o is the sugar concentration in feed (g/l); S is the sugar concentration at column height H (g/l); and H is the column height (m). Similarly, material balance on product concentration over a differential height can be expressed as:

$$FdP = q_{sp}\overline{X}dV = q_{sp}\overline{X}Adz \tag{5}$$

where, dP is the difference between ethanol concentration over the differential height (g/l). On integrating Equation (5):

$$F\int_{P_o}^{P} dP = q_{sp}\overline{X}A\int_{0}^{H} dz \tag{6}$$

$$P - P_o = q_{sp}\overline{X}\frac{A}{F}H \tag{7}$$

where, P_o is the ethanol concentration in feed (g/l); and P is the ethanol concentration at column height H (g/l). Rearranging equation (7) for calculating specific rate of ethanol formation (q_p):

$$q_{sp} = P\frac{D}{\overline{X}} \tag{8}$$

where, D is dilution rate, h^{-1}; $D = \dfrac{F}{A.H} = \dfrac{F}{V}$ and ethanol concentration in feed, $P_o = 0$.

Rate of ethanol formation or volumetric ethanol productivity (q_p):

$$q_p = q_{sp}\overline{X} = PD \tag{9}$$

RESULTS AND DISCUSSION

The bagasse pieces were chosen for immobilization of thermotolerant yeast *Kluyveromyces* sp. IIPE453 for the production of ethanol because of its high porosity to adsorb the yeast cells, easy availability, natural source and stability at high temperature. The cells were adsorbed on bagasse chips with cell loading of 120±10 mg/g of bagasse on dry basis with 60% adsorption efficiency. de Vasconcelos et al. (2004) reported Fleischmann yeast cells loading of 477 mg/g on 2 cm long dry sugarcane stalks. The average cell mass concentration in column ($\overline{X}$) was 7.5±0.5 g/l of total packed volume. Liang et al. (2008) suggested that the immobilization of yeast cells on sugarcane pieces is a result of natural entrapment into the porous structure of the support and adsorption by electrostatic forces between cell membrane and support.

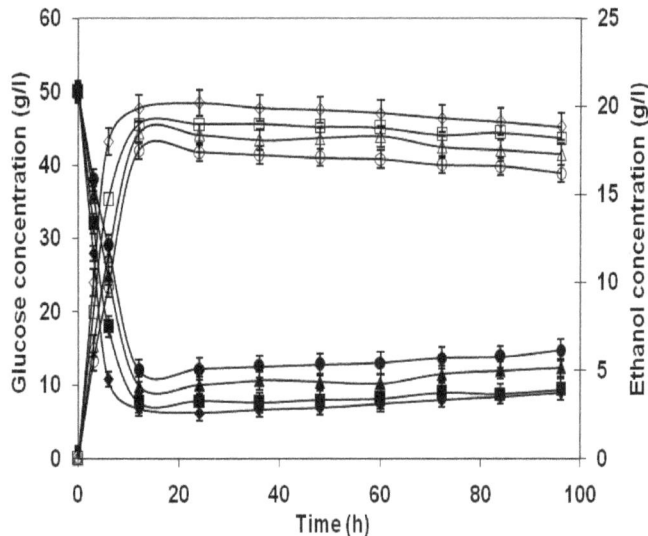

Figure 2. Ethanol fermentation by *Kluyveromyces* sp. IIPE453 immobilized in bagasse chips at different dilution rates with 50 g/l feed glucose concentration showing glucose concentrations: ($\blacklozenge$) D=0.5 h^{-1}; ($\blacksquare$) D=0.75 h^{-1}; ($\blacktriangle$) D=1 h^{-1}; ($\bullet$) D=1.25 h^{-1} and ethanol concentrations: ($\Diamond$) D=0.5 h^{-1}; ($\square$) D=0.75 h^{-1}; ($\triangle$) D=1.0 h^{-1}; ($\circ$) D=1.25 h^{-1}.

Table 1. Effect of dilution rate on different parameters in ethanol fermentation by *Kluyveromyces* sp. IIPE453 immobilized in bagasse chips.

Kinetic parameters[a]	Dilution rate (h^{-1})			
	0.5	0.75	1	1.25
S_o (g/l)	50	50	50	50
S_f (g/l)	6.8	7.6	9.9	12
P_f (g/l)	20.2	19	18.5	17.5
$Y_{P/S}$ (g/g)	0.461	0.459	0.461	0.46
q_p (g l^{-1} h^{-1})	10.1	14.25	18.5	21.87
q_{sp} (g g^{-1} h^{-1})	1.35	1.9	2.46	2.91
η (%)	87.6	84.8	80.2	76

[a]S_o = Glucose concentration in feed; S_f = Residual glucose concentration in outlet; P_f = Ethanol concentration in outlet; $Y_{P/S}$ = Ethanol yield on glucose consumed; q_p = Volumetric ethanol productivity; q_{SP} =Specific ethanol productivity; η = Sugar utilization.

Effect of dilution rate

The medium containing glucose concentration of 50 g/l was passed through the column at different dilution rates varying from 0.5 to 1.25 h^{-1} (Figure 2). At a dilution rate of 0.5 h^{-1}, 20.2±0.7 g/l ethanol concentration was obtained in the outlet with an ethanol yield of 90.2±0.2% of its theoretical yield at steady state. At a dilution rate of 0.75 h^{-1}, 19±0.55 g/l ethanol concentration was obtained in the

outlet with an ethanol yield of 89.8±0.16% of its theoretical yield at steady state. At a dilution rate of 1.0 h^{-1}, 18.5±0.6 g/l ethanol concentration was obtained in the outlet with an ethanol yield of 90.2±0.2% of its theoretical yield at steady state. At a dilution rate of 1.25 h^{-1}, 17.5±0.6 g/l ethanol concentration was obtained in the outlet with an ethanol yield of 90±0.1% of its theoretical yield at steady state.

The ethanol productivities (q_p) and specific productivities (q_{sp}) were calculated using equations 8 and 9 (Table 1). The ethanol productivity and specific ethanol productivity could be increased when the dilution rate was increased from 0.5 to 1.25 h^{-1} but the ethanol concentration and sugar utilization decreased significantly. The ethanol concentration of 20.2±0.7 g/l with 87.6±2.8% sugar utilization was achieved at a dilution rate of 0.5 h^{-1}. The ethanol concentration decreased up to 17.5±0.6 g/l with 76±2.4% sugar utilization when dilution rate was increased up to 1.25 h^{-1}. The volumetric productivity 21.87±0.75 g l^{-1} h^{-1} and specific productivity of 2.91±0.09 g g^{-1} h^{-1} was achieved at a dilution rate of 1.25 h^{-1}. The ethanol concentration was decreased due to less interaction of glucose molecule with the immobilized yeast or low hydraulic retention time. The higher ethanol productivity was achieved as compared to 1.71 and 3.7 g l^{-1} h^{-1} in a batch fermentation and continuous fermentation with cell recycle, respectively using free cells of the same strain (Kumar et al., 2009b). Thus, the ethanol fermentation using immobilized yeast is a better option.

Ozmihci and Kargi (2008a) reported the ethanol concentration of 10.5 g/l and volumetric productivity of 0.58 g l^{-1} h^{-1} with 63% sugar utilization at a dilution rate of 0.057 h^{-1} on feeding 50 g/l sugar concentration in cheese whey powder solution by *Kluyveromyces marxianus* (DSMZ 7239) in a packed column bioreactor. Yu et al. (2007) reported the volumetric productivity of 16.68 g l^{-1} h^{-1} with ~55% sugar utilization at a dilution rate of 0.3 h^{-1} on feeding 200 g/l sugar concentration using immobilized *S. cerevisiae* on sorghum bagasse. In the present study, we could achieve the highest ethanol concentration with the highest ethanol productivity and maximum sugar utilization at a dilution rate of 1.25 h^{-1} as compared to reported literature.

The ethanol yield at each dilution rate was almost same. Therefore, no effect was observed of dilution rate on ethanol yield. But Ozmihci and Kargi (2008a) reported that the ethanol yield was decreased by increasing dilution rate or increased by increasing hydraulic retention time on fermenting cheese whey powder by *Kluyveromyces marxianus* (DSMZ 7239) in a packed column bioreactor. As shown in Table 1, the volumetric productivity could be increased when dilution rate was increased from 0.5 to 1.25 h^{-1} whereas the ethanol concentration was declined consistently due to decrease in sugar utilization at high flow rate. de Vasconcelos et al. (2004) reported maximum 29.64 g l^{-1} h^{-1} volumetric

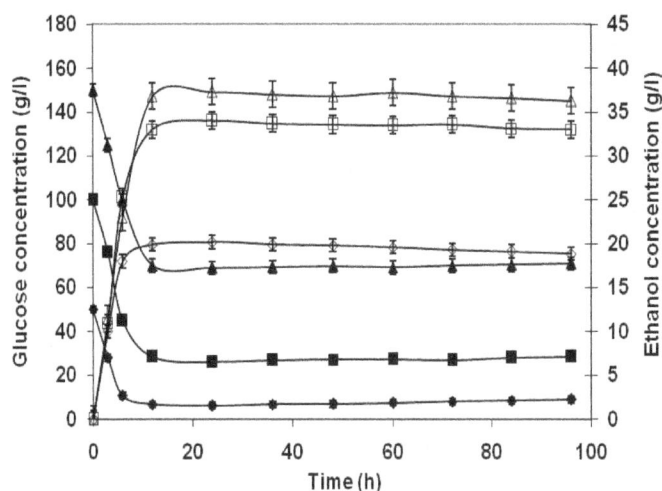

Figure 3. Ethanol fermentation by *Kluyveromyces* sp. IIPE453 immobilized in bagasse chips on varying glucose concentration at dilution rate 0.5 h^{-1} showing glucose concentrations: (♦) S_o=50 g/l; (■) S_o=100 g/l; (▲) S_o=150 g/l and ethanol concentrations: (◊) S_o=50 g/l; (□) S_o=50 g/l; (△) S_o=50 g/l.

Table 2. Effect of glucose concentration in feed on different parameters in ethanol fermentation by *Kluyveromyces* sp. IIPE453 immobilized in bagasse chips.

Kinetic parameters[a]	Glucose concentration in feed (g/l)		
	50	100	150
D (h^{-1})	0.5	0.5	0.5
S_f (g/l)	6.8	26	69
P_f (g/l)	20.2	34	37.3
$Y_{P/S}$ (g/g)	0.461	0.46	0.46
q_p (g l^{-1} h^{-1})	10.1	17	18.65
q_{sp} (g g^{-1} h^{-1})	1.35	2.27	2.5
η (%)	87.6	74	54

[a]S_o = Glucose concentration in feed; S_f = Residual glucose concentration in outlet; P_f = Ethanol concentration in outlet; $Y_{P/S}$ = Ethanol yield on glucose consumed; q_P = Volumetric ethanol productivity; q_{SP} = Specific ethanol productivity; η = Sugar utilization

productivity at a dilution rate of 0.83 h^{-1} with 74.61% sugar utilization using immobilized Fleischmann yeast cells in sugarcane stalks.

Effect of feed sugar concentration

The medium with varying glucose concentrations from 50 to 150 g/l was fed into the column at dilution rate 0.5 h^{-1} (Figure 3). At a feed glucose concentration of 100 g/l, 34±1.4 g/l ethanol concentration was obtained with an

ethanol yield of 90±0.2% of its theoretical yield at steady state. At a feed glucose concentration of 150, 37.3±1.5 g/l ethanol concentration was obtained with an ethanol yield of 90±0.2% of its theoretical yield at steady state. The ethanol productivities (q_p) and specific productivities (q_{sp}) were calculated using equations 8 and 9 (Table 2). The ethanol concentration, ethanol productivity and specific ethanol productivity could be increased when glucose concentration in feed was increased but the sugar utilization decreased considerably. The ethanol concentration of 37.3±1.5 g/l with 54±6.5% sugar utilization was achieved on feeding glucose concentration of 150 g/l as compared to 20.2±0.7 g/l with 87.5±2.8% sugar utilization on feeding glucose concentration of 50 g/l.

The volumetric productivity of 18.65±0.75 g l^{-1} h^{-1} and specific ethanol productivity of 2.5±0.12 g g^{-1} h^{-1} was achieved on feeding glucose concentration of 150 g/l. The sugar utilization decreased on increasing glucose concentration in feed due to high ratio of glucose to cell mass concentration (S/X) or sugar uptake limit. Ozmihci and Kargi (2008b) reported that the ethanol concentration increased when sugar concentration was increased in feed from 50 to 100 g/l and decreased when sugar concentration was further increased due to lower sugar utilization. They obtained the ethanol concentration of 22.5 g/l with the specific productivity of 0.075 g g^{-1} h^{-1} at feed sugar concentration of 100 g/l in cheese whey powder solution by *Kluyveromyces marxianus* (DSMZ 7239) in a packed column bioreactor.

In the present study, the ethanol concentration further increased by increasing glucose concentration. Love et al. (1998) reported maximum ethanol concentration 46 to 48 g/l on feeding glucose concentration of 100 g/l with ethanol productivity of 4.8 g l^{-1} h^{-1} at a dilution rate of 1 h^{-1} by *Kluyveromyces marxianus* IMB3 immobilized on mixed alginate and kissiris whereas Gough and McHale (1998) reported ethanol concentration of 34 g/l on feeding glucose concentration of 120 g/l with ethanol productivity of 5.1 g l^{-1} h^{-1} at a dilution rate of 1.5 h^{-1} by the same strain immobilized on alginate. In the present study, the ethanol concentration is comparable whereas volumetric productivity is much higher than the reported literature at a glucose concentration of 150 g/l. Increase in specific ethanol productivity shows increase in sugar uptake rate by the yeast on increasing glucose concentration.

The ethanol yield on feeding glucose at any concentration was almost same. Therefore, no effect was observed of glucose concentration in feed on ethanol yield. But Ozmihci and Kargi (2008b) reported that the ethanol yield was decreased by increasing feed sugar on fermenting cheese whey powder by *Kluyveromyces marxianus* (DSMZ 7239) in a packed column bioreactor. As shown in Table 2, the ethanol concentration and volumetric productivity significantly increased when feed glucose concentration was increased from 50 to 100 g/l whereas at 150 g/l feed glucose concentration slightly increase in ethanol concentration and volumetric productivity were observed. Thus, the immobilization on sugarcane bagasse chips is favorable for low feed sugar concentration.

Conclusion

Thus, ethanol fermentation at high temperature (50°C) with immobilized yeast *Kluyveromyces* sp. IIPE453 reveals that the high dilution rate is favorable to the feed with low sugar concentration. For a high sugar concentration feed, the maximum sugar utilization can be achieved either by increasing bed height or by increasing number of columns.

ACKNOWLEDGEMENTS

We thank Dr M.O. Garg, Director IIP, Dehradun for his valuable suggestion and encouragement to carry out this research work. One of the authors (Sachin Kumar) gratefully acknowledges Senior Research Fellowship awarded by Council of Scientific and Industrial Research (CSIR), India.

REFERENCES

Bakoyianis V, Koutinas AA, Agelopoulos K, Kanellaki M (1997). Comparative study of kissiris, γ-alumina, and calcium alginate as supports of cells for batch and continuous wine-making at low temperatures. J. Agric. Food Chem., 45: 4884-4888.

Banat IM, Nigam P, Singh D, Marchant R, McHale AP (1998). Review: Ethanol production at elevated temperatures and alcohol concentrations: Part I - Yeasts in general. W. J. Microbiol. Biotechnol., 14: 809-821.

de Vasconcelos JN, Lopes CE, de França FP (1998). Yeast immobilization on cane stalks for fermentation. Int. Sugar J., 100: 73-75.

de Vasconcelos JN, Lopes CE, de França FP (2004). Continuous ethanol production using yeast immobilized on sugar-cane stalks. Braz. J. Chem. Eng., 21: 357-65.

Göksungur Y, Zorlu N (2001). Production of ethanol from beet molasses by Ca-alginate immobilized yeast cells in a packed-bed bioreactor. Turk. J. Biol., 25: 265-275.

Gough S, McHale AP (1998). Continuous ethanol production from molasses at 45°C using alginate-immobilized *Kluyveromyces marxianus* IMB3 in a continuous-flow bioreactor. Bioproc. Eng., 19: 33-36.

Gunasekaran P, Raj KC (1999). Ethanol fermentation technology – Zymomonas mobilis. Curr. Sci., 77: 56-68.

Kesava SS, Panda T (1996). Ethanol production by immobilized whole cells of *Zymomonas mobilis* in a continuous flow expanded bed bioreactor and a continuous flow stirred tank bioreactor. J. Ind. Microbiol., 17: 11-14.

Valach M, Navrátil M, Horváthová V, Zigová J, Šturdík E, Hrabárová E, Gemeiner P (2006). Efficiency of a fixed-bed and a gas-lift three-column reactor for continuous production of ethanol by pectate- and alginate-immobilized *Saccharomyces cerevisiae* cells. Chem. Pap., 60: 154-159.

Verbelen PJ, Schutter DPD, Delvaux F, Verstrepen KJ, Delvaux FR (2006). Immobilized yeast cell systems for continuous fermentation applications. Biotechnol. Lett., 28: 1515-1525.

Yu J, Zhang X, Tan T (2007). An novel immobilization method of *Saccharomyces cerevisiae* to sorghum bagasse for ethanol production. J. Biotechnol., 129: 415-420.

Kocher GS, Kalra KL, Phutela RP (2006). Comparative production of sugarcane vinegar by different immobilization techniques. J. Inst. Brew., 112: 264-266.

Kourkoutas Y, Psarianos C, Koutinas AA, Kanellaki M, Banat IM, Marchant R (2002). Continuous whey fermentation using kefir yeast immobilized on delignified cellulosic material. J. Agric. Food Chem., 50: 2543-2547.

Krishnan MS, Blanco M, Shattuck CK, Nghiem NP, Davison BH (2000). Ethanol production from glucose and xylose by immobilized *Zymomonas mobilis* CP4(pZB5). Appl. Biochem. Biotechnol., 84-86: 525-541.

Kumar S, Singh SP, Mishra IM, Adhikari DK (2009a). Recent advances in production of bioethanol from lignocellulosic biomass. Chem. Eng. Technol., 32: 517-726.

Kumar S, Singh SP, Mishra IM, Adhikari DK (2009b). Ethanol production from glucose and xylose at high temperature by *Kluyveromyces* sp. IIPE453. J. Ind. Microbiol. Biotechnol., 36: 1483-1489.

Kumar S, Singh SP, Mishra IM, Adhikari DK (2010). Feasibility of ethanol production with enhanced sugar concentration in bagasse hydrolysate at high temperature using *Kluyveromyces* sp. IIPE453. Biofuels, 1: 697-704.

Liang L, Zhang Y, Zhang L, Zhu M, Liang S, Huang Y (2008). Study of sugarcane pieces as yeast supports for ethanol production from sugarcane juice and molasses. J. Ind. Microbiol. Biotechnol., 35: 1605-1613.

Love G, Gough S, Brady D, Barron N, Nigam P, Singh D, Marchant R, McHale AP (1998). Continuous ethanol fermentation at 45°C using *Kluyveromyces marxianus* IMB3 immobilized in calcium alginate and kissiris. Bioproc. Eng., 18: 187-189.

Nigam P, Banat IM, Singh D, McHale AP, Marchant R (1997). Continuous ethanol production by thermotolerant *Kluyveromyces marxianus* IMB3 immobilized on mineral kissiris at 45°C. W. J. Microbiol. Biotechnol., 13: 283-288.

Ozmihci S, Kargi F (2008a). Ethanol production from cheese whey powder solution in a packed column bioreactor at different hydraulic residence times. Biochem. Eng. J., 42: 180-185.

Ozmihci S, Kargi F (2008b) Fermentation of cheese whey powder solution to ethanol in a packed-column bioreactor: effects of feed sugar concentration. J. Chem. Technol. Biotecnol., 84: 106-111.

Rebroš M, Rosenberg M, Stloukal R, Krištofíková L (2005). High efficiency ethanol fermentation by entrapment of *Zymomonas mobilis* into LentiKats. Lett. Appl. Microbiol., 41: 412-416.

Sheoran A, Yadav BS, Nigam P, Singh D (1998). Continuous ethanol production from sugarcane molasses using a column reactor of immobilized *Saccharomyces cerevisiae* HAU-1. J. Basic Microbiol., 38: 123-128.

Shindo S, Takata S, Taguchi H, Yoshimura N (2001). Development of novel carrier using natural zeolite and continuous ethanol fermentation with immobilized *Saccharomyces cerevisiae* in a bioreactor. Biotechnol. Lett., 23: 2001-2004.

Tata M, Bower P, Bromberg S, Duncombe D, Fehring J, Lau V, Ryder D, Stassi P (1999). Immobilized yeast bioreactor systems for continuous beer fermentation. Biotechnol. Prog., 15: 105-113.

10

Application of re-refined used lubricating oil as base oil for the formulation of oil based drilling mud - A comparative study

Oghenejoboh K. M.[1], Ohimor E. O.[2] and Olayebi O.[3]

[1]Department of Chemical Engineering, Delta State University, Oleh Campus, P. M. B. 22, Oleh, Nigeria.
[2]Industries Department, Ministry of Commerce and Industry, Agbor, Delta State, Nigeria.
[3]Department of Chemical Engineering, Faculty of Engineering, Federal University of Petroleum Resources, Ugbomoro, Delta State, Nigeria.

The viability of using re-refined used lubricating oil (RULO) extracted with an aromatic selective normal methylpyrolidone (NMP) as base oil for the formulation of drilling mud was investigated. The rheological and other properties of this formulation were compared with formulations from three locally produced synthetic base oils. The synthetic base oils were Paradril® made from saturated linear ethylene polymer, Emcaid® manufactured from a blend of olefin isomers and Ty-Chem-Low Tox® made from catalytic dimerization of linear alpaolefins. RULO based mud, though alkaline in nature with a pH of 8.5 exhibits very poor filtration properties with the thickest filter cake when compared with the other formulations. It is also the least stable of the four formulations with an electrical stability (ES) of 480 volts. RULO formulation is very toxic as the cassava plant on which it was spilled survived for only 5 days compared to 15 days for Paradril®. It is therefore, not environmental friendly and may not also be cost effective as the cost of re-refining and extraction may far exceed the cost of producing synthetic base oil. RULO may not therefore be a viable alternative to existing base oils for the formulation of drilling mud.

Key words: Used lubricating oil, base oil, drilling mud, rheological properties, emulsifier, environment.

INTRODUCTION

The oil industry in Nigeria had in the past relied too heavily on imported drilling chemicals (fluids) for her drilling operations which in turn have increased their operating cost and engendered capital flight. Different types of drilling fluids are used by the oil producing companies in Nigeria due to their onshore/offshore operational nature. These drilling fluids are water-based mud (WBM), oil-based mud (OBM) and synthetic-based mud (SBM). The type used for a particular drilling operation depends on the nature and location of the oil

wells to be drilled. WBM which is made from bentonite clay with some chemicals such as potassium formate added to achieve various effects like viscosity control, shale stability, enhance drilling rate of penetration, cooling and lubricating of equipment are used mainly for drilling shallow onshore wells (Broni-Bediako and Amorin, 2010). However, oil wells are rarely shallow and sometimes complex evolving from vertical, inclined, horizontal, sub-sea to deep-sea drilling; as a result WBM becomes ineffective in accomplishing the required objective of an efficient drilling mud, therefore the use of OBM becomes imperative. OBM is a mud having a petroleum product such as diesel fuel as the base fluid. Oil-based muds are used for many reasons, some being ability to withstand greater heat without breaking down and cost environmental considerations. Other advantages of OBM over WBM are its excellent fluid loss control, no shale swelling, adequate lubrication of drill bits and good cutting carrying ability. Synthetic-based fluid is a mud with synthetic oil as the base fluid. This is most often used on offshore rigs because it has the properties of an oil-based mud, but the toxicity of the fluid fumes are much less than an oil-based fluid. This is important when men work with the fluid in an enclosed space such as an offshore drilling rig. Due to the environmental advantages of synthetic-based mud, it is more preferable by drilling companies despite its exorbitant cost. Synthetic oil based mud (SOBM) is basically water-in-oil or 'invert', emulsion. The water-in-oil emulsion itself is usually stabilized with a "primary emulsifier" (often a fatty acid salt), while the weighting material, along with drill solids which the mud acquires in use, is made oil-wet and dispersed in the mud with a "secondary emulsifier" (typically a strong wetting agent, such as a polyamide) (Broni-Bediako and Amorin, 2010; Growcock et al., 1994). For SOBM to be effective the emulsion formed must be stable, such stability is derived from the strong visco-elastic characteristics caused by the presence of asphaltenes and resins in the mixture (Akpabio and Ekott, 2013; Langevin et al., 2004). The SOBM premixes or invert emulsions are formulated to contain some amount of water (up to 30%). The amount of synthetic oil and water in the SOBM premix is referred to as the Oil-Water-Ratio (OWR). Emulsifiers are then added to emulsify the water as the internal phase and prevent the water from breaking out and coalescing into larger droplets (Huda and Nour, 2011). These water droplets, if not tightly emulsified, can water-wet the already oil-wet solids and dramatically affect the emulsion stability (Abdel-Raouf, 2011). To achieve this therefore, compounds with higher solubility in the oil phase rather than in the aqueous phase are used as emulsifiers (Dimitrov et al., 201).

To minimize the drilling industries' operating cost index, concerted efforts are ongoing to find an effective, inexpensive and ecologically safe drilling fluids that can be sourced locally in line with the current Nigerian Oil and Gas Industry Content Development Policy. To this end, the company, Skyward Resources Ltd based in Port-Harcourt, Nigeria has developed some drilling fluids chemicals such as oil mud thinner (OMT 5), oil mud wetter (OMW 5), drilling detergent (DD 3100), primary emulsifier (PEM 5) and secondary emulsifier (SEM 5) from vegetable extracts. These chemicals have been used with biodegradable plant based oil such as jatropha oil, rapeseed oil, soyabeans oil and cottonseed oil (Fadairo et al., 2012) as well as other low aromatic synthetic mineral base oil for the formulation of SOBM

that are environmental friendly and non-toxic. However, it is the belief of the authors that the production cost of drilling fluids can further be reduced by using discarded used lubricating oil as base oil for the formulation of SOBM, since plant oil is not usually available in commercial quantity. This is however, based on the fact that the used lubricating oil must meet the basic environmental requirements for such use (Nweke and Okpokwasili, 2003). Used lubricating oil is currently a source of environmental nuisance in Nigeria since it is indiscriminately dumped into rivers, soil and the environment as a result of lack of stringent enforcement of environmental laws (Oghenejoboh and Ohimor, 2012; Ogbo et al., 2009). Used lubricating oils can therefore be collected at no cost from mechanic workshops and other outlets involved in rotating machine repairs and maintenance. Used lubricating oil contains lot of impurities such as mixture of high molecular weight aliphatic and aromatic hydrocarbons as well as heavy metals acquired from engine wear and tear (Wang et al., 2000). Used lubricating oil also contains combustion products (water, un-burnt fuel, soot and carbon) as well as abrasive materials such as road dust. All these contaminants must be removed through re-refining before it can be used as base oil for the formulation of drilling mud. Re-refining of used lubricating oil involve three steps - dehydration, stripping and distillation. The dehydration step entails physical treatment in which the used oil is stored in a container for a period of time to allow water and solids to separate out of the oil followed by boiling to break water emulsion and to allow fuel diluents to evaporate from the oil. The stripping step involves normal fractionation where the bulk of the feedstock is distilled off as lubricating oil fractions. The final step in the re-refining process is the extraction process whereby a suitable solvent is used to remove all carcinogenic compounds such as poly aromatic compounds contained in the oil. This step also remove odour and colour from the oil.

In the present study, the stability and toxicity of SOBM formulated from re-refined used lubricating oil is compared with those from three commercial base synthetic base oil - Paradril® (made from saturated linear ethylene polymer), Emcaid® (made from a blend of olefin isomers) and Ty-Chem-Low Tox® (made from catalytic dimerization of linear alphaolefins) as well as results obtained with plant base oil by other workers.

MATERIALS AND METHODS

The materials used for the experiments were spent lubricating oil, soxhlet extractor fixed with 500 ml flask, distillation column, digital weighing balance, Hamilton beach mixer, mud balance, hot plate, digital thermometer, 1000 ml measuring cylinders, 500 ml measuring cylinder, 100 ml beaker, 5 ml syringes and ES-meter. Other materials used were, synthetic base oils, primary emulsifier (PEM 5) and secondary emulsifier (SEM 5) both obtained from Skyward Resources Ltd based in Port-Harcourt, organoclay, soltex,

Table 1. Viscometer reading for mud formulated from the base oils used in this work.

Dial reading (D) (RPM)	Base oil samples (lb/100 ft^2)			
	A	B	C	D
600	186	122	144	130
300	168	111	129	109
100	158	93	124	101
100	151	86	114	92
60	147	71	105	85
30	136	68	93	70
3	72	51	68	55

lime, calcium chloride, distilled water and barite.

Experimental procedure

Treatment and re-refining of used lubricating oil

Ten litres of used lubricating oil obtained from a motor mechanical workshop in Warri, Delta State of Nigeria was left in a 20 L plastic paint bucket for 5 days to allow water and solids to separate out of the oil after which the oil was decanted. Some of the decanted oil was then heated in a closed vessel immersed in a water bath maintained at 120°C for 60 min to boil off some of the emulsified water and fuel diluents. The dehydrated oil was then fractionated using a laboratory scale distillation column following the normal crude oil distillation process. The refined lubricating oil obtained as intermediate from the fractionation process is then extracted with N-methylpyrolidone (NMP) using a soxhlet extractor. The extraction step is aimed at removing unwanted aromatic contaminants present in the paraffinic lubricating oil fraction since NMP is an aromatic selective solvent. The solvent also removes colour and odour from the oil. The re-refined lubricating oil was then used as base oil for the formulation of drilling mud.

Formulation of drilling mud

175 ml of the re-refined lubricating oil and 75 ml of de-ionized water were measured into a mixing vessel using the measuring beakers. 4, 6, 6 and 2 g of organophilic clay, lime, PEM 5 and SEM 5 were then added to the mixture. 0.5 ml of brine solution prepared from 25 g of CaCl$_2$ in 100 ml of de-ionized water was added before subjecting the mixture to thorough mixing using Hamilton Beach mechanical mixer, model 936 to attain a homogeneous mixture. The formulated drilling mud was allowed to age for 24 h. The same procedure was repeated for the three synthetic base oils (Paradril®, Emcaid® and Ty-Chem-Low Tox®). For ease of identification, the base oil samples used for the drilling fluid formulation were labeled:

Sample A	re-refined used lubricating oil
Sample B	Paradril® synthetic base oil
Sample C	Emcaid® synthetic base oil
Sample D	Ty-Chem-Low Tox® synthetic base oil

Measurement of formulated fluid properties

The density, viscosity, gel strength, pH, filtered volume, filter cake thickness, electrical stability as well as the toxicity of the formulated drilling fluids were determined and compared. The density and viscosity of the fluids were measured using the method outlined by

Fadairo et al. (2012) with the values of apparent viscosity (μ_A), and plastic viscosity (μ_p) obtained from the equations developed by Amorin et al. (2011) as reproduced below.

$$\mu_p = D_{600} - D_{300} \qquad (1)$$

$$\mu_A = \frac{D_{600}}{2} \qquad (2)$$

Where D_{600} and D_{300} is the viscometer dial reading at 600 and 300 rpm in centipoises (cP) respectively.

The electrical stability of the tested drilling fluids was determined using an ES meter according to API 13B-2 procedure. Gel strength was determined using the rotational viscometer at 10 s and 10 min respectively, while the pH of the fluids was estimated by means of the pH colorimeter paper method of Fadairo et al. (2012). The API filter press was used to determine the filtered volume of the drilling fluid following the procedure of Amorin et al. (2011). The filter cake thickness of the fluids was determined using the filter paper and cake formed during the filtered volume experiment. The filter paper was thoroughly washed and placed in between two glass slides of equal diameter as the filtered paper before subjecting it to a pressure of 300 N/m^2 for 3 min. Then the slides, the filter paper and cake formed were put in an extensometer to determine the thickness of the cake formed.

To test the environmental friendliness of all the formulated fluids, 100 ml of each were spilled on 4 weeks old cassava plants and the number of days of the plants' survival was noted.

RESULTS

The result of the viscosity test is presented in Table 1, while Table 2 shows the pH, density, plastic viscosity, apparent viscosity, gel strength (10 s/10 mins) as well as the electrical conductivity of the formulated fluids.

DISCUSSION

From the results presented in Table 2 we can see that sample A (re-refined used lubricating base oil) has the highest apparent viscosity followed by sample C (Emcaid® synthetic base oil) while sample B (Paradril® synthetic base oil) exhibited the least viscosity. This result infer that re-refined lubricating base oil offers the greatest resistance to fluid flow with the least resistance offered by Paradril®

Table 2. Rheological and other properties of the formulated drilling muds.

Base oil samples	pH (-)	Density (ppg)	Plastic viscosity (μ_P) (cP)	Apparent viscosity (μ_A) (cP)	Gel strength lb/100 ft^2	ES (volts)
A	8.5	8.32	18	93	53/54	480
B	8.8	8.30	11	61	55/55	697
C	9.8	8.13	15	72	60/72	550
D	7.7	8.47	21	65	48/42	596

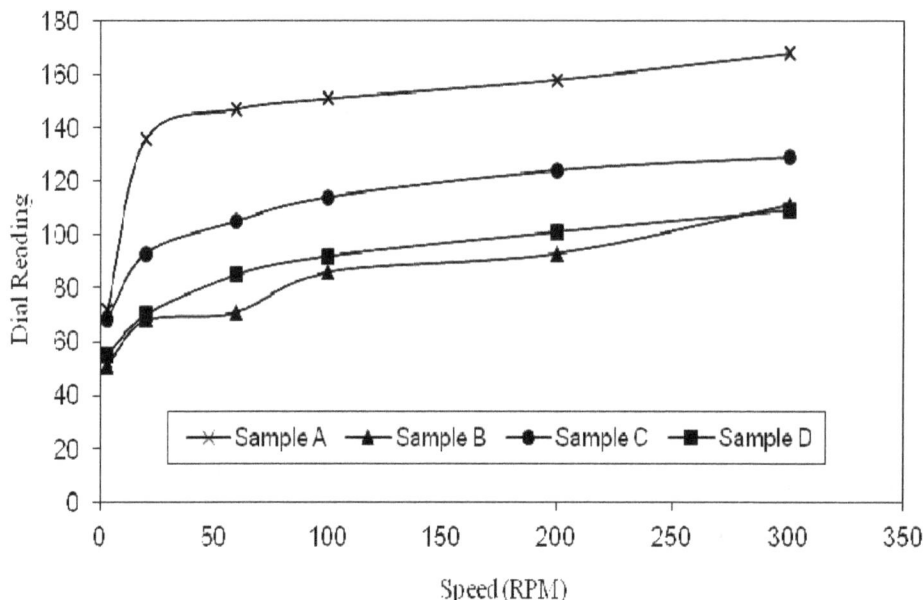

Figure 1. Viscometer plot for the formulated drilling muds.

synthetic base oil. Re-refined used lubricating base oil therefore posed the least prospect for the formulation of a good drilling fluid when compared with the three synthetic base oils used in this work since low viscosity drilling fluid lead to reduced wear in the drill string (Mitchell, 1995). However, the formulated muds from the four base oils have similar rheological behavior as they all approxi-mately exhibit the Bingham plastic model from the plots of the rotary viscometer dial reading against speed generated as shown in Figure 1. This is an indication that re-refined used lubricating oil has the potential to be used as base oil for formulating drilling mud if the viscosity is reduced by adding appropriate polymers. The formulated drilling fluid from the four base oil samples show the same range of densities, with Ty-Chem-Low Tox® synthetic base oil having the highest density of 8.47 ppg followed by re-refined used lubricating oil (8.32 ppg) and Paradril® with 8.30 ppg while Emcaid® had the least density of 8.13 ppg. According to Fadairo et al. (2012) the denser the base oil, the higher the amount of barite needed to build. From the results it is evident that Ty-

Chem-Low Tox® and re-refined used lubricating oil that have slightly higher densities will require the highest amount of barite to build.

Hydrogen ion potential (pH) is a very important parameter to consider when formulating drilling mud. Effective drilling muds are expected to be highly alkaline (that is, pH >7). This is because acidic (low pH) mud increases the corrosion of metals (pipes and casing) when it comes in contact with it. A drilling mud having a pH of between 7 and 9.5 had been reported to have the least effect on bentonite since the viscosity of such fluid remains relatively constant over a wide range of temperatures (Fadairo et al., 2012). However, a pH above 9.5 increases the mud viscosity thereby affecting the effectiveness of the drilling mud leading to complicated shale problems. As we can see from Table 2 the pH of the four formulated drilling fluids fall within the desired value, however, fluids formulated from re-refined used lubricating oil and Paradril® appear to give best hole stability and control over mud properties, since these requirements are met by fluid having a pH of 8.5 to 9.5

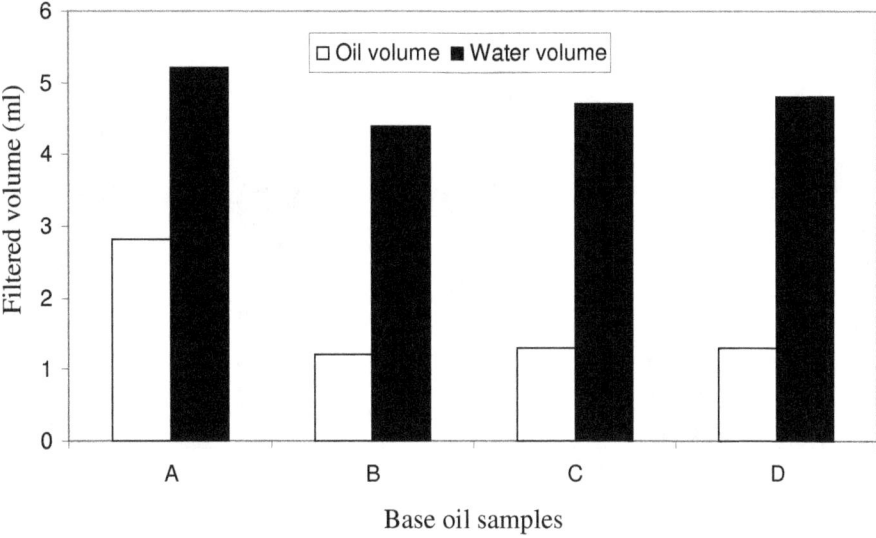

Figure 2. Filtration property of formulated drilling muds.

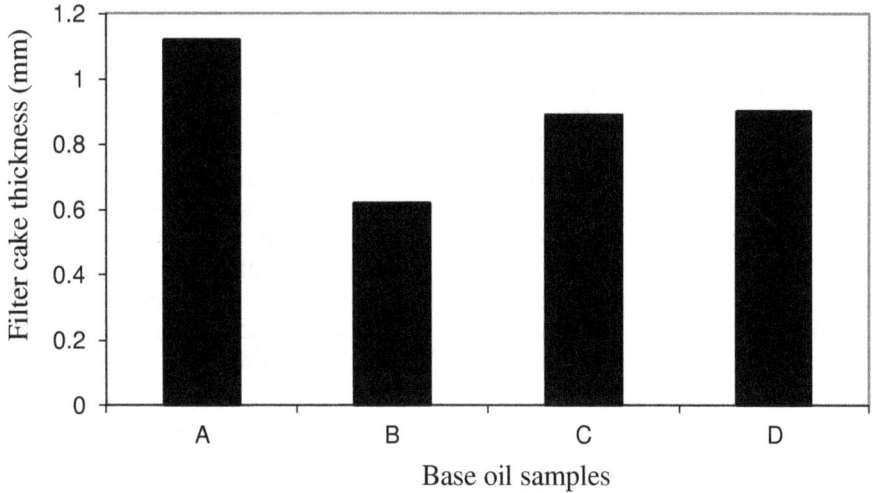

Figure 3. Filter Cake Thickness of formulated drilling muds.

(Fadairo et al., 2012).

The gel strength of the drilling fluids formulated from re-refined used lubricating oil was compared with those formulated from the three synthetic base oil used in this work. Gel strength is the ability of a drilling mud to suspend cuttings and other solid additives. From Table 2, the gel strength of mud produced from re-refined used lubricating oil and Paradril® synthetic base oil promised to be more effective than mud formulated from the other two synthetic base oils, since the shear rate of the mud remained consistent and high. High gel strength mud has the ability to suspend drill cuttings along the length of the drillpipe or bore annulus when the drilling mud circulation is stopped during pump tripping or any other secondary

operations (Shah et al., 2010). A low gel strength mud on the other hand do not efficiently suspend cuttings thereby allowing cuttings to quickly drop leading to pump shut-down, stuck pipe, hole pack-off, barite sag as well as accumulation of cutting beds. From the results, it is clear that re-refined used lubricating oil mud has excellent cutting transport capabilities even at low values of viscosity. This result is similar to that obtained by Fadairo et al. (2012) for diesel based mud.

Another factor determining the successful performance of a drilling fluid tested for, in the formulated muds was the mud filtration capacity. From Figures 2 and 3 we can see that re-refined OBM has the highest filtration rate and as a result a thicker filter cake due to its high porosity

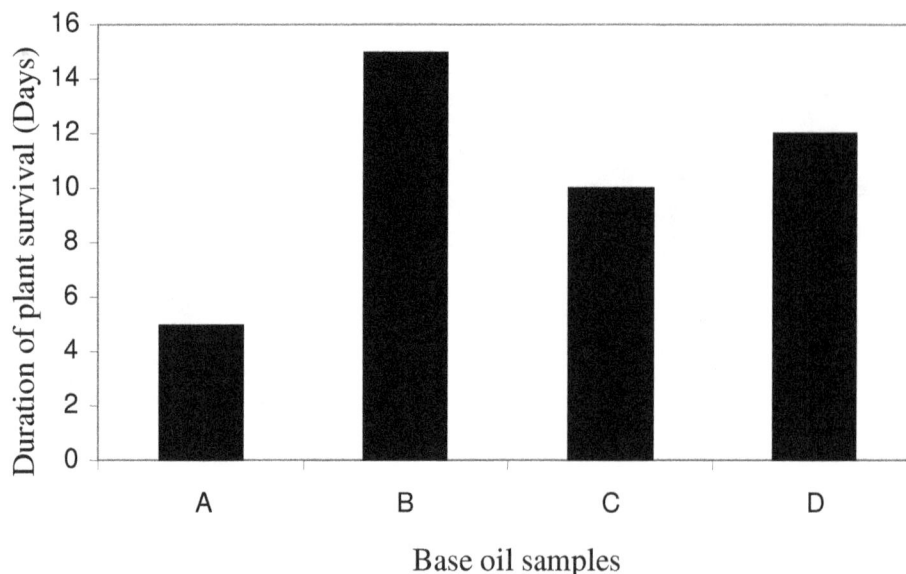

Figure 4. Survival days of cassava plants spilled with equal volume of the formulated muds

while Paradril® SOBM exhibited the lowest rate of filtration with thinner filter cake. High filtered volume is associated with thick filter cake because the cake is formed by deposition of clay particles on the walls of the hole during loss of water to the formation. So the higher the filtered volume, the thicker the filter cake and the less efficient the drilling mud. A thick cake reduces the effective diameter of the hole and increases the contact area between the tube and the cake leading to increased risk of stuck tubes (Amorin et al., 2011). Based on this result, drilling mud formulated from re-refined used lubricating oil will not be an effective fluid for drilling purposes.

Electrical stability (ES) is a vital property of oil based mud (OBM) and synthetic oil based mud (SOBM). The ES represents the stability of emulsions formed by oil and water during the formulation. A low ES mud is not conductive and therefore cannot transfer power. A good drilling mud should have an ES of between 700 and 900 V under circulation. However, an ES range of 300 to 400V is considered ideal for newly formulated mud as well as mud in storage. From the results of the ES test presented in Table 2, drilling mud formulated from the four base oils used in this study meet the specification for stable fluid, however, mud formulated from re-refined used lubricating oil exhibited the least ES value of 480 V and as such is the least stable of the formulations. The low ES value of the re-refined used lubricating oil based mud may be as a result of the low resistivity of the re-refined used oil. The resistivity of this base oil may have been reduced due to the rigorous re-processing steps it was subjected to prior to its use for the formulation. For re-refined used lubricating oil based mud emulsion to be

stable there will be need to add water and salt to the formulation, which will invariably affect the effectiveness of the mud.

The toxicity test conducted by spilling equal volume of the four formulated muds on young cassava plants show that re-refined used lubricating oil mud is the most toxic of the formulations. Cassava plant spilled with re-refined used lubricating oil based mud first showed evidence of withering after 3 days and finally died after 5 days. Cassava plants spilled with mud formulated from Paradril®, Emcaid® and Ty-Chem-Low Tox® synthetic base oils survived for 15, 10 and 12 days respectively (Figure 4). From this result it is clear that SOBM is more environmental friendly than re-refined used lubricating oil based mud. In a similar study, Fadairo et al. (2012) observed that jatropha oil based mud spilled on growing bean seedling was able to survive for 16 days before it eventually died while the same quantity of diesel oil based mud spilled on the same bean seedling survived for only 7 days before dying. Re-refined used lubricating oil based mud is therefore more toxic than all other types of base oil used for formulating drilling mud – even diesel. Toxicity of drilling mud is a function of the aromatic content of the base oil. An environmental friendly drilling mud is one with negligible carcinogenic poly-aromatic compounds. This explains why vegetable base oil mud are ecologically and environmentally friendly as seen from the drilling mud produced from the other three synthetic base oils. Though extraction of the re-refined used lubricating oil with an aromatic selective solvent (N-methylpyrolidone (NMP)) is aimed at reducing the aromatic content of the re-refined oil to non detectable level, the toxicity result shows that the re-refined used lubricating oil based mud may still contain high concentration of aromatic compounds and this may have been responsible for its high toxic nature. Based on this result, re-refined used lubricating oil does not meet the environmental conditions for the formulation of an efficient and ecologically safe oil based drilling mud.

Conclusion

The possibility of using re-refined used lubricating oil as base oil for the formulation of drilling mud had been investigated. From the results, it is clear that re-refined used lubricating oil is not a viable option neither for diesel oil based mud nor for synthetic oil based mud. Re-refined used lubricating oil based mud is very toxic and therefore fails the environmental requirement as outlined for efficient drilling mud by the Nigerian Government. The cost index for re-refined used lubricating oil based mud may also be higher than those of the synthetic oil based mud due to the combined cost of refining and extraction. As a result, re-refined used lubricating oil may not be a viable alternative to vegetable oil and other synthetic oils for the formulation of drilling mud.

REFERENCES

Abdel-Raouf ME (2011). Factors Affecting the Stability of Crude Oil Emulsions, www.intechopen.com. Assessed 3rd November 2-12. pp. 188.

Akpabio EJ, Ekott EJ (2013). Application of Physico-Technological Principles in Demulsification of Water-In-Crude Oil System. India J. Sci. Technol. 6(1):1-3.

Amorin LV, Nascimento RCA, Lira DS, Magalhães J (2011). Evaluation of the Behavior of Biodegradable Lubricants in the Differential Sticking Coefficient of Water Based Drilling Fluids. Braz. J. Pet. Gas. 5(4):197-203.

Broni-Bediako E, Amorin A (2010). Effects of Drilling Fluid Exposure to Oil and Gas Workers Presented with Major Areas of Exposure and Exposure Indicators. Res. J. Appl. Sci. Eng. Tech. 2(8):770-772.

Dimitrov AN, Yordanov DI, Petkov PS (2011). Study on the Effects of Demulsifiers on Crude Oil and Petroleum Products. Int. J. Environ. Res. 6(2):435-436.

Fadairo A, Falode O, Ako C, Adeyemi A, Ameloko A (2012). Novel Formulation of Environmentally Friendly Oil Based Drilling Mud, In New Technologies in the Oil and Gas Industry, Chapter 3, INTECH Open Science, http://dx.doi.org/105772/52136 Assessed 4th January 2013.

Growcock FB, Ellis CF, Schmidt DD (1994). Electrical Stability, Emulsion Stability, and Wettability of Invert Oil-Based Muds. SPE Drilling Completion 9(1):39-46.

Huda SN, Nour AH (2011). Microwave Separation of Water-in-Crude Oil Emulsions. Int. J. Chem. Environ. Eng. 2(1):70-71.

Langevin D, Pateau S, Hénaut I, Argillier JF (2004). Crude Oil Emulsion Properties and their Application to Heavy Oil Transportation. Oil Gas Sci. Technol.= Rev. IFP. 59(5):513.

Mitchell B (1995). Advanced Oil Well Drilling Engineering Handbook, Mitchell Engineering, 10th Edition. pp. 248-251.

Nweke CO, Okpokwasili GC (2003). Drilling Fluid Base Oil Biodegradation Potential of a Soil Staphylococcus Species, Afr. J. Biotechnol. 2(9):293.

Ogbo EM, Avwerovwe U, Odogu G (2009). Screening of four common weeds for use in phytoremediation of soil contaminated with spent lubricating oil. Afr. J. Plant Sci. 3(5):102.

Oghenejoboh KM, Ohimor OE (2012). Contamination of Soil and Rivers from Used Engine Oil: A Case Study of Choba Community in Port-Harcourt, Nigeria. Pol. Res. 32(2):131.

Shah SN, Narayan PE, Shanker H, Ogugbue CC (2010). Future Challenges of Drilling Fluids and Their Rheological Measurements, American Association of Drilling Engineers (AADE) Conference and Exhibition, Houston Texas, USA, 6 - 7 April.

Wang QR, Cui YS, Liu XM, Dong YT, Christie P (2000). Soil Contamination and Plant Uptake of Heavy Metals at Polluted Sites in China. J. Environ. Sc. Health Part A - Toxic/Hazardous Substances and Environmental Engineering 8:823-825.

11

Evaluation of different algal species for the higher production of biodiesel

Gulab Chand Shah*, Mahavir Yadav and Archana Tiwari

School of Biotechnology, Rajiv Gandhi Proudyogiki Vishwavidyalaya, Bhopal, Madhya Pradesh
(University of Technology of Madhya Pradesh), Airport Bypass Road, Gandhi Nagar, Bhopal-462 033, India.

Algae are a potential resource for biodiesel production. *Scandasmus algae* **displayed faster growth at a wide temperature range of 4 to 32°C compared to** *Chlorella vulgaris*. **As the Conventional fuels are depleting day by day, there is a need to find out an alternative fuels to fulfill the energy demand of the world. Biofuels is one of the best available resources that have come to the forefront recently. In this paper, a detailed review has been conducted to highlight different related aspects to biodiesel industry. These aspects include: biodiesel feedstocks, extraction and production methods, properties and qualities of biodiesel, problems and potential solutions of using vegetable oil, advantages and disadvantages of biodiesel, the economical viability and finally the future of biodiesel. The literature reviewed is selective and critical. Based on the overview presented, it is clear that the search for beneficial biodiesel sources should focus on feedstocks that do not compete with food crops, do not lead to land-clearing and provide greenhouse-gas reductions. These feedstocks include non-edible oils such as** *Jatropha curcas* **and** *Calophyllum inophyllum*, **and more recently microalgae and genetically engineered plants such as poplar and switchgrass have emerged to be very promising feedstocks for biodiesel production.**

Key words: Biodiesel feedstock, extraction, production, properties and qualities, problems.

INTRODUCTION

Renewable energy plays a critical role in addressing issues of energy security and climate change at global and national scales. In the US, the Federal Government passed the energy independence and security act (EISA) in 2007 which requires a gradual increase in the production of renewable fuels to reach 36 billion gallons per year by 2022. Furthermore, 28 States have passed their own mandatory renewable energy legislation (Energy Information Administration (EIA), (2009). Energy has become a crucial factor for humanity to continue the economic growth and maintain high standard of living especially after the inauguration of the industrial

revolution in the late 18th and early 19th century. According to the International Energy Agency (IEA) report (International Energy Agency (IEA) (2007) and Shahid and Jamal (2011), the world will need 50% more energy in 2030 than today, of which 45% will be accounted for by China and India.

The development of biofuels as an substitute fuel to supplement or replace conventional diesel is receiving great attention among researchers and policy makers for its numerous advantages such as renewability, biodegradability and lower gaseous emission profile. Also, concerns over increasing energy demand, continuous global warming effects, declining petroleum reserves, petroleum price hike and scarcities have raised the need to search for alternative renewable fuels (Sahoo et al., 2007; Basha et al., 2009; Refaat, 2010; Demirbas, 2009; Yang et al., 2012). The annual rate of the global

---
*Corresponding author. E-mail: gulab777@gmail.com.

$$CH_2 - O - CO - R_1 \qquad CH_3 - O - CO - R_1$$

$$CH - O - CO - R_2 + 3ROH \xrightarrow{(Catalyst)} CH_3 - O - CO - R_2 + \begin{array}{c} CH_2 - OH \\ | \\ CH - OH \\ | \\ CH_2 - OH \end{array}$$

$$CH_2 - O - CO - R_3 \qquad CH_3 - O - CO - R_3$$

Triglyceride   Alcohol   Mixture of fatty alkyl Esters   Glycerol

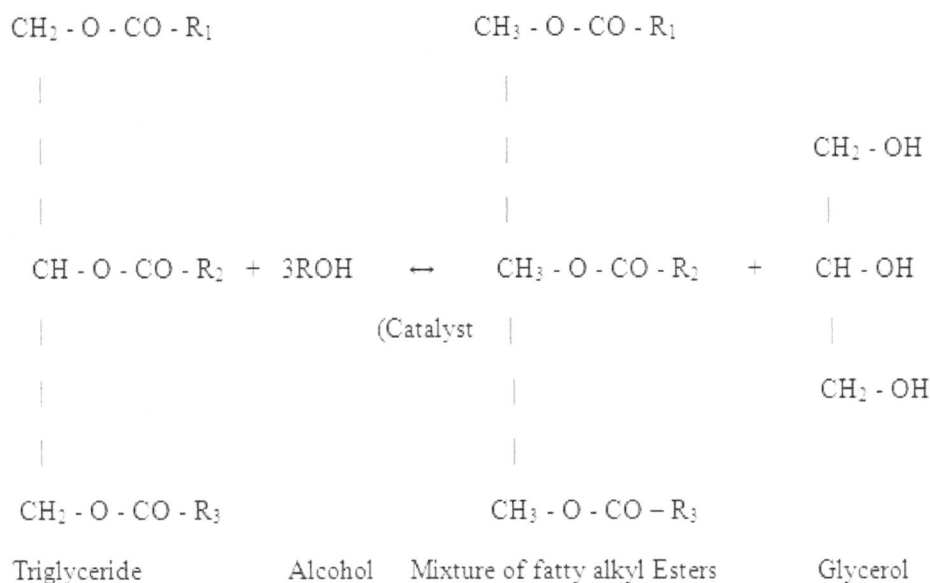

Figure 1. Transesterification of triglycerides to alkyl esters (biodiesel).

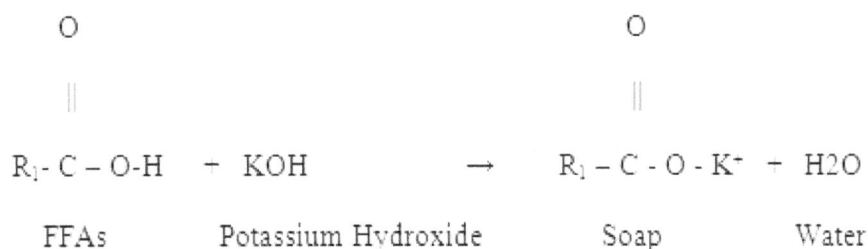

$$\begin{array}{ccccc} & O & & O & \\ & \| & & \| & \\ R_1 - C - O - H & + KOH & \longrightarrow & R_1 - C - O - K^+ & + H2O \end{array}$$

FFAs   Potassium Hydroxide   Soap   Water

Figure 2. Formation of saponified product (soap).

primary power demand is estimated to increase to a value of 1.7% and reach a value of 16,487 from 2002 to 2030 (Pandey et al., 2012).

Thus, use of renewable power is expected to get better the energy availability. Besides, renewable force is one of the most efficient routes to achieve sustainable development (Demirbas, 2009; Demirbas, 2011; Ramadhas et al., 2004; Saeid et al., 2008; Antolin et al., 2002; Bari et al., 2002; Demirbas, 2009; Kapilan et al., 2009). Transesterification is the reaction of triglycerides to fatty acid alkyl esters (FAAE) and low molecular weight alcohols such as methanol and ethanol in the presence of catalyst (Demirbas, 2011; Sharma and Singh, 2009; Demirbas et al., 2009) as shown in Figure 1. Methanol is the most favored alcohol because it is less costly and easily obtainable (Leung et al., 2010). The formed FFAs favor the formation of soap as shown in Figure 2. Saponification reaction is usually referred to as a side reaction that occurred during transesterification of fats and oils to biodiesel escalating soaps formation (Haas, 2005; Zadra, 2006).

Soap formation increases loss of methyl ester to glycerol phase, resulting to high purification costs and less biodiesel yield (Zadra, 2006). For this reason, processing low quality feedstocks containing considerable amount of water and FFAs using alkali catalyzed route requires the feedstocks to undergo pretreatment via esterification reaction using acid catalyst (H_2SO_4) as shown in Figure 3. The product (refined feedstocks) obtained with less FFAs content (≤ 3 wt.%) and water content (< 0.06 wt.%) is then used for alkali-catalyzed transesterification to produce biodiesel (Velasquez-Orta et al., 2012).

RAW MATERIALS FOR BIODIESEL PRODUCTION

Animal fats and vegetable oils are the main raw materials usually employed to produce biofuels. Also, oils from algae have shown some promise as raw materials for biofuels production. Fats and oils are primarily water-insoluble, hydrophobic substances in the plant and

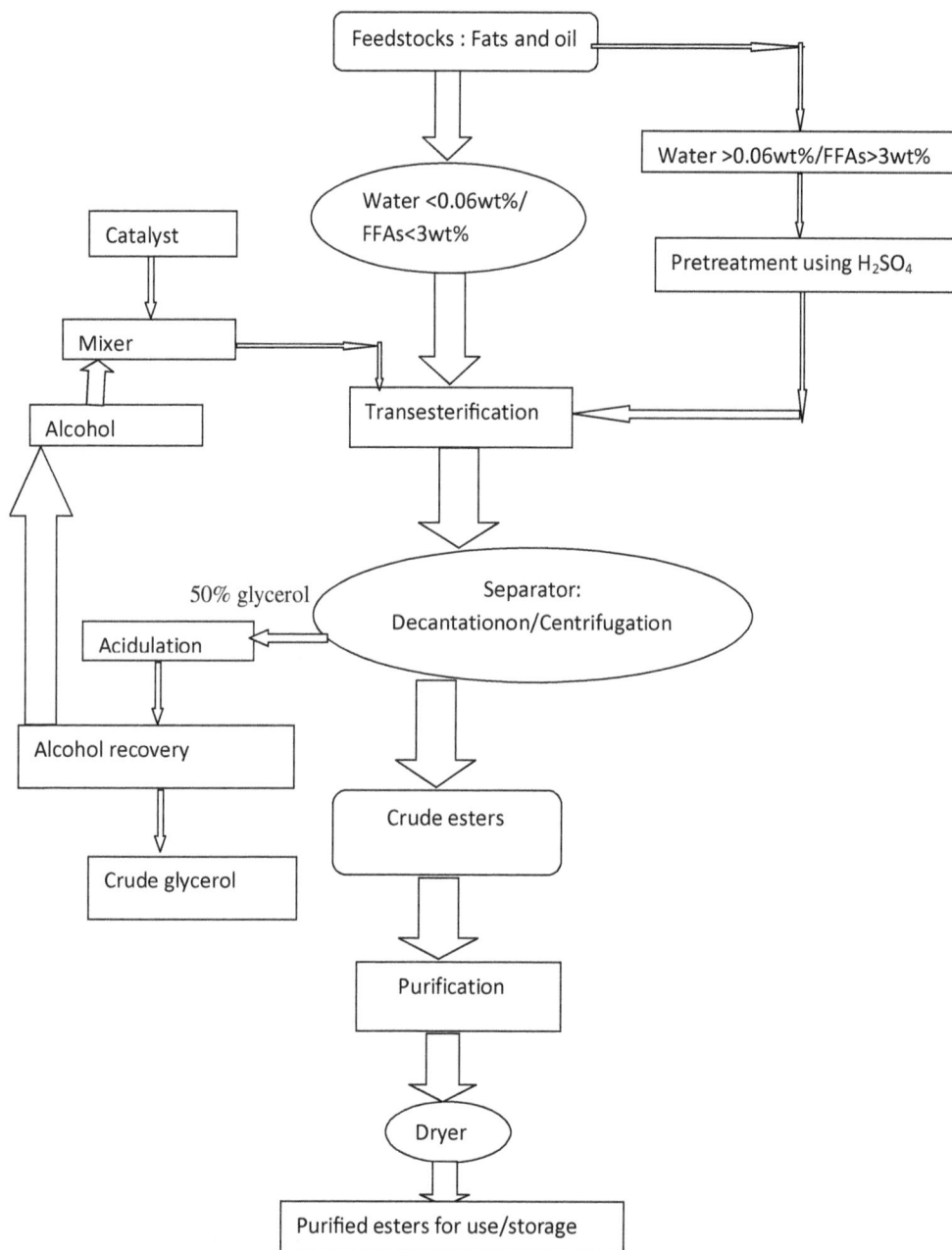

Figure 3. Schematic diagram of alkali-catalyzed transesterification for alkyl esters production.

animal kingdoms that are made up of one molecule of glycerol and three molecules of fatty acids are commonly referred to as triglycerides (Agarwal, 2007). Transesterification of triglycerides for biofuels production has been extensively studied in recent years. The raw materials being exploited commercially by the Constitute, the edible oils derived from rapeseed, soybean, palm, sunflower, coconut, and linseed (Mohibbe Azam et al. (2005). The use of refined raw materials could yield high-purity and quality biofuels with physicochemical properties comparable to diesel fuels. The major

economic factor to be considered as input costs for biodiesel production is the cost of feedstocks (Demirbas, 2010).

Algae are classified into four categories: diatoms, green algae, blue-green algae and golden algae. Microalgae are microscopic photosynthetic organisms that are found in both marine and freshwater environments (Demirbas, 2010). Oleaginous microorganisms like microalgae, bacillus, fungi and yeast are all available for biodiesel production (Meng et al., 2009). Microalgae have been recommended as good sources for biofuels production because of their advantages of superior photosynthetic efficiency, higher biomass production and faster growth rates compared to other energy yielding crops (Miao and Wu, 2006).

FACTOR EFFECTING ETHYL ESTER YIELD

Most of the points discussed here in after concern homogeneous alkaline catalysis unless specified otherwise. According to many studies focusing on the influence of reaction parameters on the ethanolysis, the main independent variables that affect ethyl ester production are (Cernoch et al., 2010):

i. The ethanol: vegetable oil molar ratio, MR,
ii. The reaction temperature, T,
iii. The amount of catalyst (weight ratio of catalyst to vegetable oil wt%), C,
iv. The purity of reactants (free fatty acid and water contents).

FEEDSTOCK'S FOR BIODIESEL

Since last few years many biodiesel production plants have been identified in India. These plants are operational as per the availability of feedstock, price of crude vegetable oils. The common feedstock for the production of biodiesel in India is palm oil, which is being imported from Malaysia and Indonesia. As palm oil is being used for edible purposes, the price of the crude palm oil fluctuates in the international market. Even though the consumption of edible oils in some countries like India is high, the availability of used cooking oil is very small. Hence, focus needs to be shifted to non-edible oilseed plants available in India and the details of such potential oilseed plant. Among these non-traditional oilseeds plants huge scope exists for the oilseeds such as *Pongamia piñnata* and *Jatropha cardus* as biodiesel feedstock. The oil from algal biomass has also attracted the attention as the future feedstock for biodiesel (Zadra, 2006).

OIL EXTRACTION METHODS

There are three main methods that have been identified for extraction of the oil: (i) Mechanical extraction, (ii) solvent extraction and (iii) enzymatic extraction. Before the oil extraction takes place, seeds have to be dried. Seed can be either dried in the oven (105°C) or sun dried. Mechanical expellers or presses can be fed with either whole seeds or kernels or a mix of both, but common practice is to use whole seeds. However, for chemical extraction only kernels are used as feed (Achten et al., 2008).

BIODIESEL PRODUCTION TECHNOLOGIES

Globally, there are many efforts to develop and improve vegetable oil properties in order to approximate the properties of diesel fuels. It has been remarked that high viscosity, low volatility and polyunsaturated characters are the mostly associated problems with crude vegetable oils. These problems can be overcome by pyrolysis, dilution with hydrocarbons blending, microemulsion, and transesterification (Canakci and Sanli, 2010; Pandey, 2008; Demirbas and Demirbas, 2007; Chauhan et al., 2010).

PROPERTIES AND QUALITIES OF BIOFUELS

Since biofuels are produced from quite differently scaled plants of varying origins and qualities, it is necessary to install a standardization of fuel quality to guarantee an engine performance without any dificulties (Pinto et al., 2005). Austria was the first country in the world to define and approve the standards for rapeseed oil methyl esters as a biofuels. The guidelines for standards and the quality of biodiesel have also been defined in other countries such as in Germany, Italy, France, the Czech Republic and the United States (Meher et al., 2010).

PROBLEMS ASSOCIATED WITH BIODIESEL PRODUCTION

The direct use of vegetable oils or blends has generally been considered to be impractical for both direct and indirect diesel engines. The high viscosity, low volatility, acid composition, free fatty acid and moisture content, gum formation due to oxidation and polymerization during storage and combustion, poor cold engine start-up, misfire, ignition delay, incomplete combustion, carbon deposition around the nozzle orifice, ring sticking, injector choking in engine and lubricating oil thickening are the major problems of using vegetable oils. In general, the problems associated with using straight vegetable oil in diesel engines are classified into short and long term probable causes and the potential solutions (Chauhan et al., 2010; Agarwal, 2007).

ADVANTAGES OF BIODIESEL

Biodiesel has 10–11% of oxygen; this makes biodiesel a fuel with high combustion characteristics (Miao and Wu, 2006; Chauhan et al., 2010). Biodiesel reduces net carbon-dioxide emissions by 78% on a lifecycle basis when compared to conventional diesel fuel and reduces smoke due to free soot (Chauhan et al., 2010; Saeid et al., 2008; Bari et al., 2002). Biodiesel is renewable, non-toxic, non-flammable, portable, readily available, biodegradable, sustainable, ecofriendly and free from sulfur and aromatic content, this makes it an ideal fuel for heavily polluted cities. Biodiesel also reduces particular matter content in the ambient air and hence reduces air toxicity. It provides a 90% reduction in cancer risks and neonatal defects due to its less polluting combustion (International Energy Agency (IEA), 2007; Demirbas, 2009; Demirbas, 2010; Miao and Wu, 2006; Chauhan et al., 2010). Biodiesel serves as climatic neutral in view of the climatic change that is presently an important element of energy use and development (Miao and Wu, 2006; Chauhan et al., 2010). Biodiesel has higher cetane number (about 60 to 65 depending on the vegetable oil) than petroleum diesel (53) which reduces the ignition delay (International Energy Agency (IEA), 2007; Demirbas, 2010; Miao and Wu, 2006; Chauhan et al., 2010).

DISADVANTAGES OF BIODIESEL

Biodiesel has 12% lower energy content than diesel, this leads to an increase in fuel consumption of about 2 to 10%. Moreover, biodiesel has higher cloud point and pour point, higher nitrogen oxide emissions than diesel. It has lower volatilities that cause the formation of deposits in engines due to incomplete combustion characteristics (International Energy Agency (IEA), 2007; Miao and Wu, 2006). Biodiesel causes excessive carbon deposition and gum formation (polymerization) in engines and the oil gets contaminated and suffers from flow problem. It has relatively higher viscosity (11 to 18 times diesel) and lower volatility than diesel and thus needs higher injector pressure (Achten et al., 2008). It can be oxidized into fatty acids in the presence of air and causes corrosion of fuel tank, pipe and injector (Demirbas, 2009; Kapilan et al., 2009).

Due to the high oxygen content in biodiesel, advance in fuel injection and timing and earlier start of combustion, biodiesel produces relatively higher NO_2 levels than diesel in the range of 10 to 14% during combustion (Demirbas, 2010). Economical viability of biodiesel is an attractive renewable energy resource. However, there are some challenges that face this vital resource (Pandey et al., 2012). These challenges include the high cost and limited availability of biodiesel feedstock beside the cheaper prices of crude petroleum. There are various factors contributing to the cost of biodiesel. These factors include feedstock prices, plant's capacity, feedstock quality, processing technology, net energy balance nature of purification and its storage, etc (Haas, 2005). However, the two main factors are the costs of feedstocks and the cost of processing into biodiesel. It has been found that the cost of feedstocks accounts for 75% of the total cost of biofuels (Miao and Wu, 2006).

FUTURE OF BIODIESEL

Biodiesel production is expanding rapidly around the world, driven by energy security and other environmental concerns. Given geographic disparities between demand and supply potential, and supply cost, expanded trade in biodiesel appears to make sense. Global potential in biodiesel production is very unclear, but in the long run it could be a substantia percentage of transport fuel demand. Currently, biodiesel can bemore effective if used as a complement to other energy sources (Kapilan et al., 2009).

CONCLUSION

Energy is an indispensable factor for human to preserve economic growth and maintain standard of living. Globally, the transportation sector is the second largest energy consuming sector after the industrial sector and accounts for 30% of the world's total delivered energy. This sector has experienced a steady growth in the past 30 years. It has been estimated that the global transportation energy use is expected to increase by an average of 1.8% per year from 2005 to 2035. Nearly all fossil fuel energy consumption in the transportation sector is from oil (97.6%). However, the expected depletion of fossil fuels and the environmental problems associated with burning them has encouraged many researchers to investigate the possibility of using alternative fuels. Among them, biodiesel seems a very promising resource.

REFERENCES

Achten WMJ, Verchit L, Mathijs Franken YJ, Singh E, Aerts VP, Muys RB(2008). Jatropha bio-diesel production and use. Biomass Bioenergy 32(12):1063-1084.

Agarwal AK (2007). Biofuels (alcohols and biodiesel) applications as fuels for internal combustion engines. Prog. Energy Combust. Sci. 3(33):233-271.

Agarwal AK (2007). Biofuels (alcohols and biodiesel) applications as fuels for internal combustion engines. Progr Energy Combust. Sci. 33:233-271.

Antolin G, Tinaut FV, Briceno Y, Castano V, Perez C, Ramirez Al (2002). Optimisation of biodiesel production by sunflower oil transesterification. Bioresour. Technol. 83:111-114.

Bari S, Lim TH, Yu CW (2002). Effects of preheating of crude palm oil (CPO) on injection system, performance and emission of a diesel engine. Renew Energy. 27:339-351.

Basha SA, Raja Gopal K, Jebaraj S (2009). A review on biodiesel production, combustion, emissions and performance. Renew. Sustain. Energy Rev. 13:1628-1634.

Canakci M, Sanli H (2010). Biodiesel production from various feedstocks and their effects on the fuel properties. J. Ind. Microbiol. Biotechnol. 35(5):431-41.

Cernoch M, Hájek M, Skopal F (2010). Ethanolysis of rapeseed oil - distribution of ethyl esters, glycerides and glycerol between ester and glycerol phases. Bioresour. Technol. 01:2071-2075.

Chauhan BS, Kumar N, Jun YD, Lee KB (2010). Performance and emission study of preheated Jatropha oil on medium capacity diesel engine. Energy 35(6):2484-2492.

Chauhan BS, Kumar N, Jun YD, Lee KB (2010). Performance and emission study of preheated Jatropha oil on medium capacity diesel engine. Energy 35(6):2484-2492.

Demirbas A (2009). Biofuels securing the planet's future energy needs. Energy Con. Vers. Manage. 50:2239-2249.

Demirbas A (2009). Biorefineries: current activities and future developments. Energy Convers. Manage. 50:2782-801.

Demirbas A (2009). Progress and recent trends in biodiesel fuels. Energy Convers. Manage. 50:14-34.

Demirbas A (2010). Importance of biodiesel as transportation fuel. Energy Policy 35:4661-4670.

Demirbas A (2010). Use of algae as biofuel sources. Energy Convers. Manage. 51:2738-2749.

Demirbas A (2011). Competitive liquid biofuels from biomass. Appl. Energy 88:17-28.

Demirbas AH, Demirbas I (2007). Importance of rural bioenergy for developing countries. Energy Convers. Manage. 48(8):2386-2398.

Demirbas F, Balat M, Balat H (2009). Potential contribution of biomass to the sustainable energy development. Energy Convers. Manage. 50:1746-1760.

Energy Information Administration (EIA) (2009). State renewable energy requirements and goals: Update through. (2008).

Haas MJ (2005). Improving the economics of biodiesel production through the use of low value lipids as feedstocks: Vegetable oil soapstock. Fuel Process Technol. 86:1087-1096.

International Energy Agency (IEA). World energy outlook 2007. Available from: http://www.iea.org/textbase/nppdf/free/2007/weo 2007.pdf; 2007 [cited 02.02.11].

Kapilan N, Ashok Babu TP, Reddy RP (2009). Technical aspects of biodiesel and its oxidation stability. Int. J. Chem. Tech. Res. 1:278-282.

Leung DYC, Wu X, Leung MKH (2010). A review on biodiesel production using catalyzed transesterification. Appl. Energy 87:1083-1095.

Meher LC, Vidya Sagar D, Naik SN (2010). Technical aspects of biodiesel production by transesterification a review. Renew. Sustain. Energy Rev. 10(3):248-268.

Meng X, Yang J, Xu X, Zhang L, Nie Q, Xian M (2009). Biodiesel production from oleaginous microorganisms. Rev. Enferm. 34:1-5.

Miao X, Wu Q (2006). Biodiesel production from heterotrophic microalgal oil. Bioresour. Technol. 97:841-846.

Mohibbe Azam M, Waris A, Nahar NM (2005). Prospects and potential of fatty acid methyl esters of some non-traditional seed oils for use as biodiesel in India. Biomass Bioenergy 29:293-302.

Pandey A (2008). Handbook of plant-based biofuels. Taylor and Francis Group. P. 87.

Pandey RK, Rehman A, Sarviya RM (2012). Impact of alternative fuel properties on fuel spray behavior and atomization. Renew. Sustain. Energy Rev. 16:1762-1778.

Pinto AC, Guarieiro LLN, Rezende MJC, Ribeiro NM, Torres EA, Lopes WA (2005). Biodiesel: An overview. J. Br. Chem. Soc. 16(6b):1313-1330.

Ramadhas AS, Jayaraj S, Muraleedharan C (2004). Use of vegetable oils as I.C. engine fuels a review: data bank. Rev. Enferm. 29:727-742.

Refaat AA (2010). Different techniques for the production of biodiesel from waste vegetable oil. Int. J. Environ. Sci. Technol. 7(1):183-213.

Saeid B, Aroua MK, Abdul Raman A, Sulaiman NMN (2008). Density of palm oil-based methyl ester. J. Chem. Eng. Data. 53:877-880.

Sahoo PK, Das LM, Babu MKG, Naik SN (2007). Biodiesel development from high acid value polanga seed oil and performance evaluation in a CI engine. Fuel 86:448-454.

Shahid EM, Jamal J (2011). Production of biodiesel: a technical review. Renew. Sustain. Energy Rev. 15(9):4732-4745.

Sharma YC, Singh B (2009). Development of biodiesel: current scenario. Renew. Sustain. Energy Rev. 13:1646-1651.

Velasquez-Orta SB, Lee JGM, Harvey A (2012). Alkaline in situ transesterification of Chlorella vulgaris. Fuel 94:544-550.

Yang CY, Li ZFB, Long YF (2012). Review and prospects of Jatropha biodiesel industry in China. Renew. Sustain. Energy Rev. 16:2178-2190.

Zadra R (2006). Improving process efficiency by the usage of alcoholates in the biodiesel production. In: IV forum Brazil-Alemanhasobrebrazil Aracatuba. 16(2012):3456-3470.

12

Jatropha oil production and an experimental investigation of its use as an alternative fuel in a DI diesel engine

Vijittra Chalatlon[1], Murari Mohon Roy[2]*, Animesh Dutta[3] and Sivanappan Kumar[1]

[1]School of Environment, Resources and Development, Asian Institute of Technology, P.O. Box 4, Klong Luang, Pathumthani 12120, Thailand.
[2]Department of Engineering, Nova Scotia Agricultural College, 39 Cox Road, Banting Building, Truro, NS, Canada, B2N 5E3.
[3]Mechanical Engineering Program, School of Engineering University of Guelph, Guelph, Ontario, Canada, N1G 2W1.

In this study, a non-edible vegetable oil was produced from *Jatropha* fruits as a substitute fuel for diesel engines and its usability was investigated as pure oil and as a blend with petroleum diesel fuel. A direct injection (DI) diesel engine was tested using diesel, *Jatropha* oil, and blends of *Jatropha* oil and diesel in different proportions. A wide range of engine loads and *Jatropha* oil/diesel ratios of 5/95% (J5), 10/90% (J10), 20/80% (J20), 50/50% (J50), and 80/20% (J80) by volume were considered. The following performance parameters were measured; brake thermal efficiency, brake specific fuel consumption and CO and CO_2 emissions. No significant change in brake thermal efficiency and brake specific fuel consumption was experienced up to J20 ratios. However, higher blends suffered from deterioration in efficiency and fuel consumption about 10 to 25%. At low load operations, CO_2 emission with blends was lower than that of diesel, whereas, at high loads, CO_2 emission became higher with a higher percentage of *Jatropha* oil in the blends. However, CO emission with blends was much higher than that of diesel; the higher the percentage of *Jatropha* oil in the blend, the higher the CO emission.

Key words: Non-edible vegetable oil, *Jatropha* oil, diesel-*Jatropha* oil blends, viscosity and heating, DI diesel engine, performance, emissions.

INTRODUCTION

Using vegetable oil in an internal combustion (IC) engine is not new. A review of literature available in the field of vegetable oil usage has identified many advantages. Vegetable oil is produced domestically which helps to reduce costly petroleum imports, it is biodegradable, non-toxic, contains low aromatics and sulphur and hence, is environment friendly. Personal safety is improved as the flash point is more than 100°C higher than that of diesel. It is usable within the existing petroleum diesel infrastructure with little or no modification to an engine (Chauhan et al., 2010). There are however some challenges. The price of vegetable oil is dependent on the feedstock price, homogeneity, consistency and

reliability of supply. Food-fuel conflict of vegetable oils is a burning issue. *Jatropha* oil is non-edible oil and thus, there is no such issue. The high viscosity of vegetable oils may cause engine problems, especially at low operating temperature. Heating the oil or blending with diesel fuel can help solve this problem. Quick (1980), Goering et al. (1981), Bari and Roy (1995) and Sapaun et al. (1996) investigated many vegetable oils in the 80's and 90's. Quick (1980) reported that over 30 different vegetable oils have been used to operate compression ignition (CI) engines since the 1900's. Initial engine performance suggests that these oil-based fuels have great potential as fuel substitutes. During the 1980's, there were many studies that tested the possi-bility of using vegetable oils as a replacement for diesel. Goering et al. (1981) studied the characteristic properties of eleven vegetable oils to determine which oils would be best suited for use as an alternative fuel source. Of the

*Corresponding author. E-mail: mroy@nsac.ca.

eleven oils tested, corn, rapeseed, sesame, cottonseed, and soybean oils had the most favorable fuel properties. Rice bran oil was attempted in a diesel engine by Bari and Roy (1995) in their study. Rice bran oil was blended with kerosene at a ratio of 1:1 (by volume) to reduce the viscosity. Performance results led to the conclusion that rice bran oil can be a very prospective substitute for diesel. Sapaun et al. (1996) reported that, in Malaysia, palm oils can be used as diesel substitutes with promising results. Performance tests indicated that power outputs were almost the same for palm oil, blends of palm oil and diesel, and 100% diesel.

Research on vegetable oil use in diesel engine is still progressing today. A study by Kalam et al. (2003) presented the results of exhaust emissions characteristics of ordinary Malaysian coconut oil blended with conventional diesel oil in a diesel engine. The results showed that the addition of 30% coconut oil with diesel oil produced higher brake power with a net reduction in exhaust emissions. The experiments were undertaken by He and Bao (2005) to test a diesel engine using oil composed of cottonseed oil and conventional diesel fuel. The experimental results showed that a mixing ratio of 30% cottonseed oil and 70% diesel was optimal in ensuring relatively high thermal efficiency of the engine. Research published by Cetin and Yuksel (2007) briefly reviewed the use of hazelnut oil as an alternative fuel in pre-chamber diesel engines, and compared it with diesel. The results showed that the hazelnut oil may be employed in most diesel operating conditions in terms of the performance and emission parameters without any modification or preheating of the fuels. An experimental investigation has been carried out to analyze the performance and emission characteristics of a CI engine fuelled with karanja oil and its blends with petroleum Diesel (K10, K20, K50 and K75) by Agarwall and Rajamanoharan (2009). The effect of temperature on the viscosity of karanja oil has also been investigated. A series of engine tests, with and without preheating have been conducted using each of the above fuel blends for comparative performance evaluation. The results of the experiment in each case were compared with baseline data of diesel fuel. Significant improvements have been observed in the performance parameters of the engine as well as exhaust emissions, when lower blends of karanja oil were used with preheating and also without preheating. Karanja oil blends with diesel (up to K50) without preheating as well as with preheating, can replace diesel for operating the CI engines, giving lower emissions and improved engine performance. Hazar and Aydin (2010) investigated raw rapeseed oil (RRO) and its blends (RRO20 and RRO50) in a DI diesel engine. The effects of fuel preheating to 100°C on the engine performance and emission characteristics of a CI engine, fueled with rapeseed oil diesel blends were clarified. Results showed that preheating of RRO lowered its viscosity and provided smooth fuel flow. It can also be concluded that preheating of the fuel has

some positive effects on engine performance and emissions when operating with vegetable oil. Chauhan et al. (2010) reported that by using a heat exchanger, preheated Jatropha oil has the potential to be a substitute fuel for diesel engines. Optimal fuel inlet temperature was found to be 80°C considering the brake thermal efficiency, brake specific energy consumption and gaseous emissions. A comparable engine performance and emissions are reported by Yilmaz and Morton (2011) using preheated peanut, sunflower and canola oils in two DI diesel engines. No (2011) reviewed seven non-edible vegetable oils including Jatropha oil as an alternative fuel for diesel engine. Jatropha oil was identified as a leading candidate for the commercialization.

Vegetable oils are considered to be suitable for Thailand due to its agricultural economy, and can help alleviate the problem of under-priced agricultural products. Thailand is blessed with many feedstocks, suitable for vegetable oil production such as palm, Jatropha, coconut and sunflower. These crops can be used to produce vegetable oil for usage in the agricultural sector; to decrease the dependence on imported oil and to help stabilize the price of agricultural products. The use of non-edible vegetable oils compared to edible oils is very significant because of the tremendous demand for edible oils as food, making them too expensive to be used as fuel at present. The scientific name of Jatropha is Jatrophs curcas L. and Jatropha oil is one such kind of non-edible vegetable oil. Not only does Jatropha have a yield of well over 200 gallons of oil per acre per year, eleven times that of corn (Khan et al., 2009), it also increases the fertility of the land on which it is grown so that it can potentially be used for food crops in subsequent years. Jatropha is a perennial which can grow in arid conditions on any kind of ground, and does not suffer in droughts or require irrigation. Therefore, unlike the common biofuel crops of today (corn and sugar); they are very easy to cultivate, even on poor land.

Jatropha is fast growing, begins yielding oil in the second year and continues for forty to fifty years (Yarrapatruni et al., 2009). Optimal yields are obtained from the sixth year. Singh and Padhi (2009) investigated Jatropha oil and its methyl ester to find out their suitability for use as petro-diesel. Different properties of Jatropha oil were experimentally determined and compared with theoretical equations developed in the study. The study suggested that Jatropha oil can be used as a source of triglycerides in manufacture of biodiesel cost-effectively. Pramanik (2003) investigated Jatropha oil in a diesel engine. Acceptable thermal efficiencies of the engine were obtained with blends up to J50. Forson et al. (2004) examined the performance of Jatropha oil blends in a diesel engine. The most significant conclusion from the study was that the J2.6 produced maximum values of brake power and brake thermal efficiency as well as minimum values of specific fuel consumption. Agarwal and Rajamanoharan (2009) and Hazar and Aydin (2010)

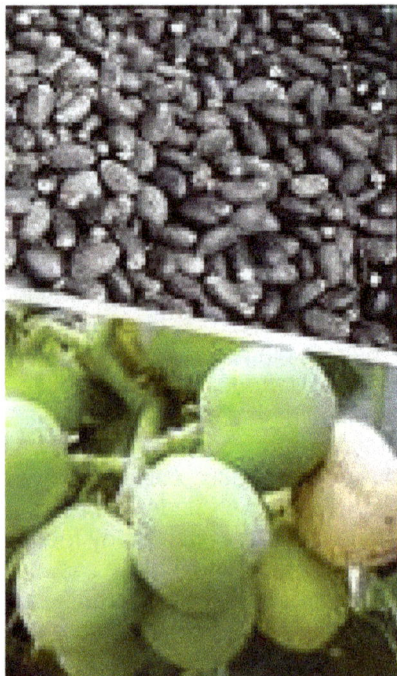

Figure 1. *Jatropha* fruits and seeds.

indicated that the successful use of *Jatropha* oil is a function of engine type, and percentage of *Jatropha* oil in the blends.

This study was therefore undertaken at the Asian Institute of Technology (AIT), Thailand to gather information on the behavior of a diesel engine when operated with *Jatropha* oil and its blend with diesel at different proportions. The objective of this study was to identify the best *Jatropha*-diesel blends, for which the engine performance is similar or better than that of diesel fuel.

MATERIALS AND METHODS

Extraction of *Jatropha* oil

It is reported by Gubitz et al. (1999) that a dry seed of *Jatropha* contains about 55% crude oil by weight. However, the maximum amount of oil that can be extracted from a given sample of the seed depends on the method of extraction and the quality of the feed stock. Two main methods of extracting the oil have been identified. They are the chemical extraction method using solvent extraction with *n*-hexane and the mechanical extraction method using either a manual seed press or an engine driven-expeller. Kpikpi (2002) has reported that solvent extraction with *n*-hexane could produce about 41% yield by weight of oil per kg of the *Jatropha* seed. Foidl and Eder (1997) reported that the dry seed of *Jatropha* would yield about 30 to 38% by expeller. However, in their study in Nicaragua, 30.8% of crude oil by weight was extracted from 12,782 tons of dry weight of *Jatropha* using an engine driven-expeller. Forson et al. (2004) used a ram press and reported that 32 kg of unshelled seeds yielded 6.88 kg of oil representing about 21.5% of crude oil by weight per kg of the unshelled dry weight of the *Jatropha* seed. In the present study, a simple mechanical cracking machine and

screw-press available at the AIT, Thailand was used for the oil extraction process. The harvested *Jatropha* fruit was dried for three to four days before cracking. About 150 kg of fruit was fed into the cracking machine. 41.8 kg of seeds was obtained after cracking (Figure 1). *Jatropha* seeds were then pressed by screw-press resulting in a yield of 11.71 kg *Jatropha* oil and 30.09 kg press cake. This means that *Jatropha* oil represented about 28% of crude oil by weight per kg of the *Jatropha* seed. On the assumption that a dry seed of *Jatropha* contains about 55% of oil, the efficiency of the mechanical extraction process used was estimated to be 51%, whereas, a value of more than 90% is reported for extraction with *n*-hexane. This suggests that the mechanical screw-press used, needs some form of improvement.

A cost analysis from cultivating *Jatropha* to obtain the final product of *Jatropha* oil has been performed in this study. The cost of *Jatropha* sprout (tree), land preparation, maintenance and labor was considered. *Jatropha* sprout costs 30 to 60% of the total cost. *Jatropha* sprout price can vary from 3 to 10 baht (1 US$ = 31 baht) per tree. *Jatropha* oil production cost was 14.67 baht/l when the sprout price was 3 baht/tree and increased to 24.64 baht/l when the sprout price increased to 10 baht/tree. The petro-diesel price selected by Thai Government Ministry of Energy was then 30 baht/l. *Jatropha* oil price was therefore lower than that of diesel for all sprout prices.

Chemical and physical properties

The chemical and physical properties of *Jatropha* oil and diesel were tested according to ASTM Standard (D975/D6751) by PTT Research and Technology Institute and PTT Public Company Limited.

RESULTS AND DISCUSSION

The results of chemical and physical properties of *Jatropha* oil and diesel are shown in Table 1. The results show that the higher heating value of *Jatropha* oil is about 85% to that of diesel and the pour point is very similar. However, the flash point and the kinematic viscosity of *Jatropha* oil are several times higher than diesel. The lower heating value of *Jatropha* oil is an indication that a higher oil consumption than that of diesel is needed for similar power output. Similar low pour points of diesel and *Jatropha* oil suggest that *Jatropha* oil can be used at low temperatures just like diesel. Due to a higher flash point,

Jatropha oil is safer for storing and transporting when compared to diesel. Kumar et al. (2003) reported that the cetane number of *Jatropha* oil is about 15% lower than that of diesel, therefore, higher *Jatropha* oil blends or pure *Jatropha* oil combustion might be less efficient. *Jatropha* oil's viscosity is about ten times that of diesel, leading to problems of fuel atomization and proper combustion. Therefore, *Jatropha* oil's viscosity reduction will be described in the next sections.

Reducing *Jatropha* oil's viscosity

High viscosity is a major problem when using vegetable oil as an alternative fuel for diesel engines. In the present

Table 1. Properties of *Jatropha* oil and diesel.

Properties	*Jatropha* oil	Diesel
Higher heating value (MJ/kg)	38.36	45.0
Lower heating value (MJ/kg)	37.65	42.5
Flash point (ºC)	> 180	64
Pour point (ºC)	-3	-2
*Cetane number	40-45	45-55
Sulphur content (% wt.)	0.0002	0.033
Density (kg/m^3, at 15ºC)	916.87	845
Kinematic viscosity (cSt, at 40ºC)	39.07	3.137

*Kumar et al. (2003).

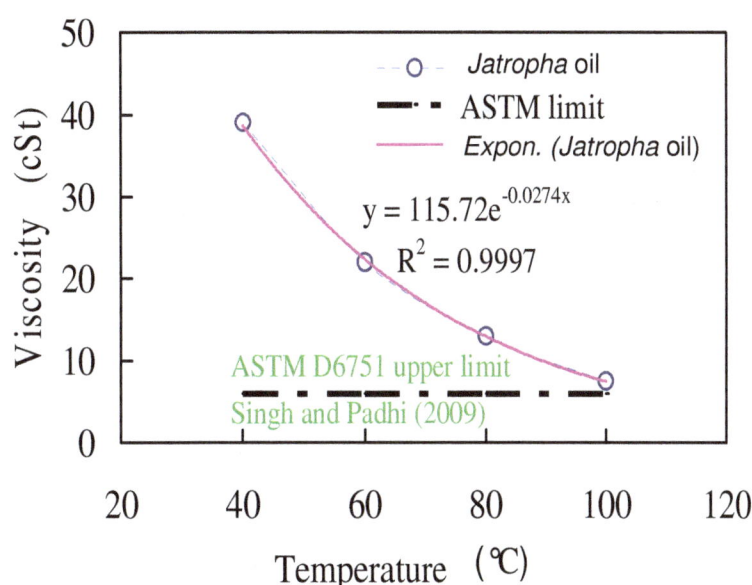

Figure 2. Relationship between viscosity and temperature of *Jatropha* oil.

investigations, viscosity was reduced by heating and blending the oil with diesel fuel. The viscosity of *Jatropha* oil was measured by a calibrated glass capillary viscometer using ASTM D445 method at different temperatures in the range of 40 to 100ºC. The results are shown in Figure 2. The viscosity of *Jatropha* oil decreased remarkably with increasing temperature and it became close to ASTM limits (ASTM D6751) for viscosity of biofuels when the temperature 100ºC or more. Figure 2 also shows a good relationship between viscosity and temperature presented by an exponential equation, y = 115.72 $e^{-0.0274x}$, where y is kinematic viscosity and x temperature.

Agarwal (1998) and Sinha and Misra (1997) reported that dilution or blending of vegetable oil with other fuels like alcohol or diesel fuel would bring the viscosity close to specification range. Therefore, *Jatropha* oil was blended with diesel in varying proportions (J5, J10, J20,

J50 and J80) with the intention of reducing its viscosity close to that of diesel fuel. The viscosity of various blends of *Jatropha* oil and diesel was also evaluated (Figure 3). The viscosity of *Jatropha* oil decreased after blending. The viscosity of J5 and J10 was within the ASTM limit. The viscosity reduction of J20 is about 85%, and very close to ASTM limit. For the three *Jatropha* oil blends (J5, J10 and J20), corresponding viscosity was found to be 3.94, 4.63, and 6.5 cSt at 40ºC, respectively.

Engine test procedure

Constant speed engine tests were carried out using *Jatropha*-diesel blends, pure *Jatropha* oil and diesel. The performance of the engine was evaluated in terms of brake thermal efficiency, brake specific fuel consumption and exhaust gas temperature. The make of the engine

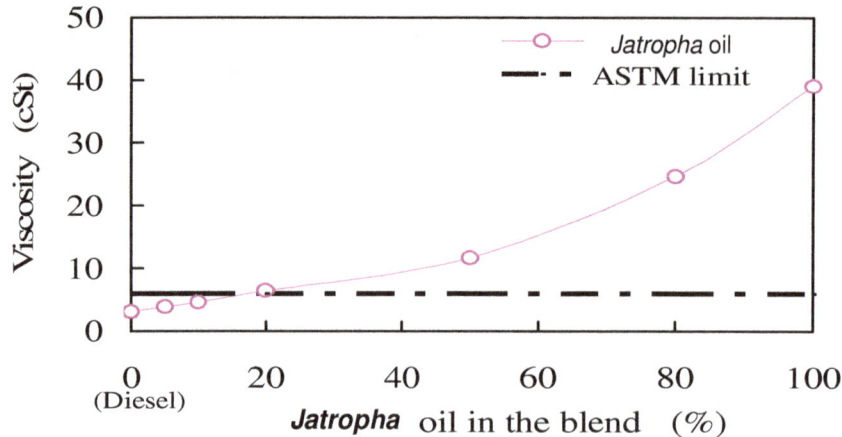

Figure 3. Effect of blending diesel on the viscosity of *Jatropha* blends.

Table 2. Engine specifications.

Specifications	Descriptions
Engine type	4-stroke DI diesel engine
Number of cylinders	Three
Bore × Stroke	91.4 × 127 mm
Swept volume	2500 cc
Compression ratio	18.5:1
Rated power	36.6 kW at 2250 rpm
Start of injection	20° BTDC
Injection pressure	20 MPa

used in this study was Perkins. Engine specifications are given in Table 2. The engine speed was kept constant at maximum brake torque (MBT) speed of 1500 rpm. Fresh lubricating oil was filled in the oil sump before starting the experiments. Loads were measured by electric dynamometer. Seven load conditions (from no load to full load) are reported. The fuel supply system was modified by adding an additional three-way, hand operated, two-positional directional control valve which allowed rapid switching between the diesel oil used as a standard and the test fuels. The engine was started with diesel and once the engine warmed up, it was switched over to *Jatropha* oil or blends. After concluding the tests with *Jatropha* oil or blends, the engine was again switched back to diesel, before stopping the engine until the *Jatropha* oil or blends was purged from the fuel line, injection pump and injector in order to prevent deposits and cold starting problems. Stop watch and fuel level indicator are used to measure the fuel consumption rate. An orifice meter-inclined manometer arrangement was used to measure the intake air flow rate. Thermocouples with temperature indicator were installed to measure intake air and exhaust gas temperatures. A gas analyzer was used to measure the CO_2 and CO emissions in

exhaust gases.

At every test point, data samples were taken at least thrice, but a single point (average of data) is used to present the results graphically. Similar conditions were maintained for all tests to allow direct comparison of the results. The standard deviations of important performance and emission parameters are approximately: 0.25% for brake thermal efficiency, 0.015 kg/kWh for brake specific fuel consumption, 0.0025% for CO and 0.2% for CO_2.

Engine performance and emissions

The variation of brake thermal efficiency of the engine with various blends and *Jatropha* oil is shown in Figure 4 and compared with the brake thermal efficiency obtained with diesel. The brake power is increased from 0 kW for no load up to about 16.75 kW for full load with different fuels. The maximum thermal efficiency obtained with diesel fuel is slightly higher than 30%. J5 showed slightly higher thermal efficiency than diesel. J10 and J20 showed similar thermal efficiency, but J50 and higher blends showed 3 to 5% less thermal efficiency than

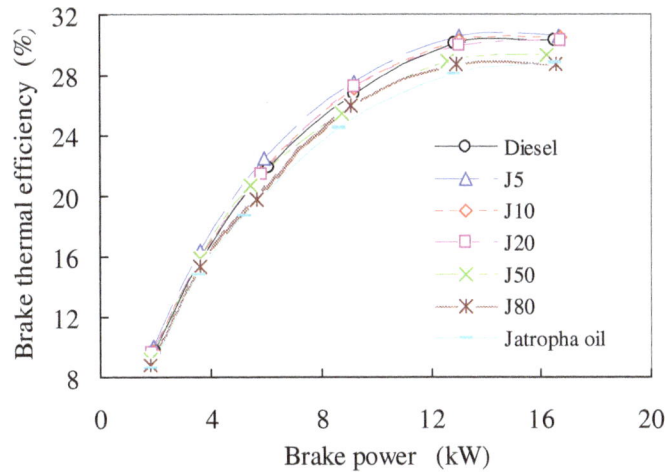

Figure 4. Brake thermal efficiency of the engine with various fuels.

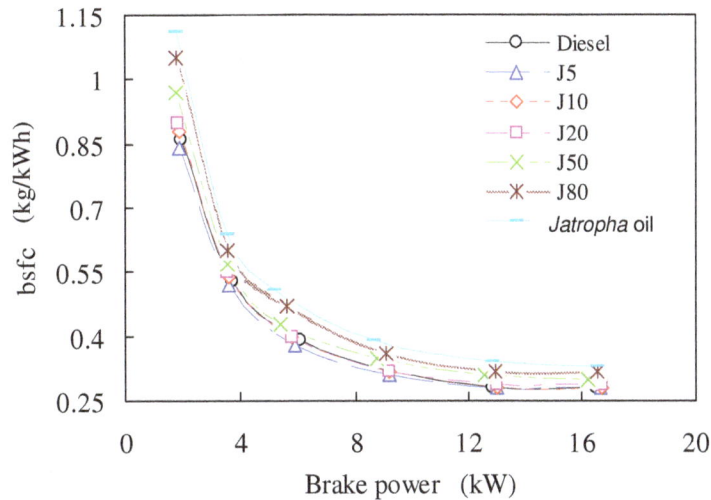

Figure 5. Brake specific fuel consumption of the engine with various fuels.

diesel fuel. The observation is that the higher the *Jatropha* oil in the blends, the higher the reduction in the thermal efficiency. The reasons might be explained as follows. Due to very high viscosity and low volatility of *Jatropha* oil, higher *Jatropha* oil blends suffer from worse atomization and vaporization followed by inadequate mixing with air. The consequence is inefficient combustion. This suggests that high fuel injection pressure and improved volatility might be helpful for better combustion with higher thermal efficiency for higher *Jatropha* blends.

Figure 5 shows the variation of brake specific fuel consumption (bsfc) of the engine with various fuels. The bsfc is decreased from 0.87 kg/kWh at low load to 0.28 kg/kWh at full load for diesel fuel. The best bsfc obtained with diesel fuel is 0.28 kg/kWh at full load operation. Better bsfc is obtained for the J5 than diesel. This blend shows about 3% less bsfc in average than diesel fuel.

Diesel and J5 have almost the same heating value. However, J5 has some oxygen present, which might have favored better combustion than the diesel only operation. The deterioration in bsfc up to J20 is 1.5 to 3.4%. J50, J80, and pure *Jatropha* oil show average bsfc deterioration of about 10, 15 and 25%, respectively. In the case of pure *Jatropha* oil, deterioration in bsfc at no or low load (up to about 30% load) condition is about 30%, whereas the load above that, the bsfc deterioration decreased to about 20%. A similar trend (the lower the loads, the higher the deterioration in the bsfc) is found for J80 and J50 blends. At low load conditions, the cylinder temperatures are low. Due to poor volatility of pure *Jatropha* oil, low cylinder temperature at low load conditions might not favor proper combustion. It seems that blending of *Jatropha* oil with kerosene instead of diesel, which was investigated by one author of this paper using

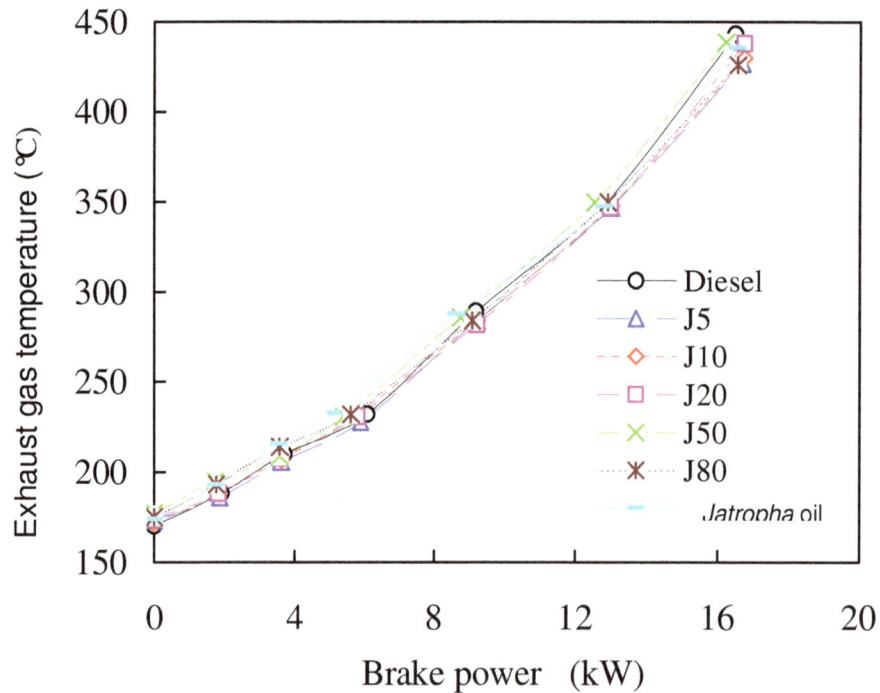

Figure 6. Exhaust gas temperatures with various fuels.

rice bran oil (Bari and Roy, 1995), might have better volatility that could improve combustion and bsfc. Furthermore, a higher amount of *Jatropha* oil can be blended without deterioration in efficiency and bsfc. Lower heating value (which was used to calculate engine efficiency) of *Jatropha* oil (37.65 MJ/kg) is about 11% less than diesel (42.5 MJ/kg) in terms of mass basis. When it is converted in volume basis (for example energy content per liter), *Jatropha* oil has only about 4% less energy content than diesel. This suggests that if specific fuel consumption is measured by volumetric basis (l/kWh), which is logical for liquid fuels as they are traded as volume basis, there will be no reduction in fuel consumption up to J20 as compared to diesel.

Figure 6 shows the variation of exhaust gas temperature with load in the range of 0 to 16.75 kW for diesel, pure *Jatropha* oil and various blends. The results show that the exhaust gas temperature increased with increase in brake power in all cases. The highest value of exhaust gas temperature of 443°C was observed with diesel, whereas, the corresponding value with *Jatropha* oil was found to be 436°C. There was no significant difference in exhaust gas temperature among various blends and pure diesel and *Jatropha* oil. The average exhaust gas temperature with different blends is more than 250°C. This high temperature exhaust gas can be used to preheat the *Jatropha* oil or blends for improved performance.

Figure 7 shows CO_2 emissions at various engine loads for different fuels. The increase in CO_2 emissions at lower load conditions is not sharp. However, at engine loads

above 50% there is a steep increase in CO_2 emissions for all fuels. At lower load operations, *Jatropha* oil blends, especially higher ones, produced less CO_2 than diesel, indicating a less efficient combustion. This is very consistent to that described in Figure 5, that is, low cylinder temperature at low load conditions does not favor proper combustion of *Jatropha* oil and higher blends. At higher load conditions (80% or higher), J50 and *Jatropha* oil produced about 20% higher CO_2 than diesel. This is thought to be partially due to improved efficiency at higher load conditions and mostly due to much higher fuel consumption of *Jatropha* oil and higher blends than diesel at this condition.

Figure 8 shows CO emissions at various engine loads for different fuels. The general trend is that CO gradually decreased up to about 80% load and then increased sharply at full load operation for all fuels. J5 produced about 50% more CO than diesel throughout the operation. J10 and higher blends produced about double the CO when compared to diesel. Less CO was expected with *Jatropha* oil and blends than diesel due to *Jatropha* oil's inherent O_2 content, which might have helped combustion (especially at high loads), where diesel operation was starved for oxygen. Excess air factor (λ) is measured to determine if the engine is starved for oxygen at high load operations.

Figure 9 shows excess air factor of different fuels tested at various engine loads. The excess air factor is decreased from 5.75 at no load to 1.75 at full load for diesel fuel. To calculate excess air factor, it is necessary to know the stoichiometric air/fuel (A/F) ratio of different

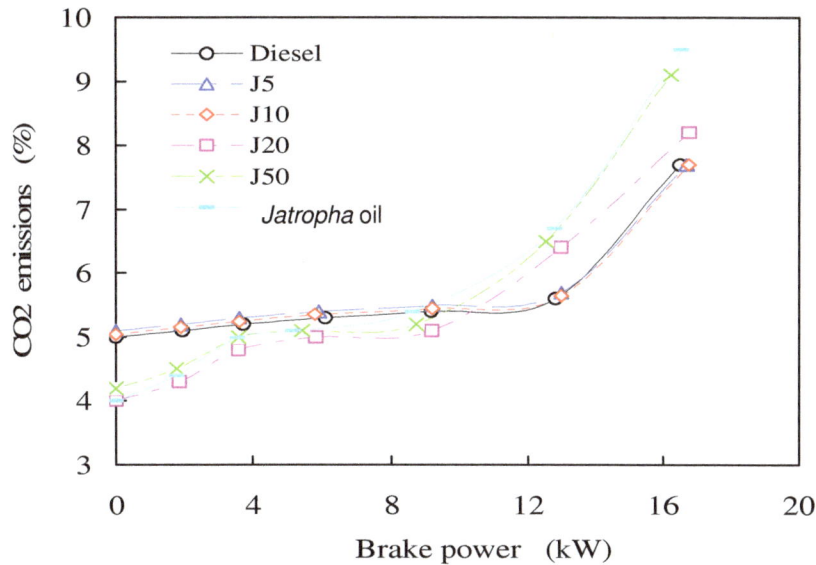

Figure 7. CO_2 emissions with various fuels.

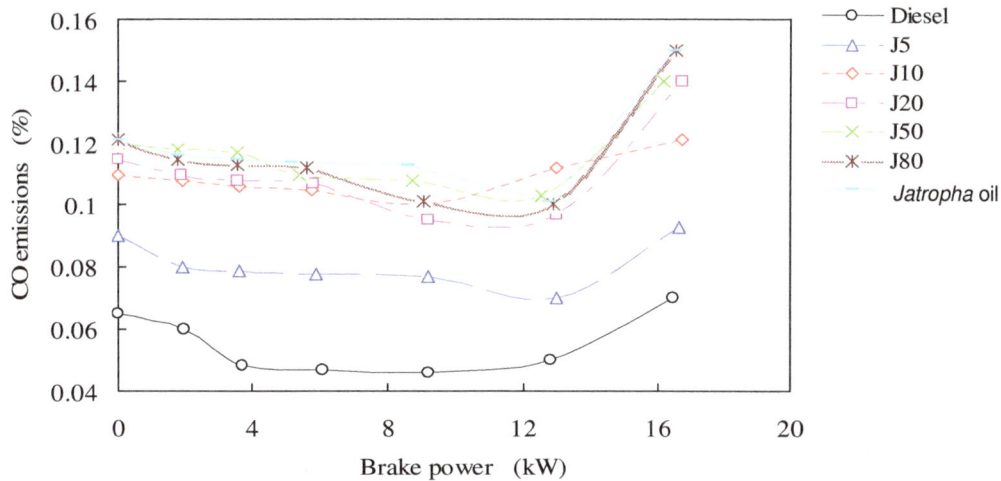

Figure 8. CO emissions with various fuels.

fuels. The conventional value of stoichiometric A/F ratio for diesel is taken as 14.7. The stoichiometric A/F ratio of *Jatropha* oil is calculated as 12.1 based on the ultimate analysis of *Jatropha* oil presented by Agarwal and Agarwal. (2007). The lowest excess air factor obtained with diesel fuel is 1.75 at full load operation. This means that there is still 75% excess air. Excess air factor of *Jatropha* blends is higher than diesel at almost all load conditions. At full load operation with *Jatropha* oil and blends, there is about 80% excess air. The amount of excess air with *Jatropha* oil and blends is higher than diesel, but much higher CO is produced with those fuels. This is possibly a result of poor spray atomization and improper local air/fuel mixture formation with *Jatropha* oil and blends.

Optimum proportion of *Jatropha* oil in the blends

Root mean squared error (RMSE) was used for comparative assessment of engine performance between diesel and *Jatropha* oil blends. The formula used is given as:

$$RMSE = \sqrt{1/n\sum_{i=1}^{n}(D_i - J_i)^2}$$

Where: D_i is the performance parameters of diesel for i^{th} value of load, J_i is the performance parameters of

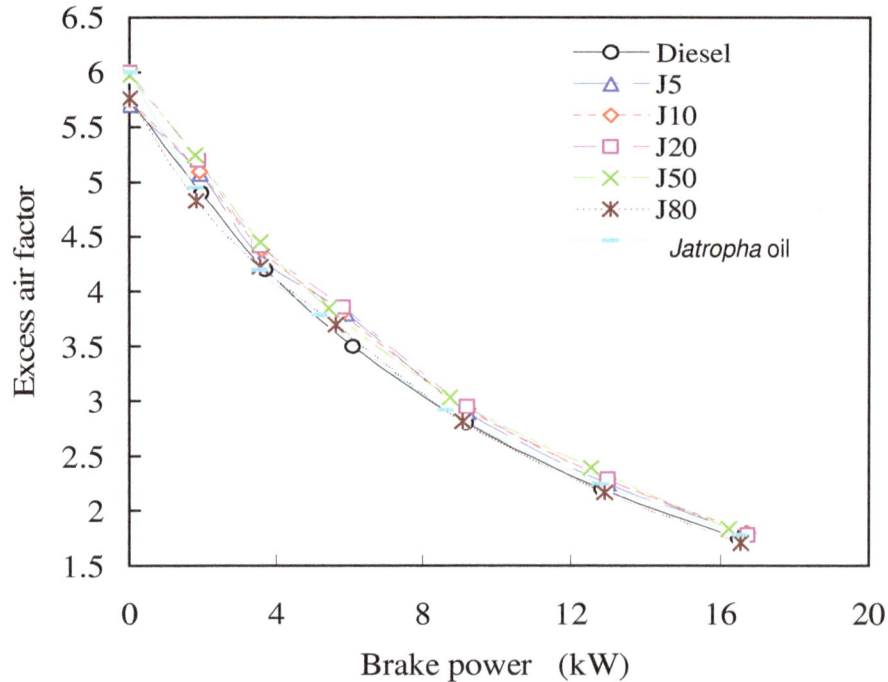

Figure 9. Excess air factor with various fuels.

Table 3. Root mean squared errors between diesel and various *Jatropha* oil blends.

	J5	J10	J20	J50	J80	*Jatropha* oil
Root mean squared error (in brake thermal efficiency, %)	0.0001	0.0001	0.0001	0.0002	0.0004	0.0006
Root mean squared error (in bsfc, kg/kWh)	0.0012	0.0012	0.0015	0.0085	0.0111	0.0144

Jatropha oil and blends for the corresponding i^{th} value of load, n is the number of loading, here, 6.

Table 3 shows the RMSE between diesel and different *Jatropha* oil blends. The results show that J5 to J20 have the lowest value of RMSE of brake thermal efficiency (0.0001), while the lowest value of RMSE of brake specific fuel consumption is 0.00012 for J5 and J10. This means that *Jatropha* oil blends up to 10% have an engine performance (both efficiency and fuel consumption) very close to that of diesel fuel. In addition, the ASTM limit for viscosity should be taken into consideration. J10 also satisfies the ASTM standard of viscosity. In conclusion, the results obtained show that blends up to J10, are optimum for diesel substitute without any engine modifycation. This is much better than that reported by Forson et al. (2004), where J2.6 was recommended. Forson et al. did not test *Jatropha* oil blends in the range of 2.6 and 20%. J5 showed the best results (even better than diesel) in this study. However, up to J20, the engine performance is comparable with diesel operation. Although, results up to J20 are promising, durability tests are required for long term usage of these oil blends in the engine.

Yilmaz and Morton (2011), No (2011) and Haldar et al. (2009) reported the results of engine performance and emissions for peanut, sunflower and canola as edible oils and *Jatropha*, karanja, mahua, linseed, rubber seed, cottonseed, neemoils and Putranjiva as inedible oils. They showed that the performance and most of the emissions (except NOx) of 10 to 20% oil blends are very similar or better than that of petro-diesel. This study shows similar and comparable results up to J20, with J5 shown to perform better than petro-diesel.

CONCLUSIONS AND RECOMMENDATIONS

Jatropha oil production from *Jatropha* fruits, its fuel characterization, temperature-viscosity relations and an experimental investigation were conducted to explore the performance of *Jatropha* oil and blends as a diesel substitute. The results obtained suggest the following conclusions:

1. A good relationship between viscosity and temperature

of *Jatropha* oil was established. It was found that heating the *Jatropha* oil to about 100 °C is adequate to reduce the viscosity to within close range of the ASTM limits of viscosity for biofuels. Viscosity of *Jatropha* blends (up to J20) was also found to be close to the ASTM limit.

2. From the experimental results obtained, *Jatropha* oil is found to be a promising alternative fuel for CI engines. It can be used for blending up to J20 without any reduction in thermal efficiency. J5 showed better thermal efficiency and bsfc than diesel. From the analysis of optimum proportion of *Jatropha* blend, up to J10 showed very good performance.

3. Emissions of CO_2 with *Jatropha* oil blends up to moderate loads are lower than that with diesel fuel. When the load is 50% or higher, J50 and higher blends produced about 20% higher CO_2 than diesel.

4. Emissions of CO with *Jatropha* oil and blends are much higher than diesel at all loading conditions, although the engine always ran leaner with *Jatropha* oil and blends than diesel. This suggests improving the local air/fuel mixture when operated by *Jatropha* oil and blends.

Two recommendations are proposed for further work which will also include the study of NOx and PM emissions:

1. Engine test with pure *Jatropha* oil at higher fuel injection pressures (up to 100 MPa or more) with a smaller nozzle hole diameter (as small as 0.1 mm) to improve atomization. This also improves the in homogeneity of local air fuel mixture.

2. Dilution (blending) of *Jatropha* oil with kerosene instead of diesel. This would improve volatility. Moreover, higher percentage of *Jatropha* oil blending is possible within the ASTM limit of viscosity due to the lower viscosity of kerosene than diesel.

REFERENCES

Agarwal AK (1998). Vegetable oils versus diesel fuel: development and use of biodiesel in a compression ignition engine. TIDE, 8(3):191-204.

Agarwal AK, Rajamanoharan K (2009). Experimental investigations of performance and emissions of Karanja oil and its blends in a single cylinder agricultural diesel engine. Appl. Energy, 86(1): 106-112.

Agarwal D, Agarwal AK (2007). Performance and emissions characteristics of *Jatropha* oil (preheated and blends) in a direct injection compression ignition engine. Appl. Therm. Eng., 27: 2314-223.

Bari S, Roy MM (1995). "Prospect of rice bran oil as an alternative to diesel", Proc. of the Fifth Int. Conf. on small engines, their fuels and the environment. The Universty of Reading, UK, pp. 31-36.

Çetin M, Yüksel F (2007). The use of hazelnut oil as a fuel in pre-chamber diesel engine. Appl. Therm. Eng. 27: 63-67.

Chauhan BS, Kumar N, Jun YD, Lee KB (2010). Performance and emission study of preheated *Jatropha* oil on medium capacity diesel engine. Energy, 35: 2484-2492.

Foidl N, Eder P (1997). Agro-industrial exploitation of *Jatropha*. In: Gubitz GM, Mittelbach M and Trabi M. editors. Bio-fuels and industrial products from *Jatropha*. Graz: Dbv-Verlag, pp. 88-91.

Forson FK, Oduro EK, Hammond-Donkoh E (2004). Performance of *Jatropha* oil blends in a diesel engine. Renew. Energ. 29(7): 1135-45.

Goering CE, Schwab AW, Daugherty MJ, Pryde EH, Heakin AJ (1981). Fuel properties of eleven vegetable oils. ASAE Paper, No. 81-3579.

Gubitz GM, Mittelback M and Trabi M (1999). Exploitation of the tropical oil seed plant *Jatropha* Cucas L., Bioresour. Technol., 67: 73-82.

Haldar SK, Ghosh BB, Nag A (2009). Studies on the comparison of performance and emission characteristics of a diesel engine using three degummed non-edible vegetable oils. Biomass Bioenerg. 33: 1013-1018.

Hazar H, Aydin H (2010). Performance and emission evaluation of a CI engine fueled with preheated raw rapeseed oil (RRO)-diesel blends. Appl. Energy, 87(3): 786-790.

He Y, Bao YD (2005). Study on cottonseed oil as a partial substitute for diesel oil in fuel for single-cylinder diesel engine. Renewable Energy, 30: 805-813.

Kalam MA, Husnawan M, Masjuki MH (2003). Exhaust emission and combustion evaluation of coconut oil-powered indirect injection diesel engine. Renewable Energy, 28: 2405-2415.

Kpikpi WM (2002). *Jatropha* as vegetable source of renewable energy. Paper Presented at ANSTI Sub-network Meeting on Renewable Energy, pp. 18-22.

Kumar MS, Ramesh A, Nagalingam B (2003). An experimental comparison of methods to use methanol and *Jatropha* oil in a compression ignition engine, Biomass Bioenerg., 25: 309-318.

No SY (2011). Inedible vegetable oils and their derivatives for alternative diesel fuels in CI engines: A review. Renewable Sustain. Energy Rev., 15: 131-149.

Pramanik K (2003). Properties and use of *Jatropha* oil and diesel blends in compression ignition engine. Renewable Energy, 28: 239-248.

Quick GR (1980). Developments in use of vegetable oils as fuel for diesel engines. ASAE Paper, No. 80-1525.

Sapaun SM, Masjuki HH, Azlan A (1996). The use of palm oil as diesel substitute. Power Energy (Part A) 210: 47-53.

Singh RK, Padhi SK (2009). Characterization of *Jatropha* oil for preparation of biodiesel. Nat. Prod. Radiance, 8(2): 127-132.

Sinha S, Misra NC (1997). Diesel fuel alternative from vegetable oils. Chem. Eng. World, 32(10): 77-80.

Yarrapathuni VHR, Ram Voleti RS, Pereddy NR and Alluru VSR (2009). *Jatropha* oil methyl ester and its blends used as an alternative fuel in diesel engine. Therm. Sci., 13(3): 207-17.

Yilmaz N, Morton B (2011). Effects of preheating vegetable oils on performance and emission characteristics of two diesel engines. Biomass Bioenerg., 35: 2028-2033.

13

Fabrication and electrical study of PEM fuel cell based on nano crystalline PEO based conducting polymer electrolyte system

Kambila Vijay Kumar

Centre for Nano Technology, FED, K. L University, Guntur-522 502(A.P), India. E-mail: vijaynanopklu@gmail.com.

Solid conducting polymer electrolyte system, nano crystalline polyethylene oxide (PEO) complexed with sodium bicarbonate ($NaHCO_3$) salt was prepared by sol gel-technique. Several experimental techniques such as IR, composition dependence conductivity and temperature dependence conductivity in the temperature range of 303 to 368 K and transport number measurements were employed to characterize this polymer electrolyte system. The conductivity of the ($PEO+NaHCO_3$) electrolyte was found to be about 3 times larger than that of pure PEO at room temperature. The transference data indicated that the charge transport in these polymer electrolyte systems is predominantly due to Na^+. Using this nano crystalline polymer electrolyte system, a three PEM fuel cell stack have been fabricated and their open circuit voltage and I-V characteristics were studied for various gas concentrations. The open circuit voltage was found to be 1.10 V. A number of other cell parameters associated with the cell were evaluated and are reported in this paper.

Key words: Complex, conducting polymer electrolyte, ionic conductivity, transport number, polymer electrolyte membrane (PEM) fuel cell, current-voltage characteristics.

INTRODUCTION

The two primary concerns of scientists today are energy and environment with the rapid depletion of fossil fuels. There is an urgent need to look for clean alternate source of energy. Fuel cells satisfy both these criteria. It is high time that we realize the need for cutting edge technologies which satisfy both conditions, that is, being able to match the present day demand and also being reliable. One such convincing technology is "Fuel cell technology". The necessity for this technology has been widely accepted globally in the energy and power sectors of the many emerging technologies such as biomass fuels, liquid-bio fuels, geothermal energy, Photo-voltaic, hydro-electric power, solar thermal-electricity, tidal-energy, wind-energy, fuel cell technology has been much considered comparative with other technologies and if developed, has more advantages because it has the practicability of being used in many mobile applications and hence this seems to be the futuristic technology

(Sossina et al., 2003). Solid polymer electrolyte membrane fuel cells are promising candidates as power generators for zero emission vehicles in place of conventional combustion engines. These cells use polymer electrolyte membrane (PEM) such as nafion as electrolyte. At present, conducting nano crystalline crystalline polyethylene oxide (PEO) complexed with sodium bicarbonate based PEM are fairly developed/fabricated and have been successfully used in H_2/O_2, fuel cell. Solid electrolyte PEM requires water to maintain their ionic conductivity. The absorption, diffusion coefficient and electro-osmotic drag of water and proton/ionic conductivity in PEMs and inter related and strongly affected by the cell operating conditions (Yeo, 1983; Verbrugge and Hill, 1990; Bernardi, 1990; Dahr, 1994; Apple and Foulkes, 1989). This problem gets aggregated when the fuel cell is operated at higher temperature (above 70°C) water content in PEM

developed so far has been indirectly managed by humidifying the reactant gases. The electrochemical performance of polymer electrolytes has been improved by the addition of inorganic fillers (Blomen, 1994). The resulting composite polymer electrolytes discharged enhanced conductivity, mechanical stability and improved interfacial stability towards electrode materials (Berers et al., 1997; Andrew and Larwine, Vishnu et al., 2003; Elroy and Nuttall, 2001).

In recent years, an attempt to investigate the possibility of fabricating PEM fuel cell based on other polymer electrolytes (Berers et al., 1997). Some researchers studied and reported on polymer electrolyte membrane fuel cells based on nafion/nafion + silica + phospotungstic (Sossina et al., 2003) etc. The present work examines a new conducting polymer electrolyte system, namely nano crystalline PEO based ion conducting polymer electrolyte system, that is, $PEO+NaHCO_3$. Using this solid electrolyte, a PEM fuel cell stack has been fabricated and its i-v characteristics studied.

EXPERIMENTALS

Polymer films (thickness $\cong$ 100 to 150 μm) of pure PEO (Aldrich ~ 6 × 10^5) and various compositions of complexed film of nano crystalline PEO with $NaHCo_3$ salt was prepared in wt. ratios (90:10); (80:20) and (70:30) by a sol-gel technique using methanol (water free) as solvent. The infrared (IR) spectra in these films have been recorded with the help of JASCO FT / IR-53000-spectrophotometer in the range 400 to 4000 cm^{-1}. The dc conductivity has been measured using the lab made conductivity setup (Sreepathi et al., 1995) in the temperature range 303 to 368 K using a Keithley electrometer (Model 614).

Solid state polymer electrolyte membrane (PEM) fuel cell stack was fabricated in the configuration electrode /solid polymer electrolyte / cathode. The readings were taken with the high purity cylinder gases and the specification design of the PEM fuel cells was reported. The details about the fabrication of the fuel cells are given elsewhere (Elroy and Nuttall, 2001; Swarna and Shyam, 2003; Rai, 2002). Finally the I-V characteristics of these fuel cells were monitored for a constant load of 0.5 Ω.

RESULTS AND DISCUSSION

The variation in dc conductivity (σ) as a function of $NaHCO_3$ composition in PEO at room temperature (R_T) that is, 303 and 368 K are given in Figure 1. The conductivity of pure PEO is approximately 10^{-10} Scm^{-1} at room temperature and its value increases sharply to 10^{-7} Scm^{-1} on complexing with 10 wt% $NaHCO_3$. The increase in conductivity becomes slower on further addition of $NaHCO_3$ to the polymer. This behavior has been explained by various researchers, who have studied

PVP and PEO based polymer electrolyte in terms of ion association and the formation of charge multipliers (Srivastava et al., 1992; Yuankang and Zhusheng, 1986; Scrosati, 1988; Reddy and Rao, 1998; Chabagno and Duclot, 1979; Narasaiah et al., 1995; Stevens and Mellander, 1987)

The complex of the polymer PEO and salt $NaHCO_3$ were confirmed by using IR spectroscopy. The IR spectra of pure PEO, PEO complexed with $NaHCO_3$ were recorded with the help of JASCO FT/IR-5300 spectrophotometer in the range 400 to 4000 cm^{-1} and are shown in Figure 2. The intensity of the aliphatic C-H stretching vibrations band observed around 2950 cm^{-1} in PEO was found to decrease with the increase in the concentration of $NaHCO_3$ salt in the polymer. The width of the C-O stretching band observed at around 1100 cm^{-1} in PEO also showed an increase in the salt concentration in the polymer. Also the appearance of new peaks along with changes in existing peaks (and/or their disappearance) in the IR-spectra directly indicates the complexation of PEO with $NaHCO_3$. This IR data clearly establishes the complexation of $NaHCO_3$ with different weight ratios of the polymer PEO.

The variation in conductivity as a function of temperature for pure PEO with different compositions of ($PEO + NaHCO_3$) polymer electrolytes over the temperature range of 303 to 368 K is shown in Figure 3. The conductivity versus temperature (log σT versus 10^3/T) plots follows the Arrhenius nature throughout, but with two different activation energies (Table 1) above and below melting point (T_m) of the polymer. In region- I (that is, below T_m), the conductivity of pure PEO increases slowly with temperature up to 65°C. At 65°C, there is a sudden increase in conductivity. In region- II (that is, above T_m), the conductivity again increases with temperature. The calculated conductivity (σ) at room temperature, at 368 K and activation energies (E_a) for pure PEO and ($PEO+NaHCO_3$) electrolyte systems are given in Table 1. The conductivity of pure PEO is ~ 6.78 × 10^{-8} Scm^{-1} at room temperature and its value increases sharply to ~10^{-7} Scm^{-1} on complexing with 10 wt % of $NaHCO_3$. The increase in conductivity becomes slower on further addition of $NaHCO_3$ to the polymer. This behavior has been explained by various researchers, who have studied PMMA and PEO-based polymer electrolyte, interims of ion association and formation of charge multiples (Ramalingaiah et al., 1996; Scrosati, 1988; Hashmi et al., 1992; Sreepathi et al., 1994, 2000). The conductivity increases with temperature in pure PEO and in all the compositions of the ($PEO+NaHCO_3$) polymer electrolyte system. The ionic conductivity in the polymer complexes may be interpreted on the basis of a hopping mechanism between coordinating sites, local

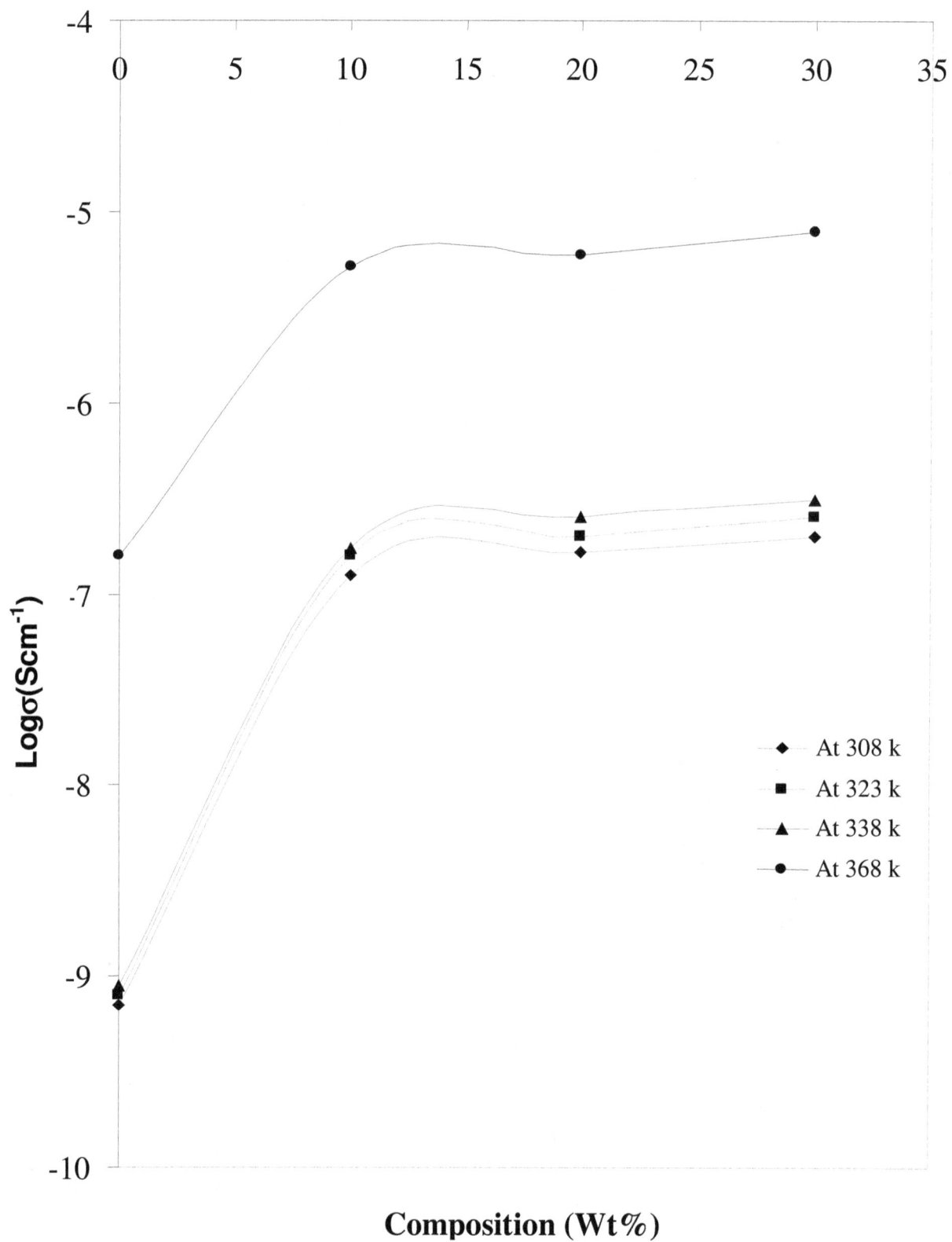

Figure 1. Concentration dependence of (PEO + NaHCO$_3$).

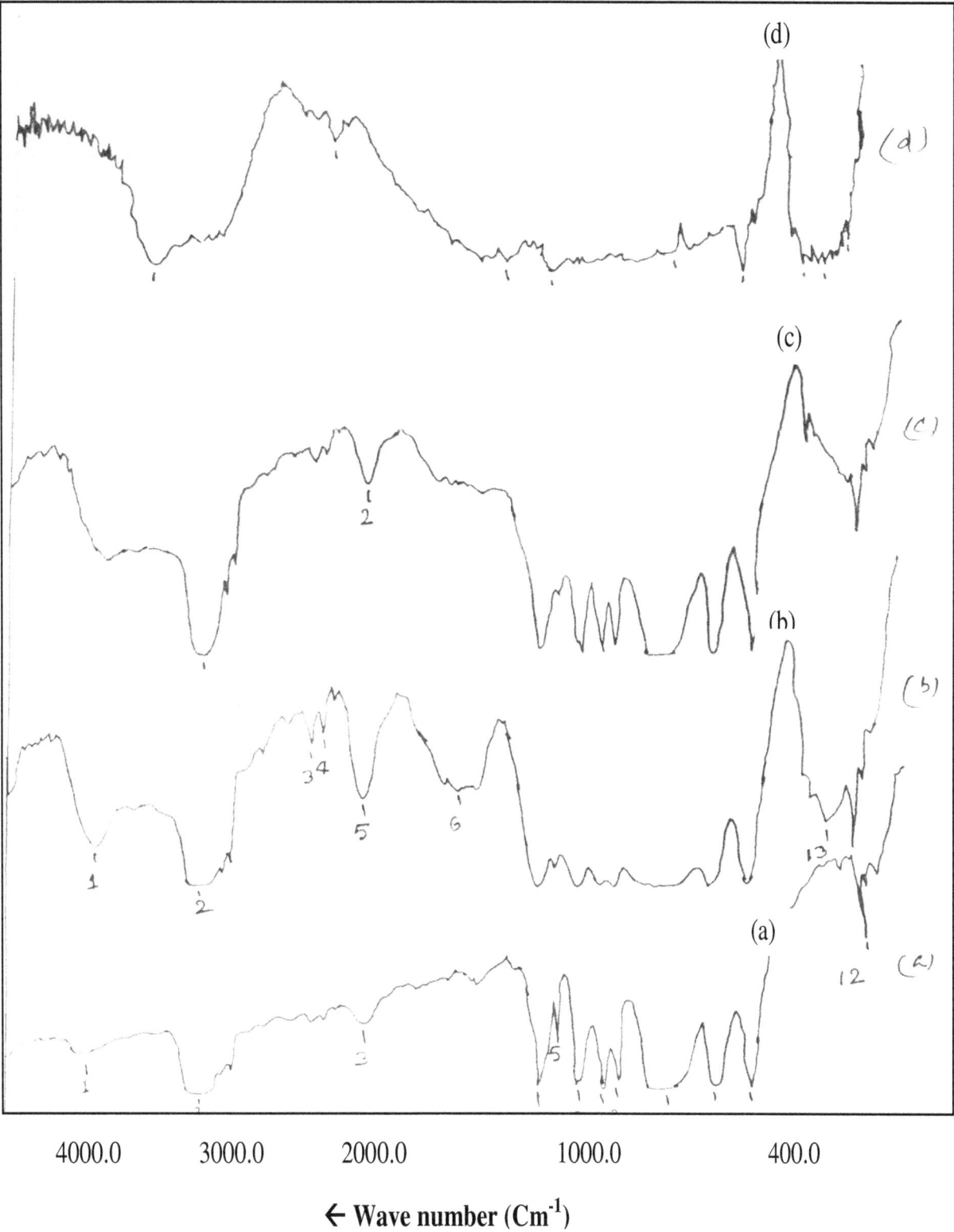

Figure 2. IR-spectra of (a) pure PEO; (b) PEO + NaHCO₃ (90:10); (c) PEO+NaHCO₃ (70:30), and (d) NaHCO₃.

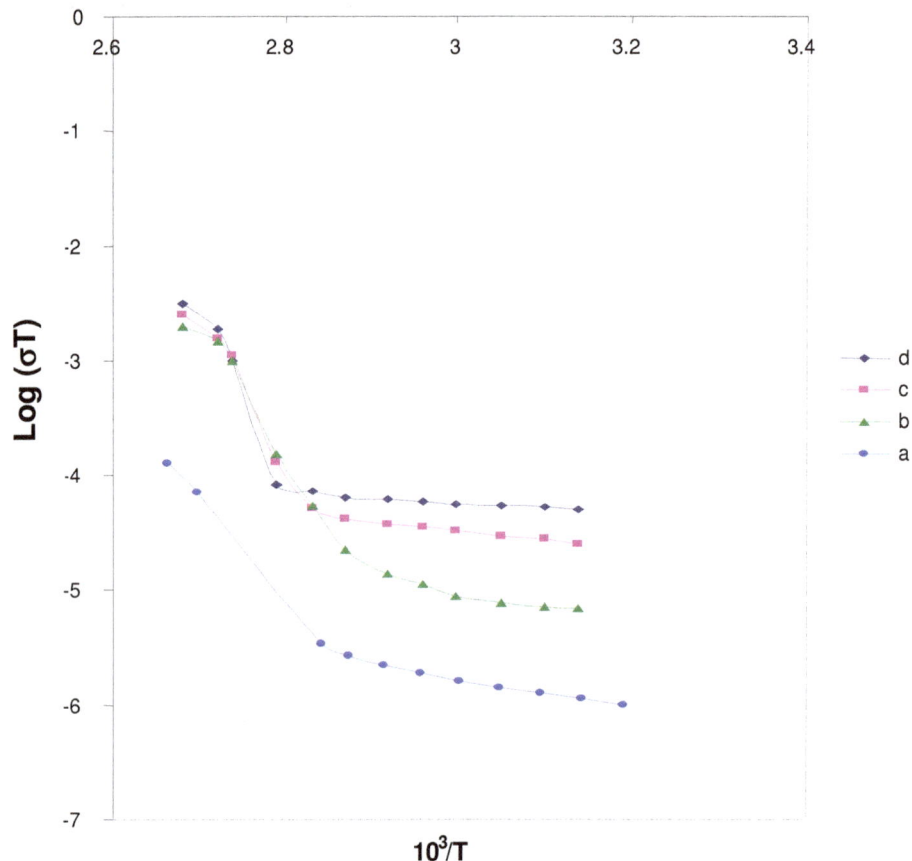

Figure 3. The temperature dependence of D.C. conductivity (a) pure PEO, (b) PEO + NaHCO₃ (90:10); (c) PEO +NaHCO₃ (80:20), and (d) PEO + NaHCO₃ (70:30).

Table 1. D. C. conductivity and activation energies of (PEO + NaHCO₃) electrolyte system.

Film	Conductivity at 303 K (R_T) (Scm⁻¹)	Conductivity at 368 K (Scm⁻¹)	Activation energies (eV)	
			Region-I	Region –II
Pure PEO	6.78×10^{-10}	1.58×10^{-7}	0.34	0.59
(PEO + NaHCO₃) (90:10)	1.55×10^{-7}	5.11×10^{-6}	0.32	0.48
(PEO + NaHCO₃) (80:20)	1.62×10^{-7}	5.91×10^{-6}	0.25	0.33
(PEO + NaHCO₃) (70:30)	1.65×10^{-7}	6.33×10^{-6}	0.19	0.28

structural relaxations and segmental motions of the polymer chains; these are essential to assure high ionic conductivity of the electrolyte (Sreepathi et al., 1994, 2000; Vijaya et al., 2007). The basic design of the cell, which consists of a solid electrolyte ion-exchange membrane, electro catalysts and gas feed fuels is illustrated in Figure 4.

The two electrodes, which consist of the electro catalyst and a plastic material for water-proofing the electrode, are in the form of fine metallic wire screens. They are bonded on either side of the electrolyte layer. The wire screen material is titanium or platinum. Metallic current collectors are ribbed onto each electrode. The hydrogen compartment of the cell is enclosed; the hydrogen gas enters this compartment through a small inlet and circulates throughout the ribbed current collectors and distributes itself evenly over the electrode. On the opposite side, oxygen or air enters the

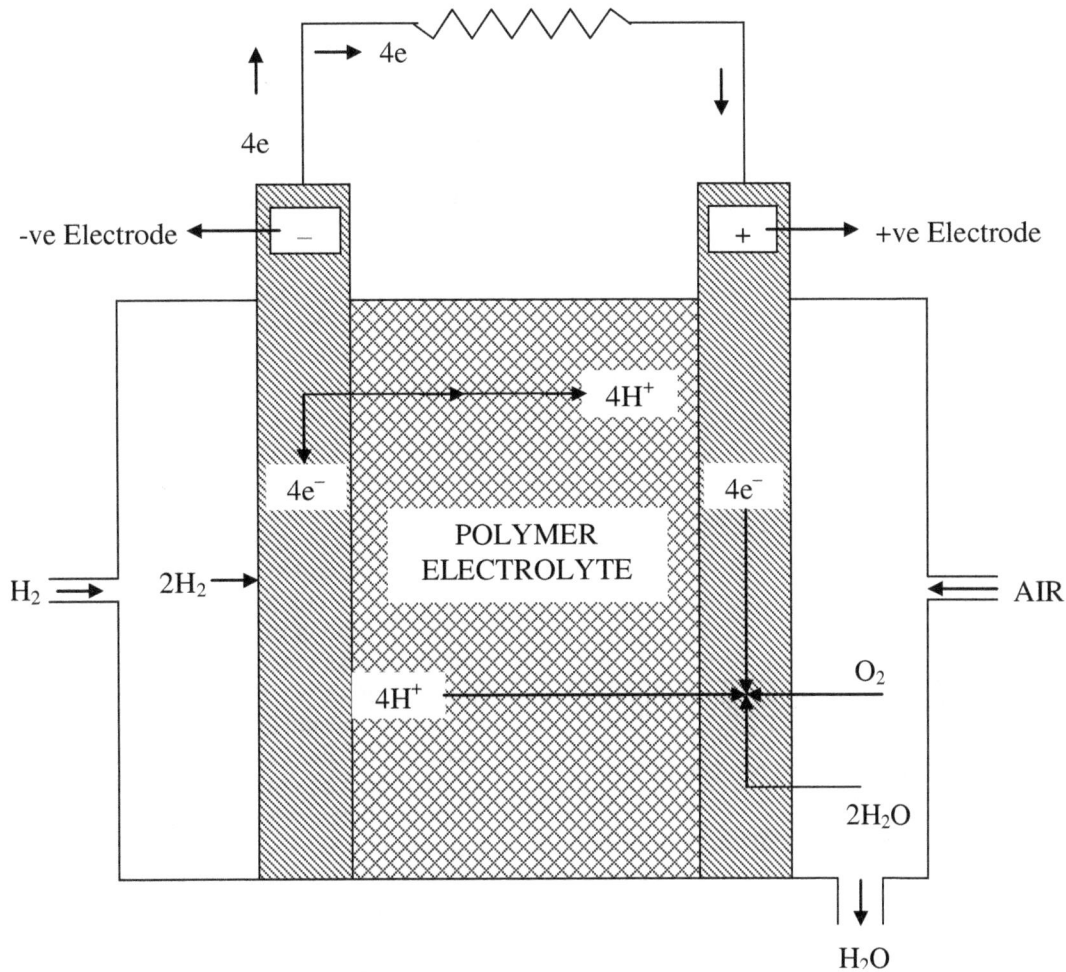

Figure 4. Polymer electrolyte membrane fuel cell (PEMFC).

compartment, coolant tubes run through the ribs of the current collectors. On the oxygen side, the current collectors also hold wicks which absorb water, the product of fuel-cell reaction and carry it over by capillary action. The water leaves the cell through an exit from the oxygen compartment. Oxygen is prevented from leaving its compartment by the inclusion of a differential pressure water-separation system. The distinctive feature of this cell is that it uses a solid polymer electrolyte in the form of an ion-exchange membrane. The membrane is non-permeable to the reactant gases, hydrogen and oxygen, which thus prevents them from coming into contact. The membrane is however; permeable to hydrogen ions (Vishnu et al., 2003) which are the current carriers in the electrolyte .The desired properties of polymer electrolyte membrane are high ionic conductivity, low permeability of fuel and oxidant, very low electronic conductivity, high resistance to its oxidation or hydrolysis and mechanical

stability (Proceedings from 3rd International Fuel Cell Conference, Nagoya Tuting, Nov. 30th to Dec 3rd 1999, Japan).

The ion-exchange membrane electrolyte is acidic and the current carriers in solution are hydrogen ions. The hydrogen ions are produced by the reaction at anode according to,

$$2H_2 \rightarrow 4H^+ + 4e^-$$

These ions are then transported to the cathode through the electrolyte and the electrons reach the cathode via the external circuit. At the cathode, oxygen is reduced producing water as represented by:

$$O_2 + 4H^+ + 4e^- \rightarrow 2H_2O$$

This cell operates at about 40 to 70°C. The single cell1

Table 2. Specifications for single cell 1.

Electrode	20% carbon anode and cathode
Polymer electrolyte	$(PEO+NaHCO_3)$ (90:10)
Type of graphite rates	Porous 1 no. and non-porous 1 no.
Concentrating $NaBH_4$	10%
Fuel	H_2 and O_2 (from cylinders)

Table 3. Specifications for single cell 2.

Electrode	20% carbon anode and cathode
Polymer electrolyte	$(PEO + NaHCO_3)$ (80:20)
Type of graphite rates	Porous 1 no. and non-porous 1 no.
Concentrating $NaBH_4$	15%
Fuel	H_2 and O_2 (from cylinders)

Table 4. Specifications for cell 3.

Electrode	20% carbon anode and cathode
Polymer electrolyte	$(PEO+NaHCO_3)$ (70:30)
Type of graphite rates	Porous 1 no. and non-porous 1 no.
Concentrating $NaBH_4$	20%
Fuel	H_2 and O_2 (from cylinders)

Table 5. Specifications for the three PEM fuel cells stack.

Electrode	20% carbon electrodes no. 6
Electrolytes	$PEO + NaHCO_3$ (90:10) $PEO + NaHCO_3$ (80:20) $PEO + NaHCO_3$ (70:30)
Electrolyte chamber thickness	4 mm (3)
Electrolyte temperature	65°C
Graphite plates	Non-porus (4) Mono-polar - 2 no. Bi-polar – 2 no.
Concentration of sodium borohydride	30%
Hydrogen and oxygen	Completely running on cylinder Gases

with the configuration anode / $(PEO+NaHCO_3)$ (90:10)/ cathode was assembled with the specifications and the first readings were taken with the cylinder gases. The specifications of the single fuel cell with the polymer electrolyte are given in Table 2. The resistor, 0.5 Ω were added in series to the circuit to gain an increase in voltage. Subsequently it was achieved. The open circuit voltage was found to be 0.95 V.

The single cell 2 and single cell 3 with the configurations anode/$(PEO+NaHCO_3)$ (80:20)/ cathode and anode/$(PEO+NaHCO_3)$ (70:30) (Tables 3 and 4) were assembled with the specifications. The specification design of the PEM fuel cell2, cell3 was given in Tables 3 and 4, respectively. The observations were taken with the cylinder gases. The resistor 0.5 Ω was connected in series to the circuit. It is observed that, when the cell current increases, cell voltage slightly decreases (Proceedings from 3rd International Fuel Cell Conference, Nagoya Tuting, Nov. 30th to Dec 3rd 1999, Japan) up to 0.61 V due to concentration of sodium borohydride.

After testing successfully with the single cell, a three cell stack was assembled (that is, the three single cells are connected in series). The material require for this assembly is similar to that of the single cell. But the quantity of the material is increased. The material required for assembly and its quantity is given in Table 5.

For the three cell stack, we used non-porous, bi-porous graphite plates. The three design aspects of the PEM fuel cell system under fabrication are (1) the hydrogen on demand generation system (that is, hydrogen reformer system) and (2) the air scrubber system (that is, the fuel cell assembly). The incorporation of the aforementioned three designs completes the fuel cell system (Stomehart, 1996). Using this conducting polymer electrolyte thin film systems, a three fuel cell stack have been fabricated and I-V characteristics studied and the results were subsequently summarized.

The basic design of the cell consists of a solid polymer electrolyte ion-exchange membrane, electro catalysts and gas fuel tubes. Hydrogen was generated from sodium borohydride ($NaBH_4$) and air was scrubbed using the initial setup of the air scrubber. The fuel cell system was run completely on generated gases. The distinctive feature of this cell is that it uses a solid electrolyte in the form of an ion-exchange membrane. The membrane is non-permeable to the reactant gases hydrogen and oxygen, which thus prevents them from coming into contact. The membrane is however, permeable to hydrogen ions (Journal of Industry Report, Global News on the Advances and Applications of the Fuel Cell Technology, Dec. 2000, Vol. 1, No. 12) which are the current carriers in the polymer electrolyte. This cell operates at around 40 to 70°C.

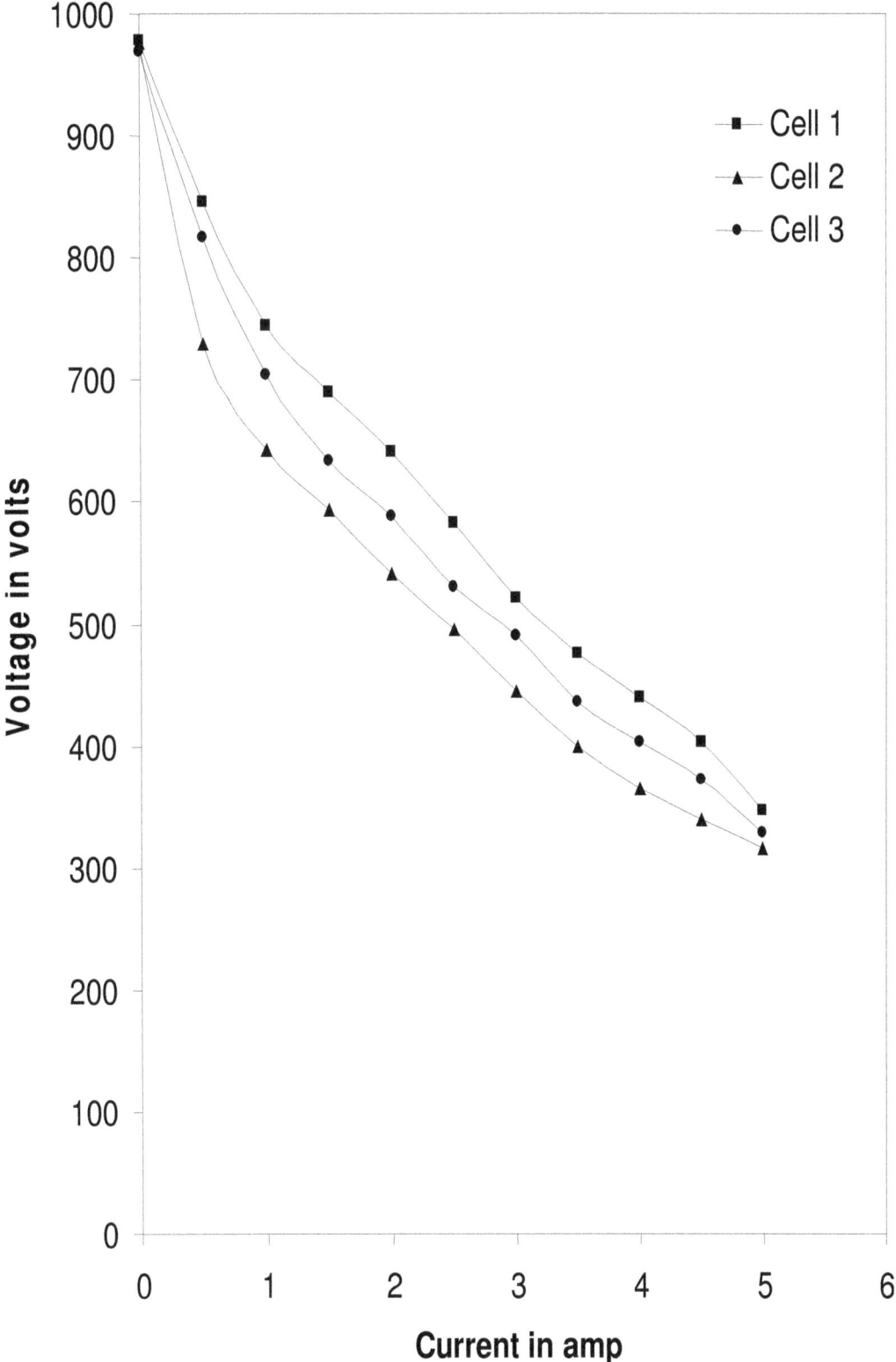

Figure 5. I-V Characteristics of a single cell 1, 2 and 3 of a (PEO + NaHO₃) based PEM fuel cell.

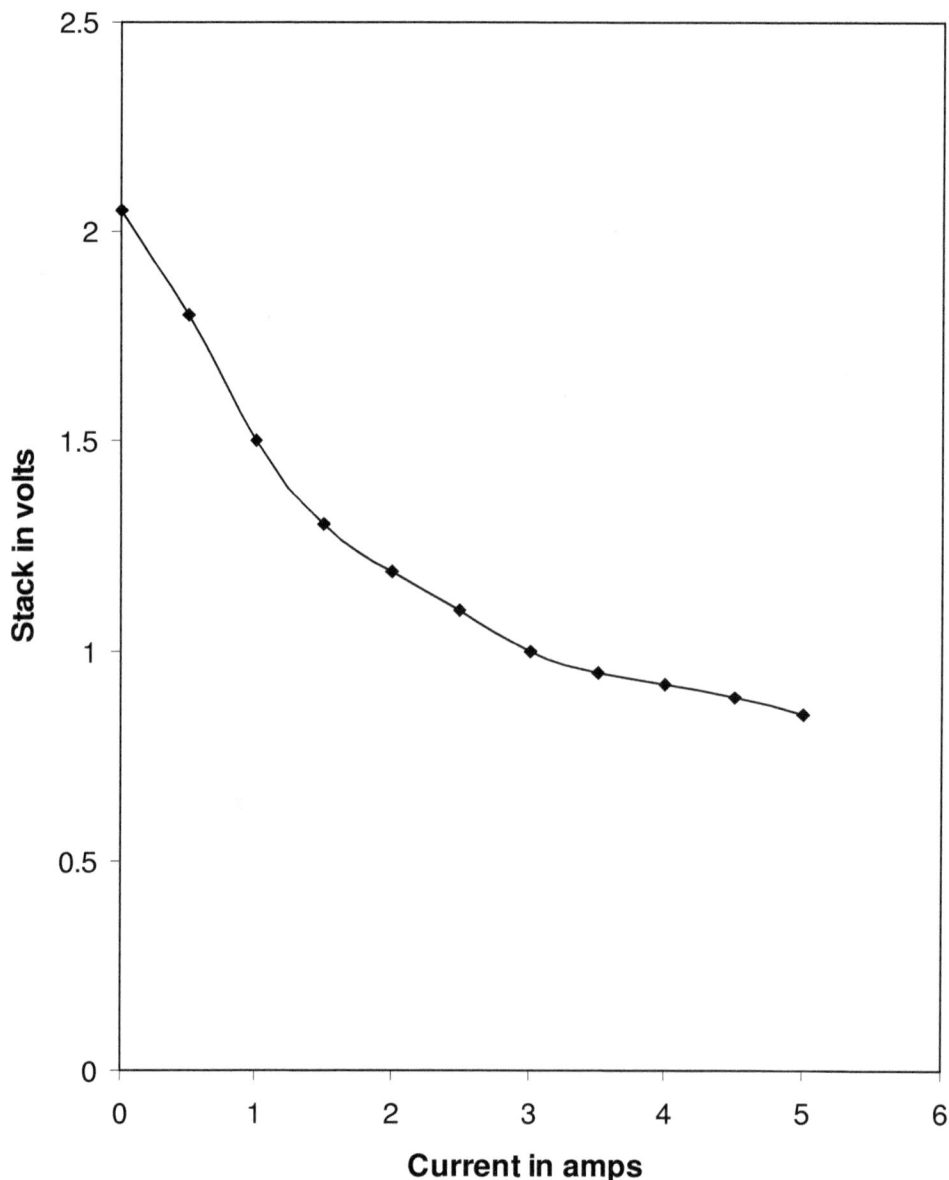

Figure 6. I to V characteristics of a three PEM fuel cell stack.

Polymer electrolyte fuel cells are assembled in the configuration (anode) / (polymer electrolyte) / (Cathode) and their I-V characteristics was studied. The main function of the electrolyte in PEMFC is to carry the O^{2-} from the cathode to the anode and also it separates the fuel from the oxidant inside the cell. The high ionic electrolytes are independent of oxygen partial pressure. In PEMFC the (PEO+NaHCO$_3$) electrolyte system is typically chosen as the base line material. And also most of the polymer electrolyte systems have good mechanical strength, fracture toughness, better ionic conductivity and lesser chemical reactivity with the cathode, anode and inter connectors (Proceedings from 3rd International Fuel Cell Conference, Nagoya Tuting, Nov. 30th to Dec 3rd 1999, Japan).

The current voltage characteristics of the single cell1, single cell2 and single cell3 are shown in Figure 5. From the figure, it is observed that, when current of the cell increases, the developed voltage of the cell gradually decreases.

The three fuel cell stack, and I-V characteristics were also studied and it is shown in Figure 6. The voltage of the fuel cell stack gradually decreases with increase in current. The open circuit voltage of these fuel cells has been measured for different compositions of various polymer electrolyte systems. It is observed that the open circuit voltage was found to be in range 0.95 to 1.1 V. it may be due to used electrolyte membrane type, electro catalysts and concentration of sodium borohydride.

Conclusion

1. The fuel cell stack is found to be better performance as compared to a single cell fuel cell.
2. (PEO+NaHCO$_3$) three cell stack wattage is found to be 4.25 watt with a stack voltage of 0.85 V. The fuel cell characteristics like I-V characteristics, open circuit voltage, power density etc. mainly depends on nature of electrolyte, the generated gases and its purity. But it is observed that the volt-ampere characteristics, efficiency of these fuel cells are very low. Further work has to be done to improve the fuel cell efficiency by proper choosing of solid polymer electrolytes along with the electrodes.
3. The improve power density, current density of the fuel cell mainly depends on the nature of catalyst used in the fabrication. Among all the electrodes, platinum is a very good catalyst but, its cost is very high. Therefore studies of other catalyst materials in the place of platinum are needed.

ACKNOWLEDGEMENTS

The author thanks Er. Koneru Satyanrayana, Chancellor, K. L. University Guntur (A.P) for his encouragement and support and also the author is highly grateful to Dr. G. L. Datta, Vice Chancellor K. L. University. Finally, the author thanks Prof. N. Venkat Ram, Dean, F. E. D and Department of Physics, K L University for his constant support.

REFERENCES

Andrew W, James L (2003). Fuel cell systems explained. John Wiley and Songs, Ltd., ISBN No. 0-471-49026-I.
Apple AJ, Foulkes FR (1989). Fuel cell Handbook. Van Nor stand Reinhold, New York, USA.
Berers D, Wagner N, Von Bradke M, (1997). Irt. Hydrogen Energy., 23: 57.
Bernardi DM (1990). Electrochemical modelling of lead/acid batteries under operating conditions of electric vehicles. J. Electrochem. Soc., 137: 3344.
Blomen LT (1994). Fuel cell systems, Plenum Publishing Corporation, New York.
Chabagno JM, MJ Duclot (1979). Fast Ion transport in Solids. Vashista P, Mundy JN, Shenoy GK, Eds., North Holland, New York, p. 131.

Dahr HP (1994). Perfluorinated Ionic Polymers for PEFCs (Including Supported PFSA). Fuel Cell Semin., p. 85.
Elroy JFMC, Nuttall LJ (2001). Status of solid polymer electrolyte fuel cell technology and potential for transport and applications by General Electric company, Direct energy conversions programmes.
Hashmi SA, Chandra A, Chandra S, Chowdari BVR (1992). Solid State Ionics: Materials and Application World Scientific, Singapore, p. 567
Narasaiah EL (1995). Study of a new polymer electrolyte (PVP+KYF4) for solid-state electrochemical cells." J. Power Source, 55: 255-257.
Proceedings from 3rd International Fuel cell conference, Nagoya Tuting, Nov. 30th to Dec 3rd (1999), Japan.
Rai GD (2002). Non-conventional Energy sources. Khanna Publishers, New Delhi, pp. 561-589.
Ramalingaiah S, Reddy DS (1996). Conductivity and discharge characteristic studies of novel Polymer electrolyte based on PEO complexed with Mg(NO$_3$) salt. Mater. Lett., 29: 285-289.
Reddy MJ, Rao UVS (1998). Transport Studies of Poly (ethylene oxide) based polymer.
Scrosati B (1988). Solid state Ionic Devices." Edited by Chodary BVR, Radha S, Krishna, World Scientific Publishing Co., Singapore, p. 113.
Sossina M, Haile F (2003). Materials for Fuel Cells. Mater. Today.
Sreekanth T, Jaipal Reddy M, Subba RUV (2000). Polymer electrolyte system based on (PEO+KBrO3) - Its application as an electrochemical cell. J. Power Sources, 3994: 1-5.
Sreepathi R, Jaipal SR, Laxmi NM, Subba RUV (1995). Polymer electrolyte system based on (PEO+KBrO3)-its application as an electrochemical cell. Mat. Sci. Eng., 33: 173.
Sreepathi RS, Rao KVS, Shareefuddin UV, Subba R, Chandra S (1994). 'Ionic Conductivity and battery characteristic studies on PEO+AgNO$_3$ polymer electrolyte." Solid State Ion., 67: 331-334S.
Srivastava PC (1992). Solid-state Ionics: Materials and Applications, Edited by Chowdar BVR, World Scientific, Singapore, pp. 561-565.
Stevens JR, Mellander BR, (1987). Conducting Polymers. Edited by Alcacer L, Reidel, Dordrecht, p. 95.
Stomehart P (1996). Polymer-silica composite membranes for direct methanol fuel cells. Proc. Fuel Cell Semin., p. 591.
Swarna R, Shyam KT (2003). Thesis – Design and fabrication of alkaline fuel cell stack.
Verbrugge MW, Hill RF (1990). Radiotracer method for simultaneous measurement of cation, anion and water transport through ion-exchange membranes. J. Electrochem. Soc., 137: 886.
Vijaya K, Venkata R, Reddy CH (2007). Characterization and electrical studies of polyethylene oxide based thin film polymer electrolytes.' Int. J. Acta Ciencia Indica, 37(4): 473.
Vishnu PB, Jai S, Prakash, Ramya K, Dhathathreyan KS (2003). Solid State Ion. Devices Sci. Technol., pp. 118-129.
Yeo RS (1983). Sorption and transport behavior of perfluorinated ionomer membranes in concentrated NaOH solution. J. Electrochem. Soc., 130: 533.
Yuankang H, Zhusheng C (1986). Materials for Solid State Ionics. In: Chowdari BVR, Radha KS (Eds),World Scientific,Singapore, p. 333.

14

Biofuel: Boon or bane

Krishna Bolla*, S. V. S. S. S. L. Hima Bindu N and M. A. Singara Charya

Department of Microbiology, Kakatiya University, Warangal – 506009, A.P, India.

Biofuels are going to play an extremely important role in meeting India's energy needs. The country's energy demand is expected to grow at an annual rate of 4.8% over the next couple of decades. Most of the energy requirements are currently satisfied by fossil fuels – coal, petroleum-based products and natural gas. Domestic production of crude oil can only fulfill 25 to 30% of national consumption. In fact, oil imports during August, 2010 were valued at US $ 7795 million which was 12.4% higher than oil imports valued at US $ 6936 million in the corresponding period last year. India was the sixth largest net importer of oil in the world, importing about 70%, of its oil needs and fourth largest consumer of oil in the world.

Key words: Biofuel, ethanol, blends, biodiesel.

INTRODUCTION

Ethanol, currently produced in India by the fermentation of sugarcane molasses, is an excellent biofuel and can be blended with petrol. Likewise, biodiesel, which can be manufactured by the transesterification of vegetable oil, can be blended with diesel to reduce the consumption of diesel from petroleum. Ethanol and biodiesel are gaining acceptance worldwide as good substitutes for oil in the transportation sector. Brazil uses pure ethanol in about 20% of their vehicles and 22 to 26% ethanol-petrol blend in the rest of their vehicles. The United States and Australia use a 10% ethanol blend. With a normal production rate of 1,900 million liters a year, India is the world's fourth largest producer of ethanol after Brazil, the United States and China. Beginning 1 January 2003, the Government of India mandated the use of a 5% ethanol blend in petrol sold in nine sugarcane-producing states. The Government will expand the 5% ethanol mandate to the rest of country in a phased manner. Biodiesel production is rapidly growing in Europe and the United States. Current estimates shows production of 2.2 Mt/year in Europe, with Germany (1.1 Mt/year), France (0.5Mt/year) and Italy (0.4 Mt/year) being the leading producers. The European Union mandated that its members derive at least 2% of their fuel consumption from biofuels by 2005 and 5.75% by 2010. Biodiesel production is about 245,000 t/year in the United States.

The Government of India has developed an ambitious National Biodiesel Mission to meet 20% of the country's diesel requirements by 2011 to 2012. Since the demand for edible vegetable oil exceeds supply, the Government decided to use non-edible oil from *Jatropha Curcas* oilseeds as biodiesel feedstock. Extensive research has shown that *J. curcas* offers the following advantages: it requires low water and fertilizer for cultivation, not browsed by cattle or sheep, pest resistant, easy propagation, high seed yield and ability to produce high protein manure. The National Biodiesel Mission will be implemented in two stages: 1) a demonstration project carried out between 2003 to 2007, which will cultivate 400,000 hectares of land and yield about 3.75 tons oilseed per hectare annually. The expected annual biodiesel production from the project is 1.2 t/ha/year for a total of 480,000 tons per annum.

The Government will build a transesterification plant with a biodiesel production capacity of 80,000 t/year as part of the demonstration project; and 2) a commercialization period from 2007 to 2012 will continue Jatropha cultivation and install more transesterification plants which will position India to meet 20% of its diesel needs through biodiesel.

BENEFITS FROM THE USE OF BIOFUELS IN INDIA

Reduced emission of harmful pollutants

Ethanol and biodiesel are both oxygenated compounds containing no sulphur. These fuels do not produce sulphur oxides, which lead to acid rain formation. Sulphur

---

*Corresponding author. E-mail: bollakrishna@gmail.com.

is removed from petrol and diesel by a process called hydro-desulphurisation. The hydro-desulphurisation of diesel causes a loss in lubricity, which has to be rectified by introducing an additive. Biodiesel has natural lubricity, and thus no lubricity-enhancing additive is required. Since ethanol and biodiesel contain oxygen, the amount of carbon monoxide (CO) and unburnt hydrocarbons in the exhaust is reduced. With the introduction of ethanol in Brazil, CO emission from automobiles decreased from 50 g/km in 1980 to 5.8 g/km in 1995. The emission of nitrogen oxides (No_x) from biofuels is slightly greater when compared to petroleum, but this problem can be ameliorated by using de-No_x catalysts, which work well with biofuels due to the absence of sulphur.

One of the disadvantages in using pure ethanol is that aldehyde emissions are higher than those of gasoline, but it must be observed that these aldehyde emissions are predominantly acetaldehydes. Acetaldehydes emissions generate less adverse health effects when compared to formaldehydes emitted from gasoline engines.

Reduction in greenhouse gas emissions

The net CO_2 emission of burning a biofuel like ethanol is zero since the CO_2 emitted on combustion is equal to that absorbed from the atmosphere by photosynthesis during the growth of the plant (sugarcane) used to manufacture ethanol. This is illustrated by the following equations:

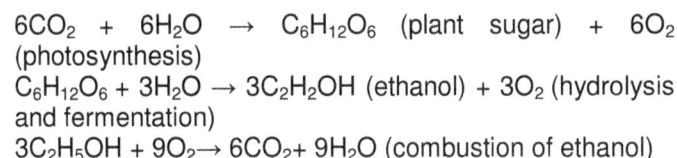

$6CO_2 + 6H_2O \rightarrow C_6H_{12}O_6$ (plant sugar) $+ 6O_2$ (photosynthesis)
$C_6H_{12}O_6 + 3H_2O \rightarrow 3C_2H_2OH$ (ethanol) $+ 3O_2$ (hydrolysis and fermentation)
$3C_2H_5OH + 9O_2 \rightarrow 6CO_2 + 9H_2O$ (combustion of ethanol)

Life cycle analysis, from well to wheels, shows that ethanol has the lowest CO_2 emission among the major transportation fuels. Biofuels contribute significantly to climate change mitigation by reducing CO_2 emissions. Biodiesel projects can qualify as CDM projects and thus bring in additional income through the sale of certified emission reductions.

Increased employment

At the beginning of the new millennium, 260 million people in India did not have access to a consumption basket, which defines the poverty line. India is home to 22% of the worlds poor. A programme that generates employment is therefore particularly welcome. The biofuels sector has the potential to serve as a source of substantial employment. The investment in the ethanol industry per job created is $11,000, which is significantly less than the $220,000 per job in the petroleum field . In

India, the sugar industry, which is the backbone of ethanol production, is the biggest agroindustry in the country. The sugar industry is the source of the livelihood of 45 million farmers and their dependants, comprising 7.5% of the rural population. Another half a million people are employed as skilled or semi-skilled labourers in sugarcane cultivation.

The first phase of the National Biodiesel Mission demonstration project will generate employment of 127.6 million person days in plantation by 2007. On a sustained basis, the program will create 36.8 million person days in seed collection and 3,680 person years for running the seed collection and oil-extraction centres.

Energy security and decreased dependence on oil imports

India ranks sixth in the world in terms of energy demand, accounting for 3.5% of the world commercial energy demand in 2001. But at 479 kg of oil equivalent, the per capita energy consumption is still very low, and the energy demand is expected to grow at the rate of 4.8% per annum. India's domestic production of crude oil currently satisfies only about 25% of this consumption. Dependence on imported fuels leaves many countries vulnerable to possible disruptions in supplies, which may result in physical hardships and economic burdens. The volatility of oil prices poses great risks for the world's economic and political stability, with unusually dramatic effects on energy-importing developing nations. Renewable energy, including biofuels, can help diversify energy supply and increase energy security.

Improved social well-being

A large part of India's population, mostly in rural areas, does not have access to energy services. The enhanced use of renewables (mainly biofuels) in rural areas is closely linked to poverty reductions because greater access to energy services can:

i) Improve access to pumped drinking water. Potable water can reduce hunger by allowing for cooked food (95% of food needs cooking);
ii) Reduce the time spent by women and children on basic survival activities (gathering firewood, fetching water, cooking, etc.);
iii) Allow lighting which increases security and enables the night time use of educational media and communication at school and home; and
iv) Reduce indoor pollution caused by firewood use, together with a reduction in deforestation.

Lack of access to affordable energy services among the rural poor seriously affects their chances of benefiting from economic development and improved living standards. Women, older people and children suffer disproportionately because of their relative dependence on traditional fuels and their exposure to smoke from cooking, the main cause of respiratory diseases.

Electricity through transmission lines to many rural areas is unlikely to happen in the near future, so access to modern decentralized small-scale energy technologies, particularly renewables (including biofuels), are an important element for effective poverty alleviation policies. A programme that develops energy from raw material grown in rural areas will go a long way in providing energy security to the rural people.

Increase in nutrients to soil, decrease in soil erosion and land degradation

In ethanol production from sugarcane, the by-products like vinasse (solid residue left after distillation) and filter cake contain valuable nutrients. Using these organic fertilizers instead of chemical fertilizers reduces the need for chemicals, which could be hazardous and avoids pollution of ground water and rivers. International Crop Research Institute for Semi-Arid Tropics (ICRISAT) compares the nutrient content of filter cake obtained from various oilseeds in biodiesel manufacture with that of commonly used fertilizers like Di-Ammonium Phosphate (DAP) and Urea and demonstrates that the filter cake is an effective fertilizer: Also the cultivation of land for sugarcane and oilseed-bearing crops contributes to a decrease in soil erosion and land degradation.

Good fuel properties

Ethanol has a research octane number of 120, much higher than that of petrol, which is between 87 and 98. Thus, ethanol blending increases the octane number without having to add a carcinogenic substance like benzene or a health-risk posing chemical like methyl tertiary butyl ether (MTBE). The energy content of ethanol is only 26.9 MJ/kg compared to 44.0 MJ/kg for petrol. This would suggest that the fuel economy (km/l) of a petrol-powered engine would be 38.9% higher than that of an ethanol-powered engine. In actuality, this difference is 30% since ethanol engines can run more efficiently (at a higher compression ratio) because of the higher octane rating. For a 10% ethanol blend the fuel economy advantage of a petrol engine is only 3%. The flammability limit of ethanol (19% in air) is higher than that of petrol (7.6%), and likewise the auto-ignition temperature of ethanol is higher than that of petrol (366 versus 300°C). Thus, ethanol is safer than petrol due to the lower likelihood of catching fire.

Ethanol's higher latent heat of vaporization and greater propensity to absorb moisture may lead to engine starting

and corrosion problems, respectively, but none of these problems have manifested in the millions of hours of running automobile engines in Brazil. Biodiesel has good fuel properties, comparable to or even better than petroleum diesel. It has 10% built-in oxygen content that helps it to burn fully. Its cetane number (an indication of its fuel burning efficiency) is 52 for biodiesel from Jatropha oil, higher than the 42 to 48 cetane number of most petroleum diesels. The esters of the long-chain fatty acids of biodiesel are excellent lubricants for the fuel injection system. It has a higher flash point than diesel, making it a safer fuel.

Other advantages are the almost zero sulphur content and the reduced amount of carbon monoxide, unburned hydrocarbons and particulate matter in the exhaust. But there are a few technical issues that need to be resolved. Biodiesel has a high viscosity at low temperatures, leading to flow problems at these temperatures. For long-term storage in hot, humid conditions, ethanol may require a biocide to prevent bacterial growth.

The other side of biofuels is not effective

The energy crisis driven by over-consumption and peak oil has provided an opportunity for powerful global partnerships between petroleum, grain, genetic engineering, and automotive corporations. These new food and fuel alliances are deciding the future of the world's agricultural landscapes. The biofuels boom will further consolidate their hold over our food and fuel systems and allow them to determine what, how and how much will be grown, resulting in more rural poverty, environmental destruction and hunger. The ultimate beneficiaries of the biofuel revolution will be grain merchant giants, petroleum companies, car companies; and biotech giants.

The biotech industry is using the current biofuel fever to greenwash its image by developing and deploying transgenic seeds for energy, not food production. Given the increasing public mistrust for and rejection of transgenic crops as food, biotechnology will be used by corporations to improve their image claiming that they will develop new genetically modified crops with enhanced biomass production or that contain the enzyme α amylase which will allow the ethanol process to begin while the corn is still in the field, a technology they claim has no negative impacts on human health. The deployment of such crops into the environment will add one more environmental threat to those already linked to GMO corn. As governments are persuaded by the promises of the global biofuel market, they devise national biofuel plans that will lock their agro-systems into production based on large scale, fuel monocultures, dependent upon intensive use of herbicides and chemical fertilizers, thus diverting millions of hectares of valuable cropland from much needed food production. There is a great need for social analysis to anticipate the food security and environmental implications of the unfolding biofuel plans.

Clearly, the ecosystems of areas in which biofuel crops are being produced are being rapidly degraded, and biofuel production is neither environmentally and socially sustainable now nor in the future. It is also worrisome that public universities and research systems are falling prey to the seduction of big money and the influence of politics and corporate power. More importantly, we need to work together to ensure that all countries retain the right to achieve food sovereignty via agro ecologically-based, local food production systems, land reform, access to water, seeds and other resources and domestic farm and food policies that respond to the true needs of farmers and all consumers, especially the poor.

REFERENCES

United Nations Conference on Trade and Development Planning Commission, Government of India. Report of the Working Group, Tenth Plan.
http://www.foodfirst.org/en/node/1662

15

Effect of properties of Karanja methyl ester on combustion and NOx emissions of a diesel engine

P. V. Rao

Department of Mechanical Engineering, Andhra University, Visakhapatnam-530003, Andhra Pradesh, India.
E-mail: prof.pvrao@gmail.com.

The aim of this work is to study the effect of properties of Karanja (*Pongamia pinnata*) methyl ester on combustion, and NOx (oxides of nitrogen) emissions of a diesel engine. The properties of the karanja methyl ester such as viscosity, density, bulk modulus, calorific value, iodine value, cetane number, saturation% and oxygen% are considered for this study. Experiments were conducted in a naturally aspirated, single cylinder, four-stroke, stationary, water cooled, constant rpm, in-line (fuel pump-pressure tube-fuel injector) direct injection diesel engine. The engine tests were conducted with karanja methyl ester (with and without preheating), and baseline fossil diesel. The peak pressures and peak heat release rates for methyl ester was slightly higher than diesel fuel. The crank angles for peak pressure of the karanja methyl ester are very close to top dead center. This is probably due to dynamic injection advance caused by their higher bulk modulus. However, the peak cylinder pressures for preheated methyl ester decreases, due to late injection and faster evaporation of the fuel. It was observed that, at full load the oxides of nitrogen emissions of karanja methyl ester are increased by 6%. A significant reduction in oxides of nitrogen emission is observed with preheated methyl ester.

Key words: Bulk modulus, combustion, diesel engine, effect, Karanja, methyl ester, NOx, properties, preheating, *Pongamia pinnata*.

INTRODUCTION

During first half of the 20th century, the exhaust gas emissions of internal combustion engines are not recognized as problem, due to lower number of automobiles (diesel engine vehicles). As the number of automobiles grew along with world population; the air pollution became an ever increasing problem. This put a major restriction on the engine design during the 1980s and 1990s. Due to technological advancements, emissions have been reduced by over 90% since the 1940s; they are still a major problem for the environment due to exponential increase of automobile population (Pulkrabek, 2003).

One of the major emissions produced by diesel engines is oxides of nitrogen (NOx). In diesel engines, the combustion process forms oxides of nitrogen (NOx) in which the nitric oxide (NO) is abundant; a small portion of the NO oxidizes into nitrogen dioxide (NO_2) at low temperatures in the presence of oxygen. The sum of NO and NO_2 is called NOx. The formation of NOx is dependent on the temperatures during the combustion, the amount of oxygen (O_2) and nitrogen (N_2) present in the charge and

the time available for them to react with each other.

NOx are the precursor pollutants which can combine to form photochemical smog. These irritate the eyes and throat, reduces the ability of blood to carry oxygen to the brain and can cause headaches, and pass deep into the lungs causing respiratory problems for the human beings. Long-term exposure has been linked with leukemia. Therefore, the major challenge for the existing and future diesel engines is meeting the very tough emission targets at affordable cost, while improving fuel economy. The focus on NOx, reductions to meet stringent emission norm 'EURO- V' may represent a major hurdle.

Different methods are being used to reduce NOx emissions from the diesel engines. One is improving the engine technology and the other is using of alternate or oxygenated fuels to increase combustion efficiency. Over the past several years, there has been increased interest in alternate fuels to control exhaust emissions. Therefore, from the point of view of protecting our surrounding environment, and human health, it is necessary to develop alternate fuels with properties comparable to fossil diesel.

Figure 1. Karanja flowers.

Figure 2. Karanja pods.

Figure 3. Karanja seeds.

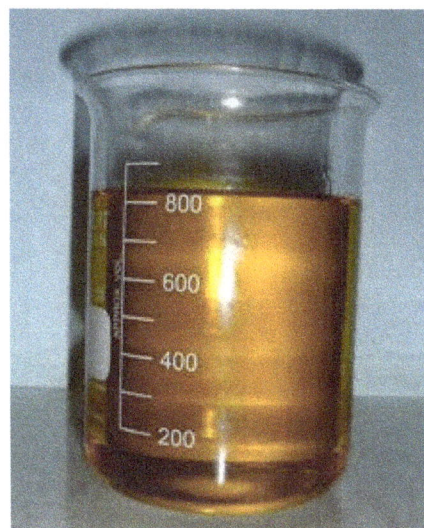

Figure 4. Karanja oil.

Vegetable oil is one such source. India has about 86 types of oilseed-bearing perennial trees (Mitra, 1963) of which Karanja seed (*P. pinnata*); mahua (*Madhuca indica*), neem (*Azadirachta)* and Jatropha (*Jatropha curcas*) are the important ones. Vegetable oils have comparable properties with diesel fuel. However, using straight vegetable oil lead to the clogging of the fuel injectors and there is a problem of starting, especially when the engine is cold. To avoid these problems, the straight vegetable oils can be chemically treated to enhance its properties, after which the vegetable oil is known as methyl ester (ME) or biodiesel. The ME is often referred to as fatty acid methyl ester (FAME).

Karanja (*Pongamia pinnata*) oil

Pongamia, a medium sized glabrous tree grown in many parts of India and Australia (Vivek and Gupta, 2003). Pongamia is popularly known as Karanja in India. Karanja belongs to the family Leguminaceae (Meher et al., 2004) and its botanical name is *P. Pinnata*. Karanja tree bears flowers (Figure 1), and then pods, which after 10 to 11 months gets matured and changes to a tan color (Figure 2) in the month of May to June. The pods are flat to elliptic in shape and contain one (or two) kidney shaped brownish red seeds (or kernels) as shown in Figure 3. The oil content of karanja seed is about 33% (Kumar and Sharma, 2008). The fresh extracted oil is yellowish orange to brown and rapidly darkens on storage as shown in Figure 4.

$$
\begin{array}{l}
\overset{\displaystyle O}{\overset{\displaystyle \|}{CH_2\text{-}O\text{-}C\text{-}R_1}} \\
| \\
| \quad \overset{\displaystyle O}{} \\
| \quad \overset{\displaystyle \|}{} \\
CH\text{-}O\text{-}C\text{-}R_2 \\
| \\
| \quad \overset{\displaystyle O}{} \\
| \quad \overset{\displaystyle \|}{} \\
CH_2\text{-}O\text{-}C\text{-}R_3
\end{array}
\;+\; 3CH_3OH \;\longrightarrow\;
\begin{array}{l}
\overset{\displaystyle O}{\overset{\displaystyle \|}{CH_3\text{-}O\text{-}C\text{-}R_1}} \\
\\
\overset{\displaystyle O}{\overset{\displaystyle \|}{CH_3\text{-}O\text{-}C\text{-}R_2}} \\
\\
\overset{\displaystyle O}{\overset{\displaystyle \|}{CH_3\text{-}O\text{-}C\text{-}R_3}}
\end{array}
\;+\;
\begin{array}{l}
CH_2\text{-}OH \\
| \\
CH\text{-}OH \\
| \\
CH_2\text{-}OH
\end{array}
$$

(KOH: Catalyst)

Triglycerides   +   Methanol   =   Mixture of Fatty Esters   +   Glycerol

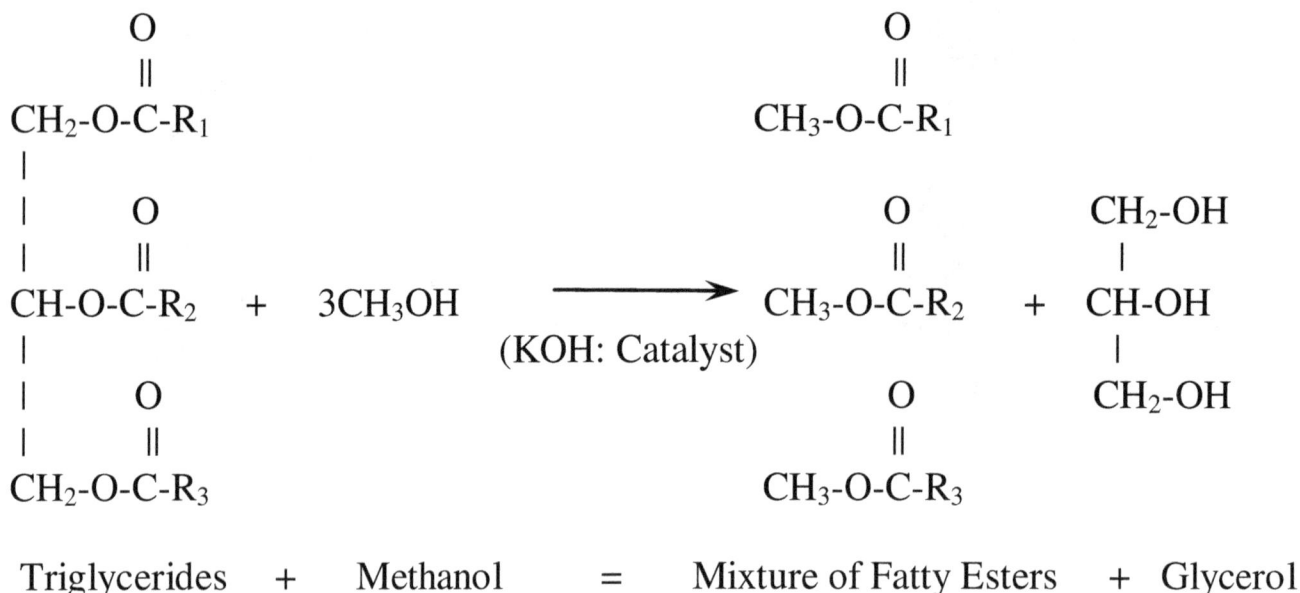

Figure 5. Mechanism of the base (KOH)-catalyzed transesterification process.

Table 1. ASTM methods and instruments used to measure physico-chemical properties of fuels.

Fuel property	Method	Instrument	Model
Density	D 1298	Hydrometer	Petroleum instruments, India
Flash and fire points	D 92	Cleveland open-cup	Petroleum instruments, India
Calorific value	D 240	Bomb CALORIMETER	Parr, UK
Kinematic viscosity	D 445	Kinematic Viscometer	Setavis, UK

Preparation of kanranja oil methyl ester

Karanja oil contains up to 20% (wt.) free fatty acids (Naik et al., 2008; Kalbande, 2008; Bryan, 2009; Sanaz et al., 2010). The methyl ester is produced by chemically reacting the karanja oil with an alcohol (methyl), in the presence of a catalyst (KOH). A two-stage process is used for the esterification (Gerpen, 2003; Aleks, 2000; Tapasvi et al., 2005; Jindal et al., 2010) of the karanja oil. The first stage (acid-catalyzed) of the process is to reduce the free fatty acids (FFA) content in karanja oil by esterification with methanol (99% pure) and acid catalyst (sulfuric acid-98% pure) in one hour time of reaction at 55°C. In the second stage (alkali-catalyzed), the triglyceride portion of the karanja oil reacts with methanol and base catalyst (sodium hydroxide-99% pure), in one hour time of reaction at 65°C, to form methyl ester and glycerol as shown in Figure 5.

To remove un-reacted methoxide present in raw methyl ester, it is purified by the process of water washing with air-bubbling. The methyl ester (or biodiesel) produced from pongamia oil is known as karanja oil methyl ester (KOME) or karanja biodiesel.

CHARACTERIZATION OF METHYL ESTER

The measurement of physico-chemical properties of the fuels were carried out according to ASTM D6751-02 (Standard Specification for Biodiesel Fuel (B100) Blend Stock for Distillate Fuels standards (ASTM, 2002)). The specifications and manufacturers of the instruments used were given in Table 1. The physico-chemical properties of karanja methyl ester are different from fossil diesel. The physical properties of biodiesel will be influenced on combustion, flow, and storage behaviors (Jo-Han et al., 2010). The independent effects of the biodiesel properties on various operational aspects of diesel engine when compared with fossil diesel are discussed.

Fatty acid composition and carbon chain

The building blocks of fatty esters, the fatty acid chain and the alcohol moiety, influence fuel properties and varying either can lead to a change in fuel properties (Knothe, 2010). As shown in Figure 6, the feed stock dependent fatty acid compositions (hydrocarbon chains)

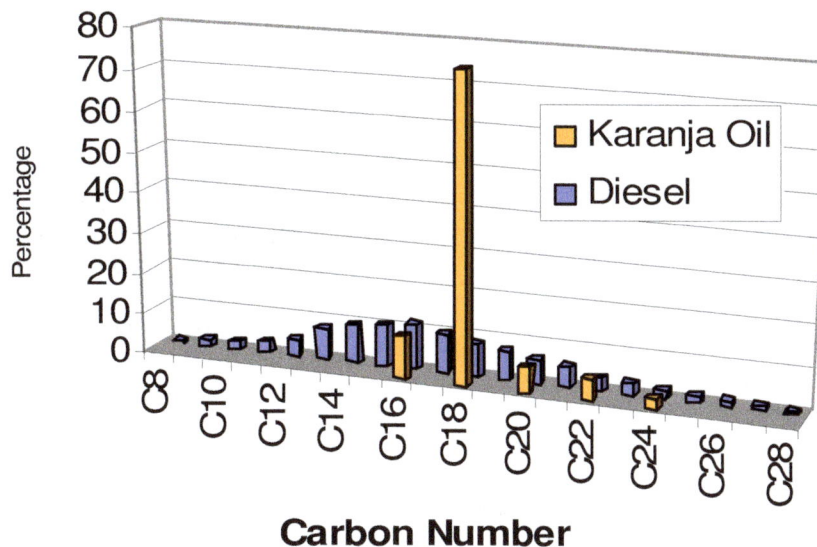

Figure 6. Carbon number distribution.

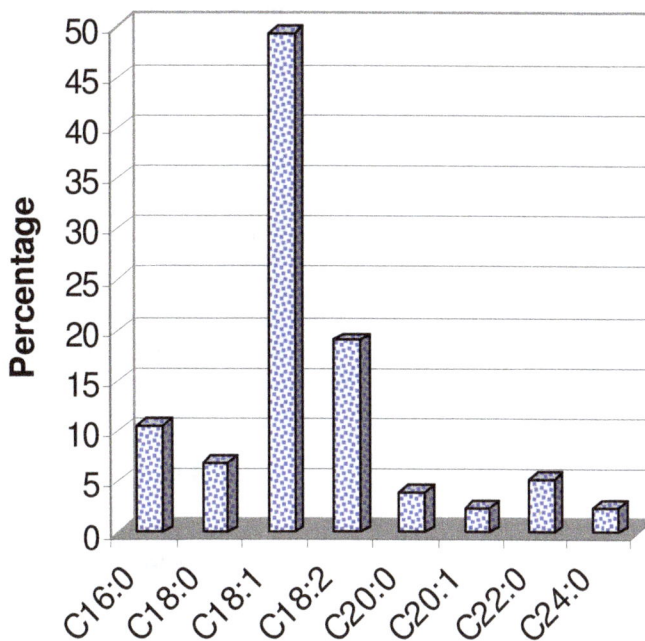

Figure 7. Fatty acids in karanja oil.

of karanja oil vary from C_{16} to C_{24}, with the long chain oleic acid ($C_{18:1}$), linoleic acid ($C_{18:2}$), palmitic acid ($C_{16:0}$), stearic acid ($C_{18:0}$), and behenic acid ($C_{22:0}$) are the highest (Kumar and Sharma, 2008; Bala et al., 2011). The amount of fatty acids present in karanja oil is oleic acid: 49.4%, linoleic acid: 19%, palmitic acid: 10.6%, stearic acid: 6.8%, behenic acid: 5.3%, as shown in Figure 7. This oil contains 29% saturated fatty acids (SFA), 52% of mono-unsaturated fatty acids (MUFA) and

19% of poly-unsaturated fatty acids (PUFA), as shown in Figure 8.

The carbon chain of diesel fuel includes both medium (C_8-C_{12}) and long (C_{14}-C_{32}) carbon chain. The hydrocarbons in diesel fuel range in size from 8 carbon atoms per molecule (C_8) to 32 carbon atoms per molecule (C_{32}) (Applewhite, 1980). The peak in the carbon-number distribution occurs at about 13 to 19 carbon atoms per molecule (C_{13}-C_{19}), as shown in Figure 6.

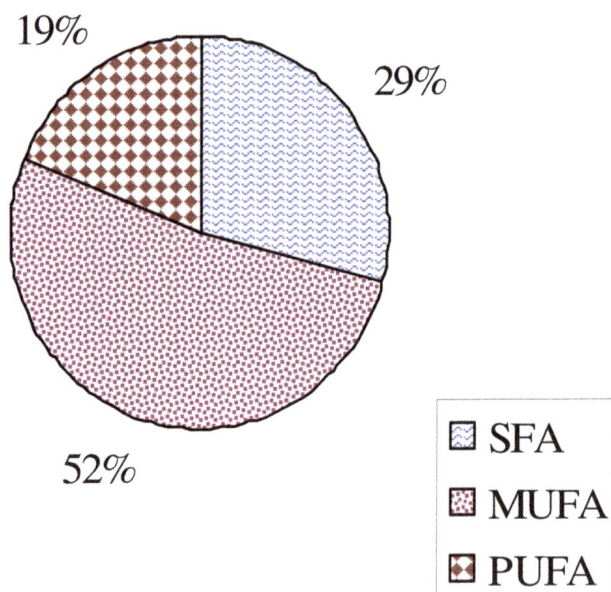

Figure 8. Saturated and unsaturated fatty acids in karanja oil.

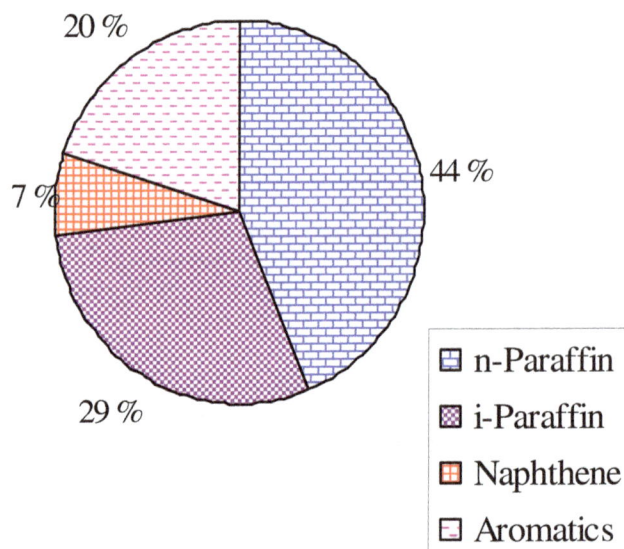

Figure 9. Composition of fossil diesel.

The fossil diesel fuel typically contains over 400 distinct types of organic compounds. Approximately 80% (vol.) of the diesel fuel contains alkanes, with the remainder (that is, 20%) comprised of aromatic molecules. Typical fossil diesel contains approximately 44% of n-Paraffin, 29% of i-Paraffin and 7% of Naphthene as shown in Figure 9. The aromatics containing multiple benzene rings are known as poly-aromatic hydrocarbons (PAHs). These aromatics include polycyclic aromatic compounds containing 2, 3, 4, and 5 fused benzene rings. Benzene

rings will act as nuclei for growth of undesired soot (Akihama et al., 2002).

Iodine value

The iodine value (IV) shows the level of un-saturation of the fuel, which means higher the percentage of unsaturation, larger will be the iodine value (Heinrich et al., 1999). Iodine value of the oil remains little affected by transesterification (Jindal et al., 2010). This is because of the fact that the process is not affecting the degree of unsaturation of the oil. The KOME with 70.8% unsaturation (29.2% saturates) has an iodine value of 81 and this value is well within the permissible limits (<120), as shown in Table 2.

Cetane number

The ignition quality of the fuel is measured by cetane number (CN) and it measures how easily ignition occurs. A fuel with a good ignition quality has a higher cetane number, where the ignition delay period between the start of fuel injection and the onset of auto ignition is short (Jo-Han et al., 2010). The higher the CN, the shorter the ignition delay time and vice versa. The CN assists in smooth combustion with lower knocking characteristics in diesel engines. The CN requirement for the engine depends on the design, size, speed, and load. The CN depends on fuel composition and influences the beginning of the process of combustion and emissions. The CN of methyl ester depends on fatty acids of feedstock (Tapasvi et al., 2005). The CN for KOME is 55.84 (Kumar and Sharma, 2008) and for fossil diesel is 48.

Oxygen and calorific value

In general, the calorific value (or heat of combustion) of methyl ester is less than that of fossil diesel due to presence of C-O and C-C bonds (Romulo et al., 2011). The fossil diesel is made up of a mixture of various hydrocarbon molecules and contains little oxygen (up to 0.3%) and small amount of sulfur (0.25%), while the methyl ester consists of three basic elements namely: carbon, hydrogen, significant amount of oxygen and negligible amount of sulfur, as shown in Table 2. The increase of O_2 in methyl ester (biodiesel) is related to the reduction of C and H_2, causes the lower value of lower calorific value (LCV) of KOME as compared to that of fossil diesel, because O_2 is ballast in fuel and 'C and H_2' are the sources of thermal energy (Gerpen et al., 2004). Therefore, the stoichiometric air-fuel ratio of KOME will be lower than fossil diesel. The KOME contains 11% of O_2; as a result the combustion efficiency of the KOME will be increased (Lebedevas and Vaicekauskas, 2004).

Table 2. Physico-chemical properties of test fuels in comparison to ASTM biodiesel standards.

Property	Units	Diesel (HC)	KOME (FAME)	ASTM D 6751-02
Carbon Chain	Cn	C8-C28	C16-C24	C12-C22
Density (ρ)	gm/cc	0.825	0.875	0.87-0.89
Bulk Modulus (β) @ 20 MPa (Boehman and Alam, 2003)	MPa	1475	1800	N A
Kinematic Viscosity (KV) @ 40°C	cSt	2.25	4.2	1.9-6.0
Cetane Number	-----	48	55.84	48-70
Iodine Value	g Iodine/100 g	38	81	120 max.
Oxygen	%	0.3	11	11
Air/Fuel ratio (Stoichiometric)	------	14.86	13.8	13.8
Lower Calorific Value	kJ/kg	42 500	38 300	37 518
Sulfur	%	0.25	0	0.05
Flash Point (open cup)	°C	66	174	130 min.
Molecular weight	------	226	281	292
Color	------	Light yellow	Yellowish orange	----

Density and bulk modulus

The values of density (ρ) and bulk modulus (β) of the methyl ester are more than that of fossil diesel as shown in Table 2. This high density compensates the lower value of the heat of combustion of the methyl ester. The bulk modulus (β) of a liquid fuel is defined as the pressure required to produce unit volumetric strain and is given by the Equation (1) (Boehman and Alam, 2003; Obert, 1973). The 'β' is a function of fuel temperature (T), pressure (p), and density. The velocity (s) of propagation of the pressure waves (or pulses) through the fuel discharge pipe is given by the Equation (2) (Giffen and Rowe, 1939; Bakar and Firoz, 2005; Brown, 1966; Rao PV, 1993). The term 'c' is the velocity of sound and 'g' is acceleration due to gravity:

$$\beta = \rho . (\partial p / (\partial \rho)_T \text{ ----------------- Equation (1)}$$

$$s = c . \sqrt{(\beta .g/\rho)} \text{ ---------------- Equation (2)}$$

Flash point

The flash point temperature of KOME is higher than that of fossil diesel, because the karanja methyl ester does not have the light fractions. Fuel with high flash point is highly coveted, as the risk associated with fuel transportation is greatly minimized, therefore the safety of KOME is ensured.

Kinematic viscosity

The kinematic viscosity (KV) influences the injection characteristics (spray pattern and depth of penetration) of

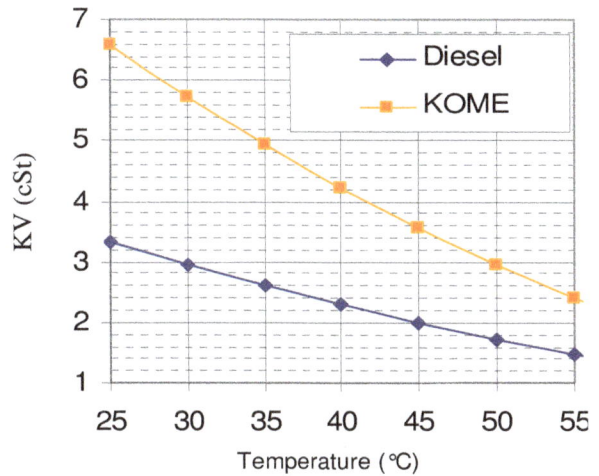

Figure 10. Kinematic viscosity versus temperature.

the fuel, and the quality of filtering. Viscosity of the liquid fuels decreases with the increase of temperature, which in turn decreases the emissions of non-combusted products. The viscosity of fossil diesel and KOME at 30°C is 3.0 and 5.8 cSt respectively, as shown in Figure 10. This high viscosity of methyl ester reduces the leakage in plunger and barrel pair of the fuel pump (Gerpen et al., 2004) and minimum viscosity limits are imposed to prevent the fuel from causing wear in the fuel injection pump. The viscosity of KOME at 50°C is 3.0 cSt and is equal to that of diesel fuel at engine room temperature of 30°C., as shown in Figure 10.

From the testing of physico-chemical properties of KOME (or Karanja biodiesel), it is observed that, these properties are meeting the specifications of ASTM biodiesel standards as shown in Table 2 and KOME was

Table 3. Test engine specifications.

Engine make and model	Kirloskar (India), AV1
Maximum power output	3.72 kW
Method of cooling	Water cooled
Bore x Stroke	80 mm x 110 mm
Compression ratio	16.5
Rated speed (constant)	1500
Fuel injection system	In-Line, Direct Injection
Nozzle opening pressure	205 bar
BMEP @1500 rpm	5.31 bar

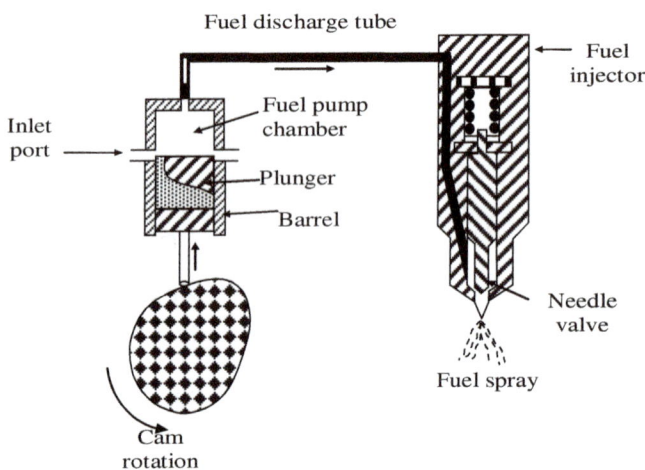

Figure 11. Traditional cam-driven, in-line fuel injection system.

Figure 12. Exhaust gas analyzer.

Table 4. Specifications of exhaust gas analyzer.

Measuring range	Precision	Resolution
0-2000 ppm	+/- 5 ppm	1 ppm

found suitable for diesel engine application.

EXPERIMENTATION, INSTRUMENTATION AND PROCEDURE

Experiments were performed on a naturally aspirated, 4-stroke cycle, single cylinder, direct injection diesel engine, with the specifications shown in Table 3.

This engine employs the traditional, cam-driven, in-line fuel injection system as shown in Figure 11. It consists of a fuel pump (jerk pump), a medium pressure tube (or fuel discharge tube) of length 585 mm (23 inch. approx.), and an injector (or atomizer). The pressure pulses (or waves) are generated by the fuel pump plunger and the 'column of fuel' is formed by the pump chamber, discharge tube, and the injector. This 'column of fuel' behaves like a stiff spring. The pressure pulses propagate through the 'column of fuel' to develop pressure at the nozzle end. When the fuel pressure reaches a pressure more than that required to open the needle valve of the injector, the needle valve is lifted to inject the fuel into the cylinder. For fuels with higher bulk modulus of compressibility, a more rapid transferal of the pressure wave takes place from pump end to the injector needle valve and the earlier needle lift causes an advanced (or early) injection. Therefore, the fuel like biodiesel (KOME), which is less compressible will inject prematurely.

This engine is tested with baseline diesel, and KOME. Diesel engine is also tested with preheated KOME (KOME_H) to find the influence of reduced viscosity on combustion, and NOx emission characteristics. For preheating KOME; heating devices were placed along the fuel discharge tube shown in Figure 11. The fuel injection was performed at a static injection timing (optimum) of 23° BTDC set for diesel fuel. The engine is allowed to warm up at constant speed of 1500 rpm, until the cooling water temperature reaches a steady state of 80°C. Eddy current dynamometer is used to measure the power (or torque). Engine brake load was varied in five equal steps (at 0, 0.93, 1.86, 2.79 and 3.72 kW), ranging from 0 to 100% of the rated power output of 3.72 kW. A computer interfaced piezoelectric sensor (water cooled) of range 0 to 145 bar was used to record the in-cylinder pressure. Apex innovations, Pune, India, software: C7112 is used to record the in-cylinder combustion pressure. Pressure signals were obtained using data acquisition system. The average pressure data from 20 consecutive cycles were used for calculating combustion pressure parameters. The net HRR and cumulative heat release (CHR) were calculated from the acquired data.

The NOx emissions of the engine were measured with the exhaust gas analyzer (make: MRU, Germany and model: Delta 1600 L) shown in Figure 12. Table 4 shows the technical specifications of the exhaust gas analyzer used for NOx measurement.

$$N_2 + O \longrightarrow NO+H \quad \text{-------- Equation (3)}$$

$$N_2 + O_2 \longrightarrow NO+O \quad \text{-------- Equation (4)}$$

$$N+OH \longrightarrow NO+H \quad \text{-------- Equation (5)}$$

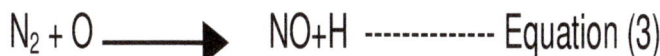

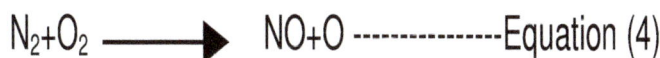

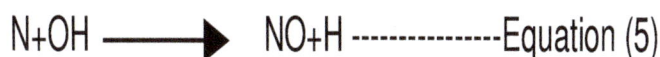

Figure 13. Reactions of Zeldovich mechanism.

Combustion and NOx formation

The combustion process in diesel engines is mainly divided into three phases. The first phase of combustion is called as ignition delay (ID), in which the tiny fuel droplets evaporates and mixes with high temperature (or high pressure) air. The delay period depends mainly on fuel cetane number (CN), and temperature of the air. The ID is also influenced by the fuel temperature. The second phase of combustion is called as period of rapid combustion or premixed combustion. In this phase the air-fuel mixture undergoes rapid combustion, therefore the pressure rise is rapid and releases maximum heat flux. In diesel engines, thermal NOx is mainly produced during premixed combustion. The third phase of combustion is called as period of controlled combustion. In this period, the fuel droplets injected during the second stage burn faster with reduced ID due to high temperature and pressure. In this third phase the pressure rise is controlled by the injection rate and the combustion is diffusive mode.

A number of fuel properties have been shown to effect the emissions of NOx (Robert et al., 2001). Fuel properties, such as density, viscosity, sound velocity, bulk modulus, cetane number, oxygen content, and so on, have significant effects on the start of injection, start of combustion, and premixed and diffusion burn peak, and over these on the emissions including NOx (Breda, 2006).

Thermal NOx refers to oxides of nitrogen formed through high temperature oxidation of nitrogen found in air during premixed combustion. The rate of thermal NOx is primarily a function of temperature and the residence time of nitrogen at that temperature. Usually at a temperatures of above 2100K, the nitrogen and oxygen disassociate and participate in a series of three principal reactions (Equations 3, 4 and 5) and this mechanism is known as Zeldovich mechanism (Heywood, 1988) shown in Figure 13.

RESULTS AND DISCUSSION

Combustion analysis

The combustion characteristics are heavily influenced by the properties of the fuel such as cetane number, calorific value, oxygen content and bulk modulus. A marked difference in the combustion characteristics are expected due to the distinct variations in fuel properties between KOME and fossil diesel. The KOME is expected to combust earlier from the shorter ignition delay caused by higher cetane number. The combustion analysis is explained in terms of peak pressure; net heat release rates (Net HRR), cumulative heat release (CHR) and exhaust gas temperatures (EGT) as follows:

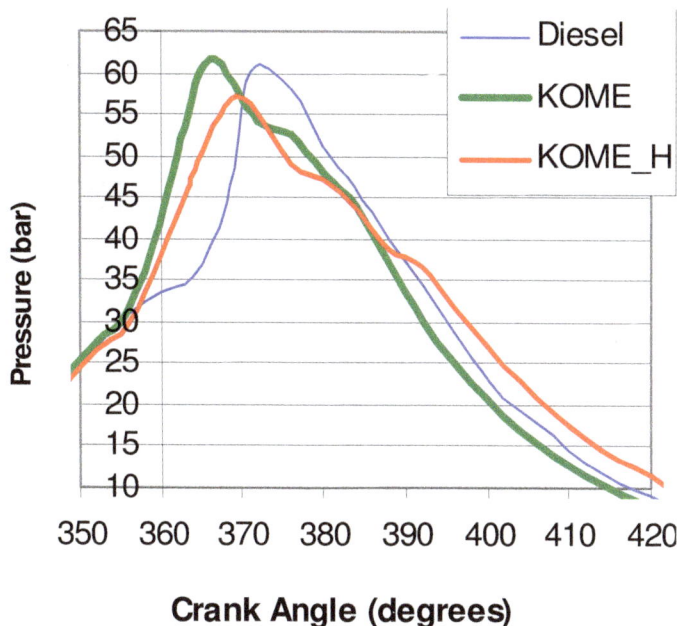

Figure 14. Cylinder pressure.

Peak pressures: Figure 14 shows the variation of cylinder pressure with respect to crank angle at maximum output of 3.72 kW. It is observed that, the KOME is burning close to TDC and the peak pressure is slightly higher than that of diesel fuel; even though the KOME is having lower value of LCV. The KOME advances the peak pressure position as compared to fossil diesel because of its higher bulk modulus and cetane number. This shift is mainly due to advancement of injection due to higher density and earlier combustion due to shorter ignition delay caused by higher cetane number of KOME. When, a high density (or high bulk modulus) fuel is injected, the pressure wave travels faster from pump end to nozzle end, through a high pressure in-line tube (Yamane et al., 2001). This causes early lift of needle in the nozzle, causing advanced injection. Hence, the combustion takes place very close to TDC and the peak pressure slightly high due to existence of smaller cylinder volume near TDC, as shown in Figure 15.

However, the peak pressures of preheated methyl ester (that is, KOME_H) are less than that of fossil diesel and KOME. When the engine is running on KOME_H, the fuel injection is slightly delayed, due to decrease in bulk modulus of KOME with the increase in fuel (methyl ester) temperature. The reasons for lower peak pressures of preheated KOME is also attributed to earlier combustion caused by short ignition delay (due to faster evaporation of KOME_H) at its preheated temperature of 50°C.

Net heat release rates (net HRR) and cumulative heat release (CHR): Figures 16 and 17, shows the net heat release rate (HRR) and cumulative heat release (CHR). A noticeable change in combustion phases were observed

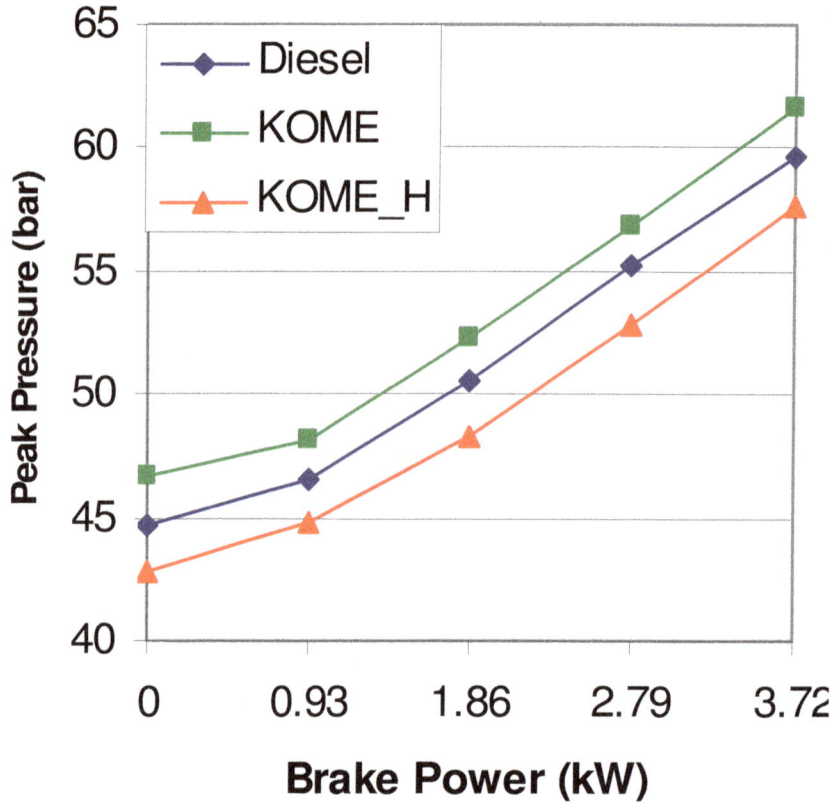

Figure 15. Peak pressure.

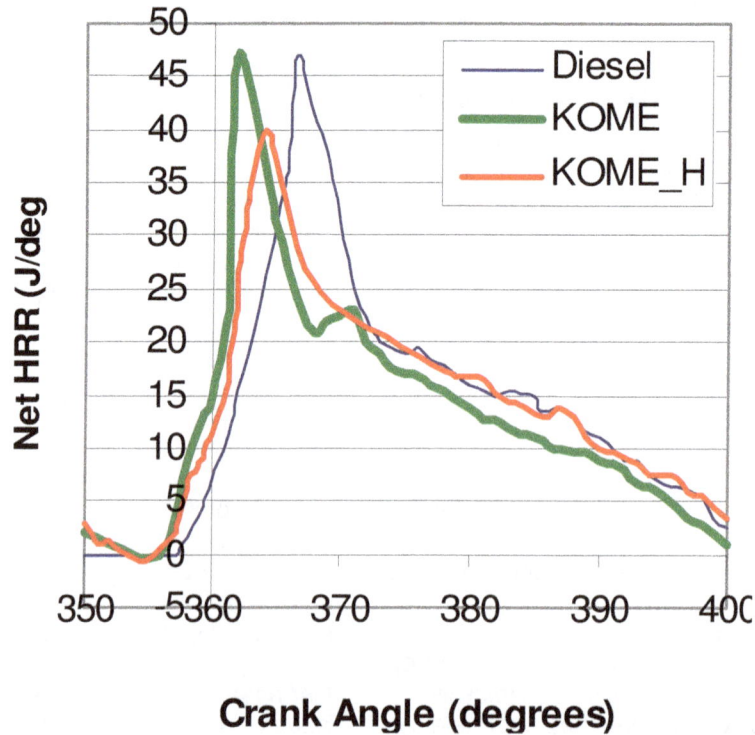

Figure 16. Net HRR at maximum output.

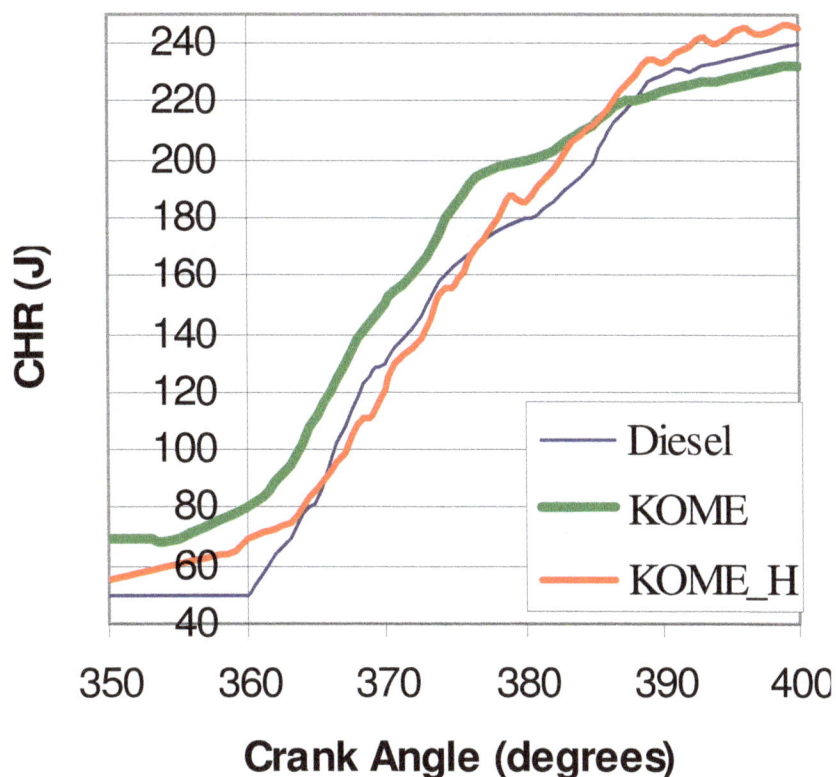

Figure 17. CHR at maximum output.

between KOME and KOME_H. The peak value of premixed combustion was more for KOME, than that of KOME_H, and the diffusive combustion phase was more for KOME_H, than that of KOME. As the fuel temperature increases, the ignition delay decreases and the main combustion phase, that is, diffusion controlled combustion increases. At the time of ignition, less quantity of air-fuel mixture is prepared for combustion. This is due to faster evaporation of the preheated methyl ester. Therefore, more burning occurs in the diffusion phase rather than in the premixed phase.

The cumulative heat release is more for KOME_H than that of KOME and this indicates that there is a significant increase of combustion in diffusion mode. This increase in heat release is mainly due to better mixing and evaporation of KOME_H, which leads to complete burning. The heat released by the KOME is less due to poor mixing of KOME with the surrounding air because of its high viscosity.

Exhaust gas temperatures: Temperature of exhaust gases, leaving the engine cylinder represents the extent of temperature reached in the cylinder during combustion. It is observed that, with increasing load the cylinder pressure increases and more of the fuel is burnt leading to an increase in temperatures as shown in Figure 18. The temperature of exhaust gases is observed to be

higher with fossil diesel as compared to KOME for entire range of power output. This is expected as the calorific value (or heat of combustion) of fossil diesel is more than that of KOME; therefore greater amount of heat is released in the combustion chamber leading to higher temperature. Also, there is an advanced combustion of KOME (Figure 16) due to its higher bulk modulus and cetane number, when compared to fossil diesel. Therefore, the heat released by fossil diesel combustion is late by few degrees and thus more heat gets exhausted. The exhaust gas temperatures of preheated KOME (KOME_H) are higher than that of KOME, which indicates the increase of diffused combustion due to high rate of evaporation and improved mixing between methyl ester and air. Therefore, as the fuel temperature increases, the ignition delay decreases and the main combustion phase (that is, diffusion controlled combustion) increases, which in turn raises the temperature of exhaust gases.

NO$_X$ analysis

Results show that for all the fuels the increased engine load promoting NOx emission as shown in Figure 19. Since the formation of NOx is very sensitive to temperature, therefore higher loads promote cylinder charge temperature, which is responsible for thermal (or

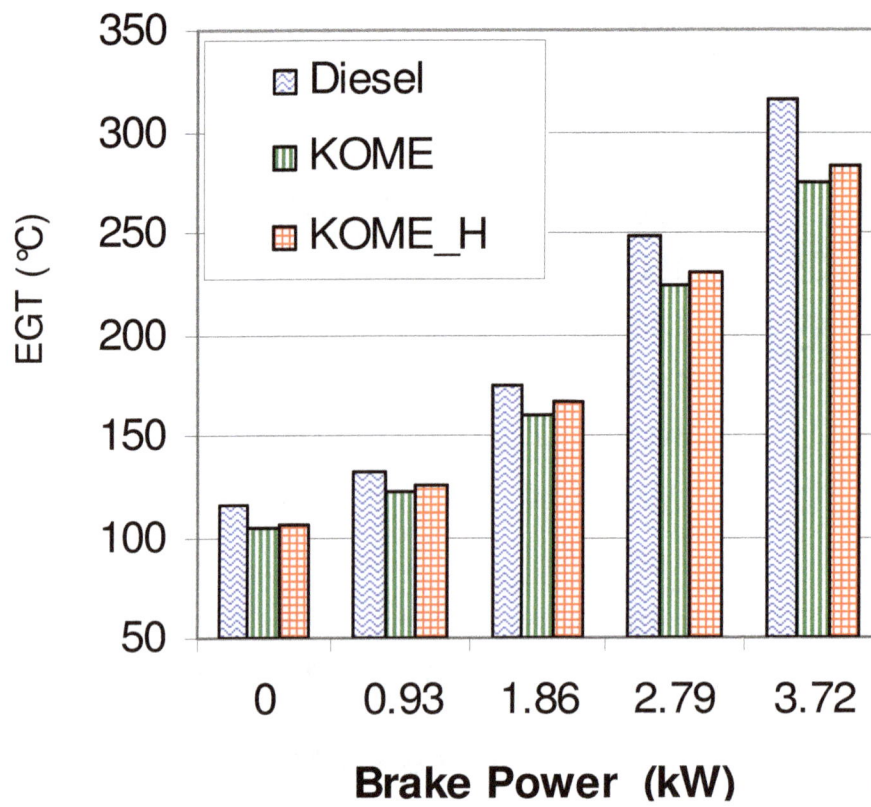

Figure 18. Exhaust gas temperatures.

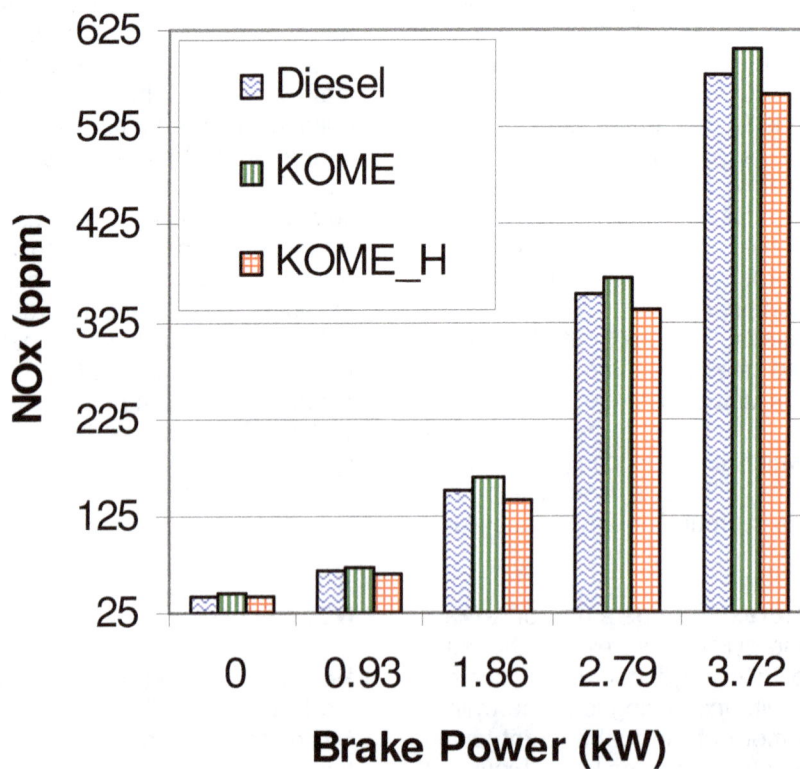

Figure 19. Oxides of nitrogen.

Zeldovich) NOx formation.

The KOME having long carbon chain (C_{16}-C_{24}) is producing more NOx than that of fossil diesel having both medium (C_8-C_{14}) as well as long chain (C_{16}-C_{28}) as shown in Figures 6 and 19. The increase in NOx emission might be an inherent characteristic of KOME due to the presence of 51.8% of mono-unsaturated fatty acids (MUFA) and 19% of poly-unsaturated fatty acids (PUFA) as shown in Figure 8. That means, the long chain unsaturated fatty acids (MUFA and FUPA) such as oleic $C_{18:1}$ and linoliec $C_{18:2}$ (Figure 7) fatty acids are mainly responsible for higher levels of NOx emission (Myo, 2008). Another reason for higher NOx levels is the oxygen (11%) present in the methyl ester. The presence of oxygen in KOME leads to improvement in oxidation of the nitrogen available during combustion. This will raise the combustion bulk temperature responsible for thermal NOx formation (Figure 13).

The production of more NOx with KOME fueling is also attributable to an inadvertent advance of fuel injection timing due to higher bulk modulus of compressibility, with the in-line fuel injection system. KOME is less compressible due to the higher bulk modulus, causing nozzle opening pressure to be exceeded prematurely. This earlier injection leading to advancement in combustion timing where a stronger premixed combustion phase (Figure 16) follows (Jo-Han et al., 2010). This in turn increases the peak in-cylinder temperature, which increases the rate of NOx formation.

The fuel spray properties may be altered due to differences in viscosity and surface tension. The spray properties affected may include droplet size, droplet momentum, degree of mixing, penetration, and evaporation. The change in any of these properties may lead to different relative duration of premixed and diffusive combustion regimes. Since the two burning processes (premixed and diffused) have different emission formation characteristics, the change in spray properties due to preheating of the KOME are lead to reduction in NOx formation. As fuel temperature increases, there is an improvement in the ignition quality, which will cause shortening of ignition delay. A short ignition delay period lowers the peak combustion temperature which suppresses NOx formation (Hess et al., 2005). Lower levels of NOx is also attributed to retarded injection, improved evaporation, and well mixing of KOME_H due to its low viscosity at preheated temperature of 50°C.

Conclusions

The present work confirms the influence of the higher bulk modulus of methyl ester (biodiesel) on injection and combustion timings of a diesel engine operated by in-line fuel injection system. The higher levels of NOx emission is attributed to the combined effect of bulk modulus, cetane number, oxygen, and unsaturated fatty acids. Decrease in premixed combustion and increase in

diffused combustion is observed with preheated methyl ester. The reduction in peak value of premixed combustion is leads to the reduction of NOx levels.

REFERENCES

Akihama K, Takatori Y, Nakakita K (2002). Effect of Hydrocarbon Molecular Structure on Diesel Exhaust Emissions. Research Report, R and D Rev. Toyota CRDL, 37(3): 46-52.

Aleks K (2000). The Two-Stage adaptation of Mike Pelly's Biodiesel Recipe, Available online: http://journeytoforever.org/biodiesel_aleks.html.

Anonymous (2002c). American Society for Testing and Materials(2002), Standard Specification for Biodiesel Fuel (B100).

Applewhite TH (1980). Kirk-Othmer Encyclopedia of Chemical Technology, 3rd Edition, John Wiley and Sons, Inc., New York, 9: 795-811.

ASTM (2002). Blend Stock for Distillate Fuels, Designation D6751-02, International, West Conshohocken, PA., pp: 1-6.

Bakar RA, Firoz T (2005). Numerical analysis of pressure pulses in the jerk fuel injection systems. Am. J. Appl. Sci., 2(5): 1003-1007.

Bala M, Nag TN, Kumar S, Vyas M, Kumar A, Bhogal NS (2011). Proximate Composition and Fatty Acid Profile of Pongamia pinnata, a Potential Biodiesel Crop. J. Am. Oil Chem. Soc., 88: 559-562.

Boehman A, Alam M (2003). Fuel formulation effects on diesel fuel injection, combustion, emissions, and emission control. Proceedings of DOE 2003, Diesel Engine Emissions Reduction Conference, Newport, RI, pp: 1-9.

Breda K (2006). Experimental Investigation of Optimal Timing of the Diesel Engine Injection Pump using Biodiesel Fuel. Energy Fuels, 20: 1460-1470.

Brown GW, McCallion H (1967-68).Simulation of an injection system with delivery pipe cavitation using a digital computer, Proc. Inst. Mech. Eng., 182(3L): 206-216.

Bryan RM (2009). Biodiesel production, properties and feed stocks: Invited Review, In vitro Cell Dev. Biol. Plant, 45: 229-266.

Gerpen JV (2003). Biodiesel production and Fuel Quality University of Idaho, Moscow, pp. 1-12.

Gerpen JV, Shanks BR, Knothe G (2004). Biodiesel analytical methods, National Renewable Energy Laboratory, p. 95.

Gerpen JV, Shanks B, Pruszko, Clements R, Knothe G (2004). Biodiesel analytical methods, National Renewable Energy Laboratory, p. 95.

Giffen E, Rowe AW (1939). Pressure calculations for oil engine fuel injection system. Proc. Inst. Mech. Eng., 142: 519-534.

Heinrich P, Manfred W, Josef R (1999), Technical Performance of Vegetable Oil Methyl esters with a High Iodine Number, 4th Biomass Conference of the Americas. BLT Wieselburg Austrlia, pp: 1-6.

Heywood JB (1988). Internal Combustion Engine Fundamentals, McGraw-Hill, Inter. New York, pp: 586-592.

Jindal S, Bhagwati PN, Narendra SR (2010). Comparative Evaluation of Combustion, Performance and Emissions of Jatropha Methyl Ester and Karanja Methyl Ester in a Direct Injection Diesel Engine, Energy Fuels, 24: 1565-1572.

Jo-Han Ng, Hoon KN, Suyin G (2010). Advances in biodiesel fuel for application in compression ignition engines, Clean Technol. Environ. Policy, 12: 459-493.

Kalbande SR, More GR, Nadre RG (2008). Biodiesel Production from Non-edible Oils of Jatropha and Karanj for utilization in Electrical Generator, Bioenerg. Res., 1: 170-178 .

Knothe G (2010). Biodiesel: Current Trends and Properties, Top Catal, 53: 714-720.

Kumar A, Sharma S (2008). An evaluation of multipurpose oil seed crop for industrial uses (Jatropha curcas L.): A review, Ind. Crops. Prod., 28(1): 1-10.

Lebedevas S, Vaicekauskas A (2004). Improvement of the parameters of maintenance of medium speed diesels applying the motor methods, Transport, XVI(6): 252-261.

Meher LC, Naik SN, Das LM (2004). Methanolysis of Pongamia pinnata (Karanja) oil for production of biodiesel. J. Sci. Ind. Res.,

63: 913-918

Mitra CR (1963), Development of Minor and Non-edible Oils, Report of the Indian Central Oilseeds Committee, Hyderabad, p. 29.

Myo T (2008). The Effect of Fatty Acid Composition on the Combustion Characteristics of Biodiesel. Doctoral Dissertation, Graduate School of Science and Engineering, Kagoshima University, Doctorate, p. 273.

Naik M, Meher LC, Naik SN, Das LM (2008). Production of Biodiesel from high free fatty acid Karanja (*Pongamia pinnata*) oil. Biomass and Bioenergy, 32:354-357.

Obert EF (1973). Internal combustion engines and air pollution, Harper International edition (3/e). Harper and Row Publishers, Inc: pp.451-453.

Pulkrabek WW (2003). Engineering Fundamentals of the Internal Combustion Engine, McGraw-HillInc., New York.

Rao PV (1993). Pressure-Time Analysis of a Unit Injector, A Thesis submitted to the Indian Institute of Science. Bangalore, India, pp: 6-9

Robert L, McCormick, Michael S, Grab ski, Teresa L, Alleman, Andrew MH, Shaine TK (2001). Impact of Biodiesel Source Material and Chemical Structure on Emissions of Criteria Pollutants from a Heavy-Duty Engine. Environ. Sci. Technol., 35: 1742-1747.

Romulo DA, Andrade, Elaine A, Faria, Amaury M, Silva, Wandallas C, Araujo, Gustavo C, Jaime, Kenia PC, Alexandre GS, Prado (2011). Heat of combustion of biofuels mixed with fossil diesel oil. J. Therm. Anal. Calorim, Springer: s10973-011-1408-x. http://www.springerlink.com/content/j30165672m014128/fulltext.pdf

Sanaz S, Ngoh GC, Rozita Y (2010). An Overview on Transesterification of Natural Oils and Fats, Biotechnology and Bioprocess Engineering 15:891-904.

Tapasvi D, Wiesenborn D, Gustafson C (2005). Process Model for Biodiesel Production from various Feedstock's, Trans. ASAE, 48(6): 2215- 2221.

Vivek, Gupta AK (2004). Biodiesel production from Karanja oil. J. Sci. Ind. Res., 63: 39-47.

Yamane K, Ueta A, Shimamoto Y (2001). Influence of Physical and Chemical Properties of Biodiesel Fuel on Injection, Combustion and Exhaust Emission Characteristics in a DI-CI Engine" Tran. Of the Jap. Soc. Mech. Eng., 32(2): 25-30.

16

A new viscosity-temperature relationship for vegetable oil

Ioana Stanciu

Faculty of Chemistry, University of Bucharest, 4-12 Regina Elisabeta Blvd., 030018 Bucharest, Romania.
E-mail: istanciu75@yahoo.com

This study proposes four relationships of dynamic viscosity temperature dependence for vegetable oils. The purpose of this study was to find a polynomial or exponential dependence between temperature and dynamic viscosity of vegetable oil, using the Andrade equation changes. Equation constants A, B, C and D were determined by fitting polynomial or exponential.

Key words: Viscosity-temperature, vegetable oil, relationship.

INTRODUCTION

Viscosity is one of the most important physical properties of a fluid system (Fasina et al., 2008). Studies on viscosity have been performed on polystyrene, protein, pineapple juice, vegetable oil, dark beer, etc. (Toth et al., 2007). Viscosity changes with shear rate, temperature, pressure, moisture, and concentration; all these changes can be modeled by equations (Dak et al., 2008; Balakrishnan et al., 2007; Severa et al., 2008).

The effect of temperature is normally fitted with the Arrhenius-type relationship that is shown subsequently (Choi et al., 2009; Thodesen et al., 2009; Saeed et al., 2009):

$$\eta = Ae^{E}a/RT \tag{1}$$

where η is dynamic viscosity (Pa.s); A is the pre-exponential factor (Pa.s); Ea is the exponential constant that is known as activation energy (J/mol); R is the gas constant (J/mol/K) and T is the absolute temperature (K).

Modeling of the temperature effect on the dynamic viscosity of oils vegetable is important and has been investigated by various researchers (Noureddini et al., 1992; Kapseu et al., 1991; Lang et al., 1992; Toro-Vazquez et al., 1993).

The two-parameter were Equations 2 and 3, and three-parameter equations Equations 4 to 7 were:

$$\ln\eta = A + B\ln T \tag{2}$$

$$\ln\eta = A + B/T \tag{3}$$

$$\ln\eta = A + B/(T+C) \tag{4}$$

$$\ln\eta = A + B/T+C/T^2 \tag{5}$$

$$\ln\eta = A + B/T+CT \tag{6}$$

$$\ln\eta = A + BT+CT^2 \tag{7}$$

where A, B and C are constants, and T is absolute temperature (K).

This article proposes four new relationships of dependence of dynamic viscosity of vegetable oils temperature. Dynamic viscosity of oils was determined at temperatures and shear rates, the 90°C and the 40°C, respectively, 3.3 to 120 s^{-1}. The purpose of this study was to find a polynomial or exponential dependence between temperature and dynamic viscosity of vegetable oil no additive using Andrade equation changes. Equation constants A, B, C and D were determined by fitting polynomial or exponential.

EXPERIMENTAL

Materials

Vegetable oils no additives used in this work are provided by a company from Bucharest, Romania.

Vegetable oils were investigated using a Haake VT 550 Viscotester developing shear rates ranging between 3 and 120 s^{-1} and measuring viscosities from 10^4 to 10^6 mPa.s when the HV_1

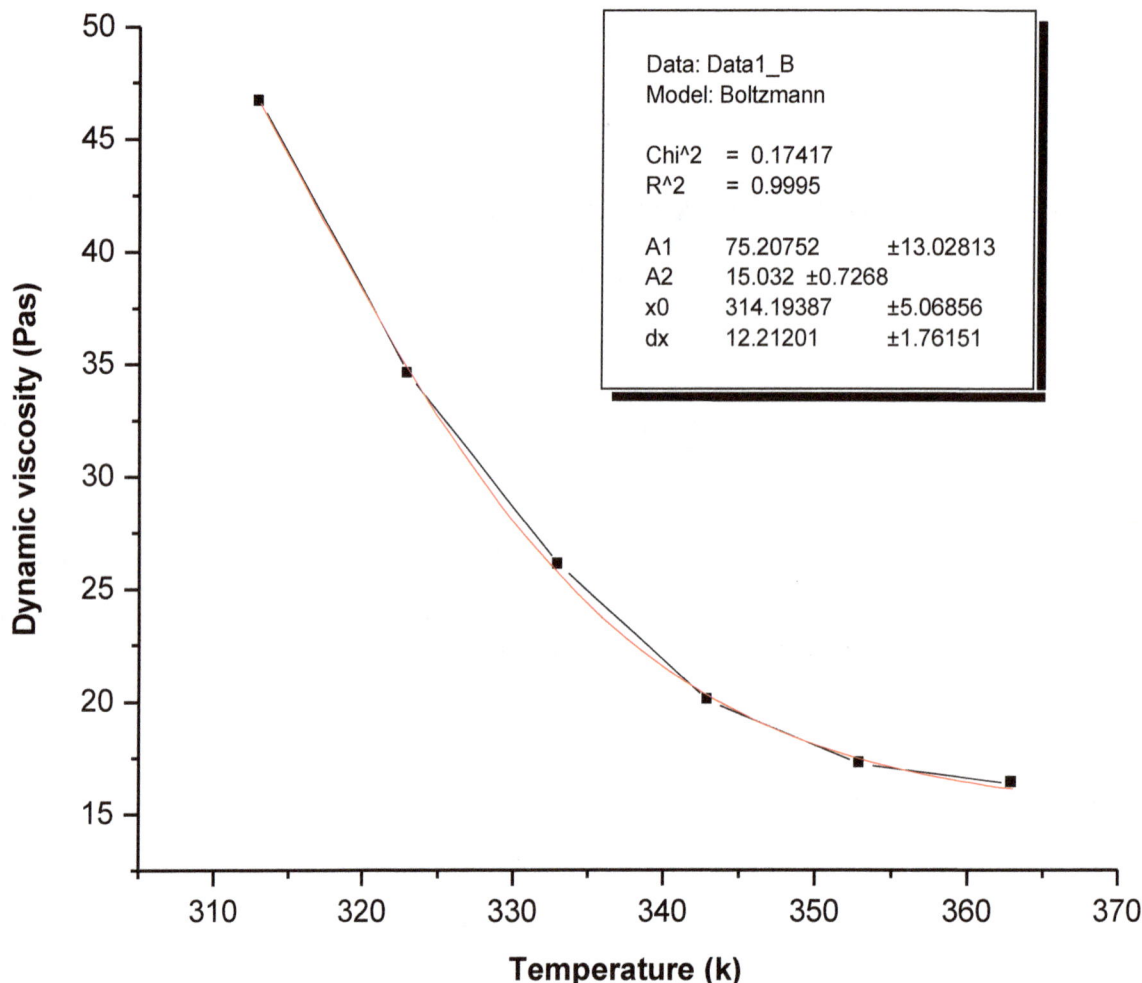

Figure 1. The viscosity dynamic of investigated oil as a function of temperature absolute of vegetable oil at shear rate 3.3 s^{-1}

viscosity sensor is used. The temperature ranging was from 40 to 90°C and the measurements were made from 10 to 10°. The accuracy of the temperature was ± 0.1°C.

RESULTS AND DISCUSSION

Figures 1 and 2 shows the dependence of dynamic viscosity temperature absolute for the same sample of vegetable oil at shear rate 3.3 and 6 s^{-1}. The behavior of vegetable oil is that the dynamic viscosity decreases with increasing temperature absolute.

This article proposes four Equations (8) to (11) temperature dependence of dynamic viscosity checked only for vegetable oils. The software Origin 6.0 was used to determine constants equation for vegetable oil. In addition, the parameters A, B, C, η_0, dT and D change with shear rate. Therefore, by imposing constant shear rate, the parameters can be determined. In order to determine the equation constants, the following steps were performed using the Origin 6.0 software: load the

non-linear regression package, input experimental data, title x-label, y-label and set the required equation, perform non-linear regression and plot experimental data and best fitted curve, calculate the mean square error and coefficient of determination and show the best fitted equation constant, mean square error and coefficient of determination.

Tables 1, 2, 3 and 4 contain the constants equations polynomial and exponential (8 to 11) and the correlations coefficient for the vegetable oil. As shown in Tables 1, 2, 3 and 4, the software found it polynomial and exponential equations applied shear rate curves of vegetable oil. The root mean square error means that experimental data is spread equation. From the results of the regression tabulated in Tables 1, 2, 3 and 4, the lowest coefficient of determination and the highest mean square error were 0.9725 and 0.9999, respectively.

$$\eta = A + BT + CT^2 \tag{8}$$

$$\eta = B + (A - B)/(1 + \exp((T-T_0)/dT)) \tag{9}$$

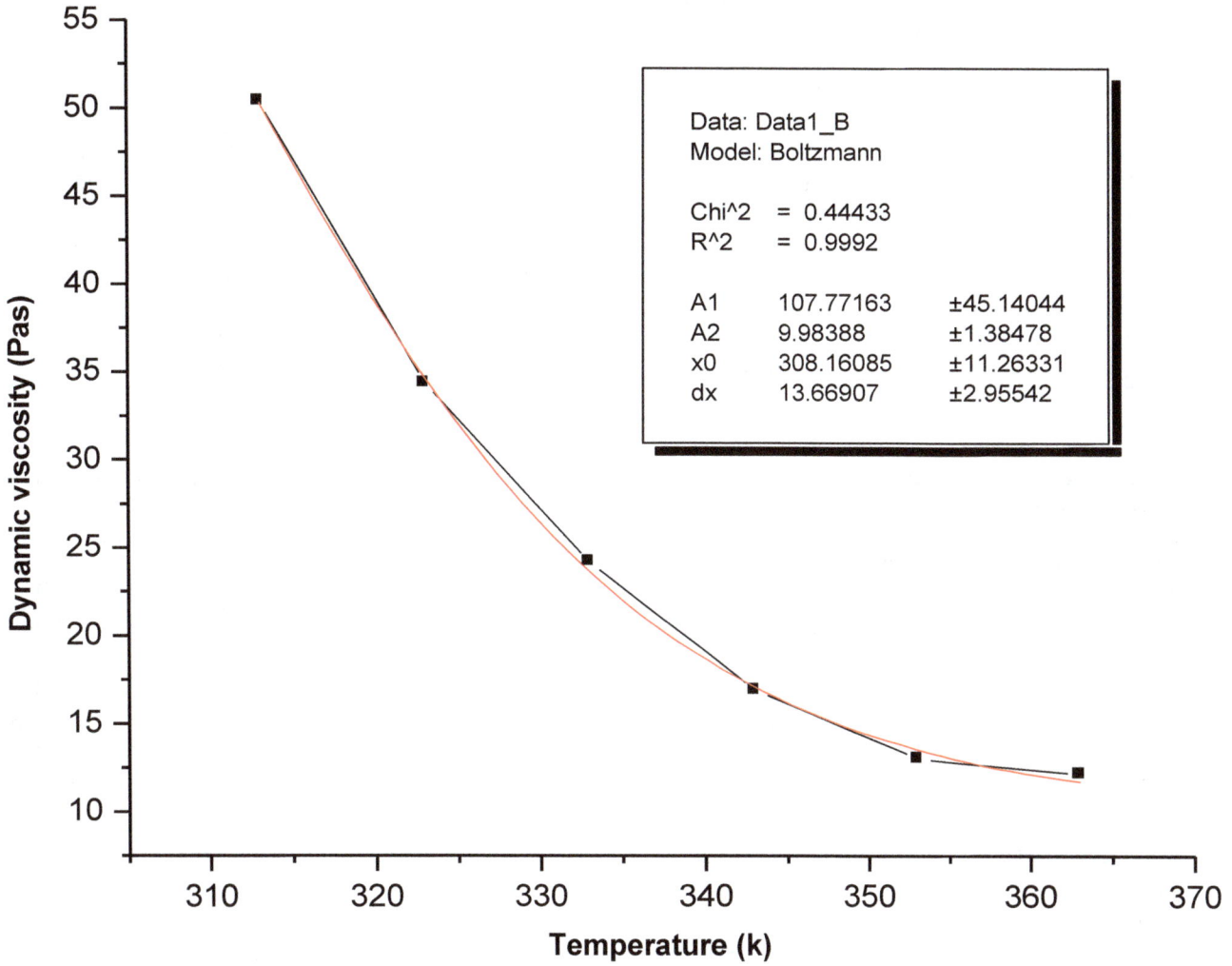

Figure 2. The viscosity dynamic of investigated oil as a function of temperature absolute of vegetable oil at shear rate 6 s^{-1}.

Table 1. The shear rate, value of parameters of the theoretical model described by Equation (8), coefficient correlation and range temperatures absolute for vegetable oil.

Shear rate (s^{-1})	Value of parameters of the theoretical model described by Equation (8)			R^2	Temp. range (K)
	A	B	C		
3.30	1836.1375	-10.1323	0.0141	0.9996	313-363
6.00	2329.3267	-12.9148	0.0180	0.9989	313-363
10.60	863.5866	-4.4560	0.0058	0.9989	313-363
17.87	1018.6742	-5.4406	0.0073	0.9991	313-363
30.00	1182.6624	-6.4256	0.0088	0.9991	313-363
52.95	1209.2660	-6.5897	0.0090	0.9989	313-363
80.00	1255.7463	-6.8596	0.0094	0.9994	313-363
120.0	856.6590	-4.5402	0.0061	0.9989	313-363

$$\eta = A + BT + CT^2 + DT^3 \qquad (10)$$

$$\eta = \eta_0 + A\exp(-T/B) \qquad (11)$$

where T is temperature absolute, A, B, C was constants vegetable oil and variation with shear rate, T_0 is temperature absolute to dynamic viscosity 50 Pa•s and

Table 2. The shear rate, value of parameters of the theoretical model described by Equation (9), coefficient correlation and range temperatures absolute for vegetable oil.

Shear rate (s^{-1})	Value of parameters of the theoretical model described by Equation (9)				R^2	Temp. range (K)
	A	B	T_0	dT		
3.30	75.2080	15.0320	314.1900	12.2120	0.9995	313-363
6.00	107.7700	9.9839	308.1600	13.6690	0.9992	313-363
10.60	252.1800	-1.3187	248.5400	37.3560	0.9990	313-363
17.87	222.4500	5.4663	262.5500	27.3870	0.9995	313-363
30.00	121.1200	7.7267	289.3600	20.2970	0.9999	313-363
52.95	98.4300	7.2421	296.2100	18.8600	0.9989	313-363
80.00	73.0500	7.5299	306.4100	16.2820	0.9992	313-363
120.0	432.8100	4.5940	238.8900	27.8120	0.9998	313-363

Table 3. The shear rate, value of parameters of the theoretical model described by Equation (10), coefficient correlation and range temperatures absolute for vegetable oil.

Shear rate (s^{-1})	Value of parameters of the theoretical model described by Equation (10)				R^2	Temp. range (K)
	A	B	C	D		
3.30	6364.1295	-50.9599	0.1367	-1.2259E-4	0.9999	313-363
6.00	10499.3346	-88.8027	0.2514	-2.378E-4	0.9996	313-363
10.60	5186.6121	-42.1897	0.1152	-1.0546E-4	0.9989	313-363
17.87	3768.3185	-29.9054	0.0798	-7.1389E-5	0.9990	313-363
30.00	6334.2189	-52.4365	0.1456	-1.3528E-4	0.9999	313-363
52.95	5618.0514	-46.0452	0.1266	-1.1644E-4	0.9997	313-363
80.00	6294.1093	-52.0333	0.1442	-1.3379E-4	0.9999	313-363
120.0	5014.4810	-40.9763	0.1124	-1.0333E-4	0.9999	313-363

Table 4. The shear rate, value of parameters of the theoretical model described by Equation (11), coefficient correlation and range temperatures absolute for vegetable oil.

Shear rate (s^{-1})	Value of parameters of the theoretical model described by Equation (11)			R^2	Temp. range (K)
	η_0	A	B		
3.30	9.2527	56.1092	2.4778	0.9996	313-363
6.00	11.7142	5.2391E7	21.7004	0.9987	313-363
10.60	12.3522	2.7673E9	16.9276	0.9799	313-363
17.87	11.3668	2.1719E9	17.1222	0.9740	313-363
30.00	10.6029	1.1579E9	17.7312	0.9880	313-363
52.95	10.1838	3.0160E9	16.8042	0.9847	313-363
80.00	9.7130	1.9394E9	17.2275	0.9869	313-363
120.0	9.6286	5.0556E10	14.7457	0.9725	313-363

width dT.

Conclusions

This study proposes four new relationships dynamic viscosity dependence of the absolute temperature no additive. Check the only vegetable oils. Equation constants were determined by exponential or polynomial beast curves obtained at different shear rates using the program Origin 6.0. The correlation coefficients thus obtained were 0.9725 and 0.9999 values between.

REFERENCES

Balakrishnan RK, Guria C (2007). Thermal degradation of polystyrene in the presence of hydrogen by catalyst in solution. Polym. Degrad. Stab., 92(8): 1583- 1591.

Choi HM, Yoo B (2009). Steady and dynamic shear rheology of sweet potato starch-xanthan gum mixtures. Food Chem., 116(3): 638-643.

Dak M, Verma RC, Jain MK (2008). Mathematical models for prediction of rheological parameters of pineapple juice. Int. J. Food Eng., 4(3): 1-17.

Fasina OO, Colley Z (2008). Viscosity and specific heat of vegetable oils as a function of temperature: 35°C to 180°C. Int. J. Food Prop., 11(4): 738-746.

Lang W, Sokhansanj S, Sosulski FW (1992). Modelling the temperature dependence of kinematic viscosity for refined Canola oil J. Am. Oil Chem. Soc., 69: 1054-1062.

Noureddini H, Teoh BC, Clements LD (1992). Viscosities of oil and fatty acids. J. Am. Oil Chem. Soc., 69: 1189- 1192.

Saeed R, Uddin F, Masood S, Asif N (2009). Viscosities of ammonium salts in water and ethanol + water systems at different temperatures. J. Mol. Liq., 146(3): 112-115.

Severa L, Los L (2008). On the influence of temperature on dynamic viscosity of dark beer. Acta Universitatis Agriculturae et Silviculturae Mendelianae Brunensis, 56(2): 303-307.

Toro-Vazquez JF, Infante-Guerrero R (1993). Regressional models that describe oil absolute viscosity. J. Am. Oil Chem. Soc. 70: 1115-1122.

Toth J, Simon Z, Medveczky P, Gombos L, Jelinek B, Szilagyi L, Graf L, Malnasi-Csizmadia A (2007). Site directed mutagenesis at position 193 of human trypsin 4 alters the rate of conformational change during activation: Role of local internal viscosity in protein dynamics. Struct. Funct. Genet., 67(4): 1119-1127.

Thodesen C, Xiao F, Amirkhanian SN (2009). Modeling viscosity behaviour of crumb rubber modified binders. Constr. Build. Mater., 23(9): 3053-3062.

Kapseu C, Kayem GJ, Balesdent D, Schu_enecker L (1991). Estimation of dynamic viscosities of vegetable oils J. Am. Oil Chem. Soc., 68: 128-133.

17

Renewable fuel saving obligation: Innovative mechanism for emission reduction

Ajay Chandak[1,2]* and Sunil Somani[3]

[1]SSVPS BSD College of Engineering, Deopur, Dhule: 424005, India.
[2]Prince Suman Foundation, Dhule, India.
[3]Medicaps Institute of Technology and Management, Pigdambar, Indore - 453 331, MP, India.

All energy requirements of human beings can be broadly fulfilled through electricity (power) and fuels. Many governments including Government of India have taken policy measures to make renewable power component as legally binding to power producers through renewable power obligations. Under these laws, power distribution companies have to buy renewable power as fixed percentage of the total power sold. Similarly in few countries, there are legally binding norms for mixing fixed proportion of bio-fuels in automotive fuels. However, there are no policies adopted by any government for cooking fuels like kerosene and Liquid petroleum gas (LPG). Current article proposes RFSO (Renewable Fuel Saving Obligation) as one innovative policy measure which provides legally binding obligations on cooking fuel selling companies rather than independent agencies who do not have any control on sale of cooking fuels. Mechanism like RFSO needs serious attention by the policy makers and needs to be included in emission reduction acts in respective countries.

Key words: Renewable fuel saving obligation (RFSO), electricity, fuels, India.

INTRODUCTION

All energy requirements of human beings can be broadly fulfilled through electricity (power) and fuels. Even though fuels constitute more than 55% of our energy requirement, majority of renewable energy development efforts are directed towards power production. Recent announcement of 'Jawaharlal Nehru Solar Mission' by GOI also has a commitment of generating 20000 MW of solar power and similar commitments are made by Gujrath government (Jawaharlal Nehru National Solar Mission, 2010; Solar Power Policy, 2009). For complying with emission reduction commitments as per provisions of Kyoto Protocol (http://unfccc.int/kyoto_protocol/items/3145.php) many governments give high priority for electricity generation through renewable sources and have made amendments in their laws accordingly. No such efforts are seen for reducing emission from fuels. Few countries have laid down norms for mixing ethanol and bio-diesel in fixed proportions with gasoline and diesel respectively (http://www.ethanolproducer.com/articles/2571/canadian-government-unveils-renewable-fuel-requirement-plans/). However such schemes have also given rise to food versus fuel debates (Kenneth and Adam, 2007). No policies are seen to reduce emission resulting from cooking fuels. In India, domestic energy consumption of kerosene and Liquid petroleum gas (LPG) is approximately 10% of the total energy mix (Bureau of Energy Efficiency, 2005). Almost 50% of population uses firewood for cooking, which in non commercial energy and is not accounted for in the energy mix (Bureau of Energy Efficiency, 2005). Pohekar et al reports very uncertain figures on contribution of different energy sources for cooking needs between 1950 to 2000 (Pohekaret al., 2005). Solar cookers, biogas or wood efficient stoves are some of the possible alternatives which can reduce consumption of conventional fuels like kerosene and LPG (Pohekaret al., 2005). In India, distribution of cooking fuels like kerosene and LPG is

---
*Corresponding author. E-mail: chandak.ajay@gmail.com.

Table 1. Maharashtra RPS (now called RPO) Operational Framework 2007.

Financial Year	Renewable purchase specifications (RPS)* (%)	Penal rate for shortfall in Rs./kwh
2006-07	3	Nil
2007-08	4	5
2008-09	5	6
2009-10	6	7

*Percentage RPS denotes minimum quantum of procurement of electricity by eligible persons from renewable energy sources.

Table 2. Targets for RPO for state of Maharashtra.

Year	Minimum quantum of purchase (in %) from renewable energy sources (in terms of energy equivalent in kWh)		
	Solar (%)	Non-solar (other RE) (%)	Total (%)
2010-11	0.25	5.75	6.0
2011-12	0.25	6.75	7.0
2012-13	0.25	7.75	8.0
2013-14	0.50	8.50	9.0
2014-15	0.50	8.50	9.0
2015-16	0.50	8.50	9.0

responsibility of petroleum ministry and these fuels are highly subsidized. Petroleum ministry has accepted obligation of replacing 5 to 10% of petrol with ethanol (Belum et al., 2005), however same ministry has not been given any obligations on reducing emission on account of selling cooking fuels. On the other hand, ministry of new and renewable energy (MNRE) have the responsibility of promoting clean technologies like solar cooking, biogas and clean wood stoves (http://www.mnre.gov.in/). There is conflict of interest in the ministries which are allowed to pollute (petroleum and natural gas) and MNRE who has responsibility to clean up. There is no accountability on either of ministries as both have independent functioning and independent targets. RFSO (Renewable Fuel Saving Obligation) is one innovative policy measure suggested by authors, which puts up responsibility of promoting green technologies for cooking fuel in a manner that targets for emission reduction by cooking fuels are also set and met on similar lines of Renewable Power Obligation (RPO). Such mechanism needs serious attention by the policy makers and needs to be included in emission reduction laws in respective countries. An integrated approach is required for promoting renewable energy and discouraging carbon based energy sources.

RPO (Renewable Power Obligation)

Renewable power obligation has been in place in many countries. In India, RPO has been introduced by many state electricity commissions and it has fixed responsibility of generating a fixed percentage of power through renewable sources. This responsibility is fixed on the power producing and distributing company. This share of renewable energy is increasing every year as shown in Table 1, applicable to state of Maharashtra. Similar provisions were made by other states as well. There are penal provisions as specified in the last column (Maharashtra RPS Operational Framework, 2007). These provisions are made by state and central electricity regulatory commissions.

Till 2010, such legal obligation used to be called RPS (Renewable Power Specifications), since 2010, the terminology used is RPO (Renewable Power Obligation). New targets are specified in Table 2 (Maharashtra Electricity Regulatory Commission - Renewable Purchase Obligation and its Compliance and Implementation of REC Framework Regulations, 2010). These targets now include solar power as a separate component along with other RE technologies. Such measures can bring in time bound promotion of renewable power and corresponding increase in share of renewable power in energy mix of the state. Credit of RPO is a tradable commodity and such mechanism brings in accountability and commitment for promoting renewable power.

Integrated energy policy of Government of India also provides high thrust on renewable power and also agrees that increase in availability of electricity will not reduce kerosene consumption (Draft Report of the Expert Committee on 'Integrated Energy Policy' prepared for Planning Commission, 2005). This policy document mentions obligations only for renewable power and not for renewable fuels. Promotional programs and policies for solar cookers, biogas and biomass are mentioned without specific targets (Draft Report of the Expert Committee on 'Integrated Energy Policy' prepared for

Table 3. Proposed RFSO Operational Framework 2011 for India.

Financial year	(RFSO)Renewable fuel saving obligation (%)	Penal rate for shortfall per kg of kerosene/LPG
20011-12	5	Rs. 50
2012-13	6	Rs. 60
2013-14	8	Rs. 70
2014-15	10	Rs. 80

Percentage RFSO denotes minimum quantum of fuel savings/supplement by eligible persons from renewable energy sources.

Planning Commission, 2005).

NEW PROPOSALS FOR RFSO (RENEWABLE FUEL SAVING OBLIGATION)

Need for RFSO

Issues of fossil fuels and renewable fuels required for cooking are dealt by too many ministries. Ministry of petroleum and natural gas is primarily responsible for providing Liquid petroleum gas (LPG) and kerosene. These fuels are highly subsidized when deployed for cooking needs. Competing technologies like solar cookers, biogas, biomass technologies etc. are dealt by Ministry of New and Renewable Energy (MNRE). MNRE has onus of cleaning up the emissions caused by ministry of petroleum and natural gas. MNRE has no control on these powerful ministries of petroleum and natural gas and for this reason MNRE cannot implement any measurable mechanisms to reduce fossil fuel consumptions substantially. These ministries have conflict of interest. One ministry is allowed to do any amount of emission and another ministry, MNRE is given a daunting task to reduce emissions by all means. It definitely makes more sense that on the lines of RPO the responsibility of reducing fossil fuel consumption or promotion of renewable fuels should be fixed on petroleum and natural gas ministry and not on MNRE. On the same lines as that of RPO, fixed percentage of cooking fuel consumption (or saving) should come from renewable sources and such responsibilities are to be fixed on the fossil fuel manufacturing/selling companies. Initially, 5% of the cooking fuel requirement should be supplemented by renewable energy gadgets like solar cookers, biogas plants and efficient wood stoves and the contribution should go up to 10% in next 4 years. Promoting right kind of technologies will be responsibility of fuel selling companies. Similar to RPO, new specifications for fuel savings need to be evolved. Authors propose 'RFSO' (Renewable Fuel Saving Obligation). This innovation of RFSO is a rational idea because it puts the responsibility of reducing the emission caused by fossil fuel burning, on the

companies, who are responsible for the emission. This is on the similar lines as that of Kyoto protocol, which has put onus of cleaning the earth on developed annex-I countries, those are responsible for global warming. One proposal of RFSO is suggested in Table 3.

It will be responsibility of the fuel selling companies, like Indian oil and Bharat Petroleum, to promote RE technologies so that RFSO targets are met. In case RFSS targets are not met then the penalties realized from these companies should go to a 'Green Fund' and MNRE/ PCRA should use this fund to achieve RFSO targets. RFSO compliance should result in 'Renewable Fuel Certificates' (RFCs) similar to 'Certified Emission Reductions' (CERs) generated under CDM. These RFCs should be a tradable commodity. Any agency, company or NGO promoting renewable energy for cooking needs can generate such 'RFC's and trade those RFCs to eligible companies and use such discounts for promoting green technologies.

Conclusions

Legislations like RPO makes it obligatory that certain portion of power has to be generated from green sources, however no such provisions are available for cooking fuels. Kerosene and LPG are promoted by petroleum ministry while green cooking technologies are dealt by MNRE and there is a conflict of interest. None of the existing ministries like MNRE and departments like PCRA, CII, TERI etc. has any accountability towards achieving measureable savings in fossil fuels for cooking needs. Authors' innovative concept of 'RFSO' (Renewable Fuel Saving Obligation) needs to be incorporated in EC-Act or relevant emission saving acts and fix onus of promoting green cooking technologies on the companies, selling fossil fuels. Companies like Bharat Petroleum, Indian Oil, Reliance petroleum limited etc. should be made accountable for the fossil fuel they sell. Measureable portion of fossil fuels need to be replaced by clean technologies like solar cookers, biogas and efficient wood stoves in time bound manner. Proposed RFSO measures can bring in measurable and targeted emission savings and more accountability for the same.

Inclusion of such RFSO measures in EC-Act or other emission reduction act with pollution control boards or with ministry of environment will give legal teeth to such measures.

Even though the paper has been discussed in the background of Indian situation, the guiding principle of fixing responsibility of clean up on the polluters is applicable in all the countries where petroleum or carbon products are used.

REFERENCES

Belum VS, Reddy S, Ramesh, P Sanjana Reddy (2005). Sweet Sorghum – A Potential Alternate Raw Material for Bio-ethanol and Bioenergy: SAT eJournal, Open Access J. published by ICRISAT, 1(1).

Bureau of Energy Efficiency (2005). 'Energy Scenario', Book I - General Aspect of Energy Management and Energy Audit. Second edition, pp. 1-6.

Draft Report of the Expert Committee on 'Integrated Energy Policy' prepared for Planning Commission, Govt. of India, and Dec. 2005. http://planningcommission.nic.in/reports/genrep/intengpol.pdf , (29), pp. 100-102.

Ethanol producer's magazine available at: http://www.ethanolproducer.com/articles/2571/ canadian-government-unveils-renewable-fuel-requirement-plans/.

Jawaharlal Nehru National Solar Mission (2010). Guidelines published by MNRE.

Kenneth G, Cassman and Adam J (2007). Liska: Food and fuel for all: Realistic or foolish? Published in Biofuels, Bioprod. Biorefin., 1: 1: 18-23; doi 10.1002/bbb.3.

Kyoto Protocol, Emission saving targets available at: http://unfccc.int/kyoto_protocol/items/3145.php.

Maharashtra Electricity Regulatory Commission (2010). Renewable Purchase Obligation and its Compliance and Implementation of REC Framework Regulations 2010, available at http://www.mercindia.org.in/pdf/Order%2058%2042/Final_MERC(RPO-REC)_Regulation_2010_English.pdf, p. 5.

Maharashtra RPS Operational Framework (2007). MEDA Order No. BP-07/RPS/06-07/001: http://www.mahaurja.com/PDF/RPS%20Final.pdf.

Official Website of Ministry of New and Renewable Energy, Government of India. http://www.mnre.gov.in/.

Pohekar SD, Dinesh K, Ramachandran M (2005). Dissemination of cooking energy alternatives in India—A review. Renew. Sustain. Energy Rev., Elsevier Publication, 9: 379–393.

Solar Power Policy (2009). Government of Gujrath, Energy and Petrochemicals Department, G.R. No. SLR-11-2008-2176-B, dt. Jan. 6, 2009.

18

Technical and economical feasibility analysis of energy generation though the biogas from waste in landfill

Fábio Viana de Abreu[1]*, Mila Rosendal Avelino[2], Mauro Carlos Lopes Souza[2] and Diego Preza Monaco

[1]Petrobras - Brazilian Oil S. A. Brazil.
[2]Program of Pos Graduation in the Mechanical Engineering, Universidade do Estado do Rio de Janeiro (UERJ), Brazil.

Biogas from wastes in landfill can reduce the dependence on fossil fuels, beyond finding solutions that are environmentally sustainable to collaborate with the energy matrix of the countries. The intensification of human and industrial activities in the last few decades has generated increase in the production of municipal solid wastes (MSW), becoming a serious problem for the society. Furthermore, the uses of large landfills in great urban centers are still common, which causes sanitary and ambient problems. Gramacho's landfill was chosen as study case, for technical and economical feasibility analysis of energy generation though the biogas from waste, had it's importance for the city of Rio of Janeiro and metropolitan region. Moreover, an ambiental concern of the contamination of the Guanabara Bay with the leachate of this landfill. The more important environmental contribution associated to this project is the reduction of greenhouse gases emissions (GHG), by means of the conversion of methane in carbon dioxide. Studies and comparative analysis was presented demonstrating when gas turbine, internal combustion engines (Otto or Diesel cycles) or other technologies of energy conversion have technical and economical feasibility for implantation of the thermoelectrical plant.

Key words: Biogas, renewable energy, landfill.

INTRODUCTION

Waste disposal in landfills can generate environmental problems such as water pollution by leachate, unpleasant odors, risks of explosion and combustion, risk of asphyxiation, vegetation damage, and greenhouse gas emissions (Popov, 2005).

According to prediction of the United Nations Organization (United Nations, 2002), the world-wide population must grow until 2050 about 40% in relation to 2002, reaching 8.9 billion people.

The Agenda 21 from ECO-92 Conference foresees the duplication of the amount of residues produced in the world until 2010, based on values of 1990 and they will quadruplicate until 2025 (United Nations, 1992).

The amounts of wastes generated by the societies are increasing in the whole world, either due to population increase or due to increment of the per capita production of residues. Additionally, current production and consumption models prioritize the use of disposable materials and products, not taking in account the necessity of maintenance of a sustainable ambient (Abreu, 2009).

Landfill gas is generated under both aerobic and anaerobic conditions. Aerobic conditions occur immediately after waste disposal due to entrapped atmospheric air. The initial aerobic phase is short-lived and produces a gas mostly composed of carbon dioxide.

*Corresponding author. E-mail: fabiovian@bol.com.br

Table 1. MSW disposal in Brazil.

Region	Total (tones/day)	Open dump (%)	Control Landfill (%)	Landfill (%)	Others (%)
North	11.067	56.7	28.3	13.3	1.7
Northeast	41.558	48.2	14.6	36.2	1.0
Southeast	141.617	9.7	46.5	37.1	6.7
South	19.875	25.7	24.3	40.5	9.5
Center-west	14.297	21.9	32.8	38.8	6.5
Brazil	228.413	21.2	37.0	36.2	5.6

Source: IBGE (2001).

Since oxygen is rapidly depleted, a long-term degradation continues under anaerobic conditions, thus producing a gas with a significant energy value that is 55% methane and 45% carbon dioxide with traces of a number of volatile organic compounds (Meraz et al., 2004; Zamorano et al., 2007). For Polprasert (1996), the biogas generated in landfills is basically composed of methane (CH_4, 55 to 65%), carbon dioxide (CO_2, 35 to 45%), nitrogen (N_2, 0 to 1%), hydrogen (H_2, 0 to 1%) and sulfidric gas (H_2S, 0 to 1%).

The anaerobic process begins after the waste has been in the landfill for 10 to 50 days. Although the majority of CH_4 and CO_2 are generated within 20 years of landfill completion, emissions can continue for 50 years or more (Popov, 2005).

In Brazil, 149,199 tons of municipal solid wastes (MSW) have been daily collected (Abrelpe, 2009). The national average daily production is 0.950 kg per capita. Table 1 shows MSW disposal in Brazilian geographical regions.

Brazilian Energy Matrix is compound of approximately 48.4% from renewable energy sources and 51.6% from non renewable ones (EPE, 2009).

Nearly 80% of electricity in Brazil originates from hydroplants, not considering thermal generation. World average for renewable generation is 15.6% (EPE, 2009). So, Brazil has one very advantageous position in facing global environmental problems.

Electricity generation in Brazil reached 463.1 TWh in 2008, or 4.2% higher than 2007 total. Main contributors are public utilities, with 89.0% of shares. From those, hydro utility plants remain as main source, even with a reduction of 1.4% in comparison to 2007.

Thermal generation increased in 63.2%, specially from natural gas (116.6%) and nuclear (13.1%) (BEN, 2009). Landfill gas (LFG) recovery and utilization have not been significantly evaluated in Brazil. A number of reasons might have contributed for this scenario, including: public regulation uncertainties, lack of financial incentives, absence of public and private investments, operational conditions of landfills, and low level of technical support.

The only full scale LFG power plant started its operation in the beginning of 2004 with an installed capacity of 20 MW (Bandeirantes Landfill/São Paulo) (Maciel and Jucá, 2005).

Bandeirantes and Sao Joao landfills were disabled in 2007 and 2009, respectively, and thermoelectric power plants were installed to burn LFG produced by the decaying waste. Eleven million tons of CO_2 eq shall be prevented from being thrown in the atmosphere by 2012, generating tradable reduced emissions certificates (RECs), part of it sold at two public auctions in the Brazilian Stock Exchange (C40 cities, 2010). Table 2 shows potentials of methane recovery and electricity generation in main Brazilian landfills

This article aims at presenting a technical and economical evaluation of energy generation from MSW at Gramacho's landfill in Brazil.

Waste-to-energy (WTE) technologies, which combust municipal solid waste to produce energy, are often not competitive, when viewed solely from a waste management or energy production perspective. However, more appropriate analysis examines the energy and solid waste management questions simultaneously (Miranda and Hale, 2005). Although their proposed strategy to include social costs is quite reasonable, and it increases the feasibility of the thermo power facility, difficulties in accounting add to lack of precise data do not allow that social costs were included in the present study.

METHODOLOGY

Gramacho's landfill

Gramacho's landfill was chosen as study case because its importance for the city of Rio of Janeiro and its metropolitan region. In 2009, the first phase of the Effluent Liquids Treatment Station, treated daily, according to *Companhia de Limpeza Urbana* (Comlurb), was completed in 960 m³ of leachate. The leachate was one of the main concerns of the ambient professionals, because of the contamination risk of the Guanabara Bay.

The Gramacho's landfill is located at the following coordinates: 22°44'46" South and 43°15'37" West, as it is showed in Figure 1.

Table 2. Potentials of methane recovery and electricity generation in main Brazilian landfills.

Municipality	Unit of treatment	Waste disposal (tones/years)	Methane recuperation (MM m³/day)	Power generation (MW average)
Duque de Caxias/RJ	Gramacho Landfill	2.258.429	484	53.8
Rio de Janeiro/RJ	CTR Gericinó	1.081.848	232	25.8
Caucaia/CE	ASMOC Landfill	1.038.670	223	24.8
Jaboatão	Muribeca Landfill	955.746	205	22.8
Belo Horizonte/MG	CTRs BR040	909.520	195	21.7
Brasília/DF	Joquei Landfill	846.669	182	20.2
Salvador/BA	Centro Landfill	828.514	178	19.7
São Paulo/SP	Bandeirantes Landfill	743.208	159	17.7
Manaus/AM	KM 19 Landfill	709.696	152	16.9
São Paulo/SP	São João Landfill	701.472	150	16.7
Curitiba/PR	Caximba Landfill	670.790	144	16.0

Source: Zanetti (2009).

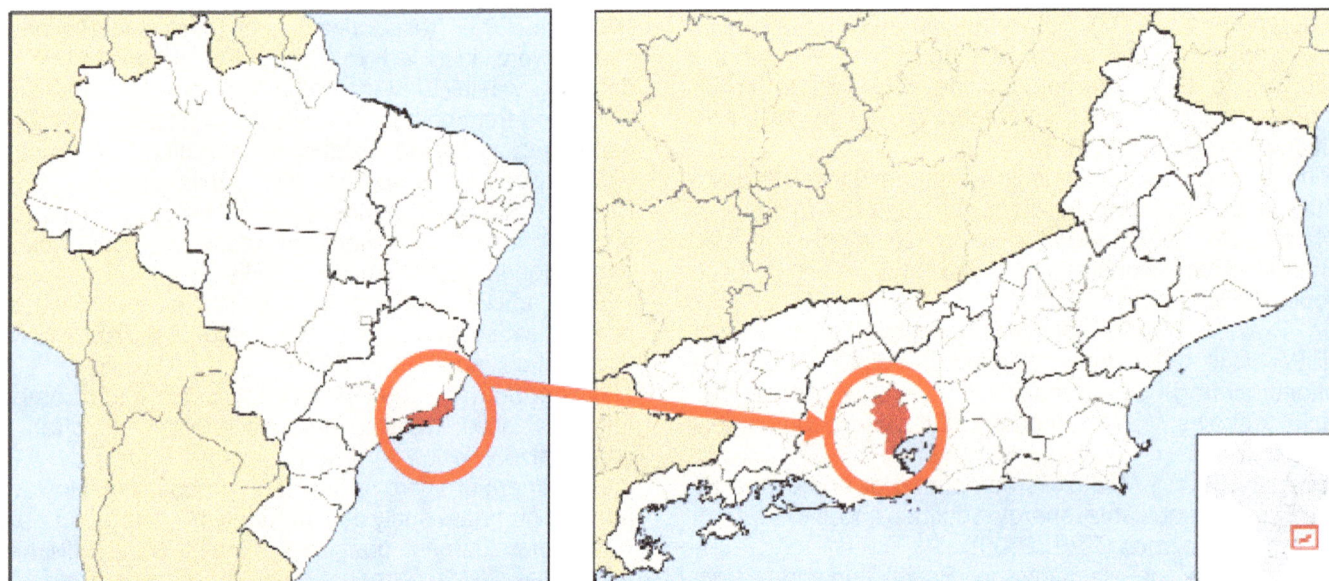

Figure 1. Gramacho's landfill localization.

Gramacho's landfill operations started as an open dump in a mangrove swamp in 1978. Initial filling was performed by pushing waste into the swamp area to fill it to a point where it was above high sea level. Subsequent fill activities consisted of haphazard dumping, waste burning, and uncontrolled scavenging. Since the beginning of the decade of 1990 it has started to receive some cares to minimize its environmental impact. In the early 1990s, the landfill operator, Companhia de Limpeza Urbana (COMLURB), began converting the open dump into a sanitary landfill. By 1996, most of the attributes of a modern sanitary landfill were in place, including controlled access, a recycling facility, well-maintained access roads, waste compaction by bulldozers, and the application of daily and intermediate cover soils. (SCS Engineers, 2005).

Table 3 shows solid waste disposal evolution in Gramacho's landfill. All waste deposited prior to 1993, during the open dump

operations, were not included in the present study. Excluding waste disposed in the open dump, it is estimated that there was more than 40 million tonnes of waste in place at the Landfill as of the end of 2010. The site currently receives approximately 250,000 tonnes per month (3,000,000 tonnes per year). The Landfill is expected to close at the end of 2011, at which time there will be more than 43 million tonnes of waste in place.

Technical solutions for energy generation in landfills

Most suitable conventional technologies for direct electric energy conversion from biogas are gas turbines and internal combustion engines, since steam turbines require a furnace e for steam

Table 3. Solid waste disposal in Gramacho's landfill.

Year	Waste disposed	Waste in place
	Tonnes	Tonnes
1993	1.646.374	1.646.374
1994	1.669.443	3.315.817
1995	1.800.209	5.116.026
1996	2.325.161	7.441.187
1997	2.414.508	9.855.695
1998	2.390.021	12.245.716
1999	2.403.311	14.649.027
2000	2.454.563	17.103.590
2001	2.417.409	19.520.999
2002	2.473.918	21.994.917
2003	2.359.715	24.354.632
2004	2.400.000	26.754.632
2005	2.400.000	29.154.632
2006	2.568.000	31.722.632
2007	2.747.760	34.290.632
2008	2.920.000	37.210.632
2009	3.000.000	40.210.632

Source: Comlurb (2010).

generation. From small to medium power generation capacities, internal combustion engines are more appropriated because of its lower cost and greater efficiency in this range. Only for higher capacities, gas turbines are competitive, and their yielding is improved when they are used in combined cycles.

Internal combustion engines are more efficient within the operation range of this project. Diesel cycle engines work on higher compression rates, requiring that biogas is fed mixed with diesel or biodiesel, which would represent an additional input to the energy facility. Moreover, in the Brazilian internal market, Otto cycle engines can be more easily adapted to operate with biogas (Abreu, 2009).

Economical analysis

The following assumptions have been considered:

(i) The economical analysis is carried out through a 15-years period.
(ii) Two financing options have been evaluated: one without financing of capital expenditures and another with a 75% financing of the initial capital expenditures.
(iii) Recipes from RECs have been included, with the selling price of US$ 17 per ton of CO_2 equivalent.
(iv) The same 8% interest tax has been adopted for the liquid present value (LPV) determination and for the financing of the loan.
(v) The loan's payment period for the initial investment is 15 years.
(vi) The payment of approximately 20% of REC recipes to the landfill proprietor for the biogas use has been considered, representing a tax of $0.43/MMBtu;
(vii) The value of biogas has a 3% annual readjustment.

For biogas generation potential calculation, it has used the model

recommended by the United States Environment Protection Agency, showed in Equation 1 (EPA, 2005).

$$Q_M = \sum_{i=1}^{n} 2 \, k \, L_o \, M_i \left(e^{-kti} \right)$$

(1)

where: QM = methane generation (m³/years); L_o = potential methane generation capacity (m³/tonnes); Mi = annual waste disposal in year i (tonnes); k = methane generation (decay) rate constant (1/years); t = time elapsed (years); i = time increment in one year.

The USEPA model requires that the site's waste disposal history (or, at a minimum, the amount of waste in place and opening date) be known. The model employs a first-order exponential decay function, which assumes that LFG generation is at its peak following a time lag representing the period prior to methane generation. The USEPA model assumes a one-year time lag between placement of waste and LFG generation. After one year, the model assumes that LFG generation decreases exponentially as the organic fraction of waste is consumed.

The Methane decay rate constant (k) is a function of refuse nutrient availability, pH, temperature and, in particular moisture content. For the Gramacho Landfill evaluation, k is 0.06 based on the degradability of the waste components (SCS Engineers, 2005).

The methane recovery potential (Lo) is the total amount of methane that a unit mass of refuse will produce given enough time, and is a function of the organic content of the waste. For the Gramacho Landfill, started with a default Lo value based on 1,140 mm of annual precipitation, and then adjusted this value based on the ratios of organic and moisture contained in U.S. waste and

Table 4. Summarizes TEP schedule, proposed by SCS Engineers (2005).

Years	Planning of TEP – Biogás
1	System of collection of gas and burning in construction
2	Beginning of the collection system and burns. Plant in construction
3	Beginning of the functioning of the energy plant; System to operate the capacity of 10 MW
4 to the 8	System with capacity of 10 MW
9 and 10	System with capacity of 7,2 MW
11 to the 15	System with capacity of 4,3 MW

Source: SCS Engineers (2005).

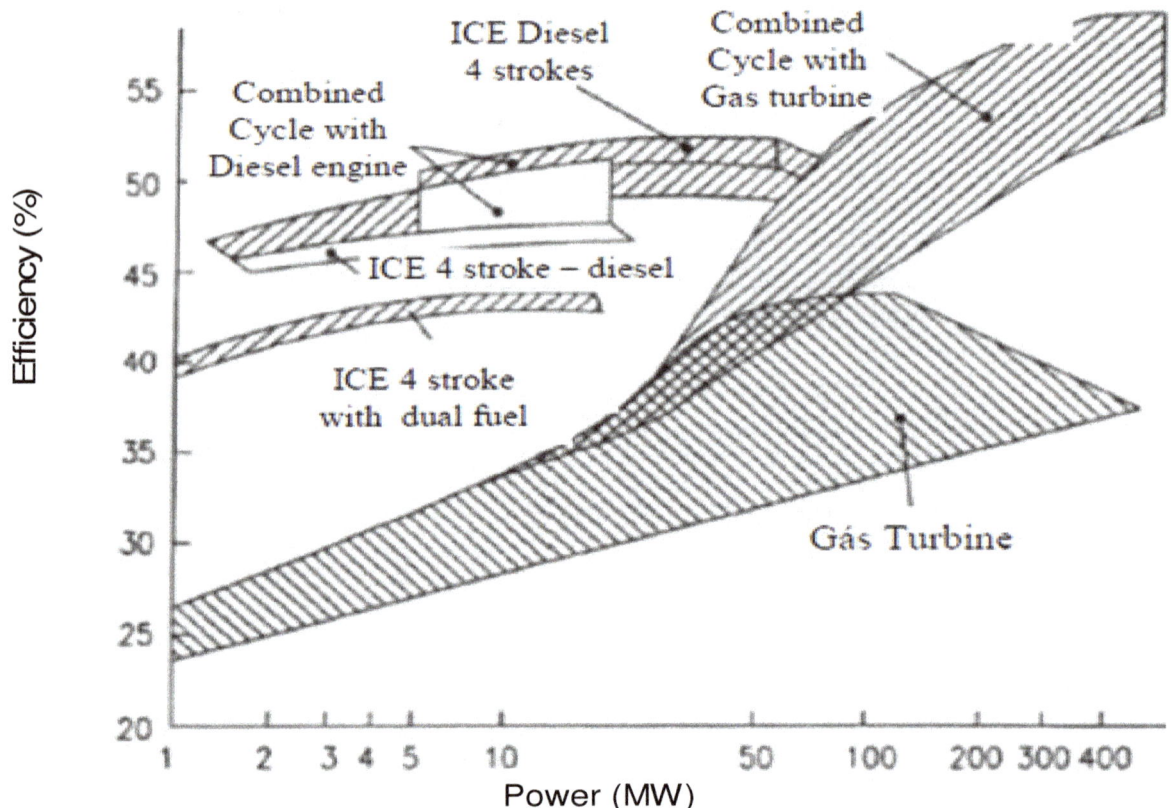

Figure 2. Efficiency comparison among diverse energy conversion technologies (Lora and Nascimento, 2004).

waste at the Landfill. The methane recovery potential for Gramacho Landfill is 84.8 m³/Mg (SCS Engineers, 2005). Table 4 summarizes TEP schedule, proposed by SCS Engineers (2005).

RESULTS

The costs of capital for the development of a biogas recovery project and those related to the operation, maintenance and regular expansion of the biogas collection system were estimated, including recurrent costs for capacity expansion of the ventilation and burning station.

Figure 2 shows the energy efficiency in function of the thermoelectric plant (TEP) capacity, for gas turbines, internal combustion engines (Otto and Diesel cycles) and combined cycles. Since Gramacho's potential power generation has been estimated at 10 MW, internal combustion engines present better performance than gas turbines form this application.

The initial cost for accomplishment of the 10 MW (bulk) TEP has been estimated in US$ 11,885,640 (Table 5) using internal combustion engines, fed with biogas, intended to attain all landfill and its own energy

Table 5. Costs of the thermoelectrial plant (TEP).

Detail	Estimated total cost ($)[1]
Plant of Energy of 10MW supplied with biogas	$9,910,875
Interconnection of 3 km	$617,500
Construction of the Plant/work in the place (including tubing)	$214,890
Measurement of biogas and equipment of register	$61,750
Engineering/contigency (10% of other costs)	$1,080,625
Total costs	$11,885,640

Table 6. Costs of biogas collection and burning system.

Detail	Estimated total cost ($)
Mobilization and management of the project	$61,750
Main tubing of gas collection	$2,779,058
Lateral tubing	$213,902
Footbridge	$58,415
Management of the condensed	$33,715
Wells of vertical draining	$398,905
Horizontal collectors	$1,200,210
Equipment of ventilation and burns (Burning)	$1,729,000
Engineering, contingency, and Initial costs of Transaction of the MDL	$689,130
Total costs	$7,164,086

consumption and to sell the exceeding energy to the electrical grid.

The costs of the biogas collection and burning system were added (cost of 7,164,086 US$ - Table 6). It was assumed that the plant will start to operate in first day of the third year of the project and will continue to operate until 15th year (in this case until 2024). So, the value of investment is US$ 19,049,726. Table 7 shows the other costs of Thermoelectrical Plant. Table 8 shows the recipes and costs of Thermoelectrical Plant. The typical payback for Thermoelectrical Plant is nine years, in this scenario actual. Then, 2014 is the year of payback of this project. Table 9 shows a summary of the results of the economic evaluation in the scenario without taking account recipes from RECs or carbon credits. Table 9 shows sensibility analysis, scenario without carbon credits.

Table 10 shows a summary of the results of the economic evaluation in the scenario of the energy plant with carbon credits, having presented a composition of financing options using the LPV and RIT. The results do not include calculations of taxes. Table 10 shows sensibility analysis, scenario with carbon credits ($17 tCO$_2$ eq. – Gramacho's adopted tax).

As demonstrated in Table 10, the economic projections of the TEP are presented attractive for financing scenarios. On the other hand, the scenario without carbon credits is not attractive.

Conclusions

Biogas energy is one of the important options which might gradually replace oil, which is facing increasing demand and may be exhausted early in this century. Brazil can depend on the biogas energy to satisfy part of local consumption.

Support for biogas research and exchange of experiences with countries that are advanced in this field is necessary. In the meantime, the biogas energy can help to save exhausting the oil wealth.

Based on results, the landfill biogas energy exploitation of Gramacho's Landfill is viable taking as reference the value of CER in $17 of ton.CO$_2$eq and any of the financing options analyzed.

The results are based on limited factors of contingency enclosed in the estimates of capital and the operation and maintenance costs. Improvements to be added in some of the used estimates in the economic evaluation, mainly the electricity sale price, can positively modify the results of this analysis.

Brazilian GHG emissions are mainly originated from hydro power plant reservoirs, forest burning and uncontrolled emissions from landfills. By employing

Table 7. Others costs of thermoelectrical plant.

Year	Annual cost O&M - thermoelectrial plant	Annual O&M of the collection system and gas of control and ampliation of costs	CDM register and annual verification	Comlurb Recipe	Payment of Garbage's participation deep
2005	-	-	-	-	-
2006	-	-	-	-	-
2007	-	-	-	-	-
2008	-	-	-	-	-
2009	-	-	-	-	-
2010	-	-	-	$741,000	$1,482,000
2011	-	$435,023	$58,986	$770,640	$1,541,280
2012	$2,010,809	$448,073	$60,755	$801,465	$1,602,931
2013	$2,071,133	$461,516	$62,578	$833,524	$1,667,048
2014	$2,133,267	$475,361	$64,455	$866,865	$1,733,730
2015	$2,197,265	$489,622	$66,389	$901,539	$1,803,079
2016	$2,263,183	$504,311	$68,381	$937,601	$1,875,202
2017	$2,331,079	$519,440	$70,432	$975,105	$1,950,210
2018	$1,715,031	$535,023	$72,545	$1,014,109	$2,028,219
2019	$1,766,482	$551,074	$74,721	$1,054,674	$2,109,348
2020	$1,819,476	$567,606	$76,963	$1,096,861	$2,193,722
2021	$1,874,061	$584,634	$79,272	$1,140,735	$2,281,470
2022	$1,930,283	$602,173	$81,650	$1,186,364	$2,372,729
2023	$1,988,191	$620,238	$84,100	$1,233,819	$2,467,638
2024	$2,047,837	$638,846	$86,623	$1,283,172	$2,566,344

Table 8. Thermoelectrical plant - recipes and costs.

Year	Recipe	Costs
2005	-	(19,160,877)
2006	-	(20,693,747)
2007	-	(22,349,247)
2008	-	(24,137,186)
2009	-	(26,068,161)
2010	22,043,968	(30,376,614)
2011	42,724,070	(37,225,110)
2012	66,862,995	(46,544,500)
2013	90,663,678	(56,609,719)
2014	114,448,897	(67,507,297)
2015	138,516,482	(79,328,396)
2016	163,144,780	(92,169,497)
2017	188,597,458	(106,133,088)
2018	213,640,347	(120,642,457)
2019	239,855,345	(136,424,834)
2020	267,472,725	(153,598,599)
2021	296,721,613	(172,290,691)
2022	327,832,772	(192,637,454)
2023	361,041,200	(214,785,521)
2024	396,588,563	(238,892,757)

Table 9. Investment analysis (scenario without carbon credits).

Value of initial investment	Percentual value of the initial investment of capital (%)	LPV	RIT*
19.160.877	100	-$36.157.454	-
4.790.219	25	-$37.221.947	-

* RIT – Return internal tax.

Table 10. Investment Analysis (scenario with carbon credits).

Value of initial investment	Percentual value of the initial investment of capital (%)	LPV	RIT (%)
19.160.877	100	$33.833.352	24.95
4.790.219	25	$32.768.859	35.40

control improvements followed by energy generation, a great amount of GHG emissions will be avoided.

Energy generation facilities shall be included in future landfill projects. A methodology for evaluation of social and environmental costs shall be added in economical evaluation of WTE.

Energy generation from landfills does not impact Brazilian Energy Matrix, although saves transmission costs since landfills are close to urban concentrations. The main advantage of implementing WTE facility is waste volume reduction, lengthening landfill useful life and technical servicing.

The following measures are being suggested to promote the growing of energy production through biogas from the waste:

(i) Simplification of the environmental licensing procedures for landfills.

(ii) Adoption of fiscal favorable instruments as, for example, "ICMS green". The cities will have these fiscal privileges case if they fit in criteria of ambient preservation and/or carry through investments in sustainable projects (as it is the case of the implantation of landfill with energy exploitation).

(iii) Dissemination of technical and economical data on construction and operation of landfill with exploitation of biogas, as well as the achieved benefits.

(iv) Establishment of special credit lines by development banks (as BNDES) with favored taxes and dedicated calls in official researching support agencies to promote the scientific initiation and technological innovation for energy exploitation from biogas in landfill.

REFERENCES

Abrelpe, Brazilian Association of Public Cleanness and Special Wastes. Solid waste in Brazil-2004 [on-line]. São Paulo: ABRELPE; 2009.

Abreu FV, Costa Filho MAF, Souza MCL (2009). Technical and economical feasibility analysis of energy generation though the biogas from waste in landfill - an alternative of renewable energy generation. In: 20th International Congress of Mechanical Engineering (COBEM), Gramado - RS, 2009, pp. 50-60.

Abreu FV, Costa Filho MAF, Souza MCL (2009). Biogás of landfill for generation of renewable and clean energy - a feasibility study technical and economic In: IX Congreso Iberoamericano de Ingeniería Mecânica (CIBIM), 2009, pp. 50-60.

Ben (2009). National Energy Balance 2009 – https://ben.epe.gov.br/downloads/Relatorio_Final_BEN_2009.pdf - Access in: August of 2010.

C4O Cities - http://www.c4ocities.org/docs/casestudies/waste/sao-paulo-landfill.pdf - Access in Maio of 2010.

Comlurb (2010). Transforming Ambient Liabilities into an Energy Resource: The garbage as power plant, Rio de Janeiro.

Epa (2005). Landfill Gas Emissions Model (LandGEM) Version 3.02 User's Guide. EPA-600/R05/047 (May 2005), Research Triangle Park, NC. U.S. Environmental Protection Agency.

Energy Source Company - EPE. Brazilian energy balance, 2009 – Base year 2008 – Rio de Janeiro : EPE, 2009.

Lora EES, Nascimento MAR (2009). Thermoelectrical Generation: plan, project and operation. São Paulo: Únit, 2004, p. 457-461.

Maciel FJ, JUCÁ JFT (2005). Gas Recovery Investigation in a Brazilian Landfill.

Miranda M, Hale B (1997). Waste not, want not: the private and social costs of waste-to-energy production. Energy Policy, 25: 587-600.

Miranda M, Hale B (2005). Paradise recovered: energy production and waste management in island environments. Energy Policy, 33: 1691-1702.

Meraz RL, Vidales AM, Domínguez A (2004). A fractal-like kinetics equation to calculate landfill methane production Fuel, 83: 73-80.

Popov V (2005). A new landfill system for cheaper landfill gas purification, Renew Energy, 30: 1021-1029.

Polprasert C (1996). Organic Waste Recyclin Technology And Management. 2nd Edition. John Wiley and Sons, P. 412.

Zamorano M, Pérez JL, Aguilar I, Ramos A (2007). Study of the energy potential of the biogas produced by an urban waste landfill in Southern Spain, Renew. Sustain. Energy Rev., 11(5): 909-922.

Zanetti AL (2009). Potential of energy exploitation of biogás in Brazil, Master in Energy Planning Program COPPE, UFRJ, Rio de Janeiro, 2009.

19

Power transfer analysis of a hybrid fuel cell/battery as portable power source generator

Joevis J. Claveria* and Akhtar Kalam

School of Engineering, Science and Health, Victoria University of Technology, Victoria, Australia.

The configuration of hybrid power system (HPS) are widely used and accepted in providing a stable and higher power capability sources in generating energy for human needs. Hybrid configuration is composed of at least two generating sources combined to produce larger supply of energy. The advantages of using hybrid configuration are distinctively beneficial. It reduces the stress on each source and decreases the internal power loss of the system. It also smoothes the hybrid source terminal voltage of the system due to load demands and create quality energy when conventional source is incapable. Moreover, in portable application, it is quicker to replenish the fuel for a fuel cell than for a battery which requires recharge. This paper presents an analysis of hybrid system consisting of- a fuel cell and battery as portable power source. The fuel cell used in this experiment is a Heliocentris ^{TM} "Constructor", a 50 W proton exchange membrane (PEM) fuel cell, a rechargeable lead acid battery pack, DC – DC converter and a variable resistive load. The experimental results show the characteristics of the hybrid configuration under different load conditions.

Key words: Hybrid power system (HPS), proton exchange membrane fuel cell (PEMFC), DC-DC converter, microcontroller, pulse width modulation (PWM), lead acid battery.

INTRODUCTION

According to the International Energy Outlook (2010), the use of renewable energy is the fastest growing source of electricity generation. The estimated total generation from renewable resources increases by 3.0% annually and the renewable share of world electricity generation grows from 18% in 2007 to 23% in 2035 (U.S. Energy Information Administration, 2010). The increase in the use of renewable resources is remarkable and accepted globally; promoting the green revolution of power energy. The boost in generating capacity using alternative generating sources like wind, hydropower, solar, bio-fuels and fuel cells are most likely to be used in supplying additional energy to the grid, power back up application and in stand-alone portable power generation.

The introduction of utilizing renewable energy resources improves the energy production, fuel economy and operating performances of a system. It shows significant environmental benefits in terms of operation and purpose. Hence, distributed generation of energy has an excellent effect in combining different alternative sources of energy in power industry to step-up the use of renewable energy.

A rewarding improvement in the generation area is the hybrid power systems. Hybrid power system (HPS) is a process of combining at least two sources providing energy in a system. Both could work at the same time or alternately generate energy in a system. Initially, HPS were designed for powering remotely located telecommu-nication stations by integrating one or two renewable energy sources with storage devices such as batteries (Maskey et al., 2010). The significance of this became popular for remote area power generation applications and a reasonable solution to cover up the energy demands of both stand-alone and grid connected consumers (Dagdougui et al., 2010). Hybrid systems are more likely to yield energy once it is needed. A concrete

---

*Corresponding author. E-mail: joevis.claveria@live.vu.edu.au.

illustration of this configuration is the fusion of a micro hydropower generator and batteries in remote location. The downfall of this arrangement arises due to insufficient amount of water to supply the micro hydropower plant during summer time. On this occasion, batteries work alone to sufficiently provide enough energy to a residential home.

Another type is the combination of wind power and batteries. The irregular blowing of wind is another concern on this design. Furthermore, the photovoltaic (PV) cell and batteries are also used. The inadequacy of sunlight, weather conditions and the intensity of sunlight that the PV absorbs are the drawback of this configuration. However, the backup batteries are continuously and perfectly providing energy to the load.

Generally, batteries play a very important role in a hybrid power system. They are often used as an energy storage unit for the DC power produced from different charging sources (that is, PV modules, wind generator, micro-hydro generator and fuel cell). Aside from this, they are also a good back-up power source of energy when primary source cannot deliver the intended supply to the load (Batteries, 2009).

A novel hybrid configuration is fuel cell and batteries. The significant aspect as presented in this paper is the production of energy from the fuel cell is continuous as long as there is uninterrupted supply of hydrogen (H_2) to the fuel cell. The use of a fuel cell in stand-alone generation is promising due to the characteristics of the fuel cell. It has: high efficiency conversion; high power density; low emission; operates quietly; simple in construction (Larminie and Dicks, 2000).

Hybrid fuel cell/battery power sources have potentially widespread uses in applications wherein the power demand is impulsive rather than constant. Such systems could provide much needed capability of high current pulses and extend the operation time of comparable in size and weight battery system. This is especially true for the military applications (Gao et al., 2004; Cygan et al., 1998).

This arrangement defies the downside of other hybrid configurations. It is not affected by weather, wind condition and circumstances of sunlight hours. Aside from this, the high efficiency of the fuel cell makes this design on top of these configurations.

Hybridization is the key in the fast growing source of electricity generation in the world, especially with the renewable energy resources that were involved in the production. It offers several advantages over a single source generation system. One of the advantages of using a hybrid power system is a continuous supply of power to the load when other source fails.

The improvement of this system is significant to economic, social, technological and environmental challenges point of view. As a result, this paper is arranged as follows. Subsequently, the components used are enumerated and the hybrid design of the fuel cell and battery is shown, after which the systems configuration and its set-up are analyzed. This is followed by verifying the viability of the experimental results and measurements of the experiment, before presenting the conclusion. The result of the study shows the capability of a hybrid fuel cell and battery to act as stand-alone portable power generator. This will provide other researchers to study and design a hybrid source using other alternative energy resources.

HYBRID CONFIGURATION

Lead acid battery

A deep-cycle battery has the ability to work for a long period of time. Lead acid is a type of a deep-cycle battery that is designed to regularly charge and discharge almost 80% of their capacity hundreds of times. It was designed to absorb and give up electricity by a reversible electrochemical reaction. Lead acid battery is a complex, non-linear device exhibiting memory effect in hybrid power systems. On the other hand, automotive starting, lighting and ignition batteries are of shallow-type and are not recommended for use for renewable systems since they only discharge small amount of their capacity (Gao et al., 2004; Saiju and Heier, 2008).

When compared with other types of deep cycle batteries, the rating of lead acid battery is average on efficiency conversion. In this research, further experiments will be done using other types of deep-cycle batteries.

Microcontroller board

A microcontroller unit (MCU or µC) works like a small computer on a single integrated circuit consisting internally of a relatively simple processor, clock, timers, input/output (I/O) ports and memory. The microcontroller board controls most of the parameters in the system, including the generation of pulse width modulation (PWM) for the electronic gate switch and the battery charge/discharge state.

DC – DC converter

The heart of the energy conversion between a generator and a load is a converter. In this paper, the experiments were conducted at Victoria University's Power laboratory that uses a DC-DC boost converter. This converter implies that the output voltage is higher than the input

Table 1. Technical data of fuel cell stack.

Denomination	Specification
Rated output power	40 W
Maximum output power	Approximately 50 W
Open circuit voltage	Approximately 9 V
Voltage at rated power	5 V
Maximum current	10 A
Current at rated power	8 A

voltage. Moreover, the addition of embedded electronic converters in renewable power system is now used for higher efficiency result. They are more efficient when compared with the conventional converters.

Resistive load

The load used in this experiment was a variable resistive load. It is linear and simpler to investigate its characteristics.

Constructor fuel cell

The constructor (Heliocentris Constructor Instruction Manual, 2008) is a 50 W, proton exchange membrane (PEM) fuel cell. Basically, a PEM fuel cell is a type of fuel cell that uses a polymer mebrane that separates the hydrogen and oxygen ions to produce electricity. The PEM fuel cell operates between 50 to 100°C. New applications are growing in the field of portable power generation, where fuel cells are compared with primary and rechargeable batteries in portable electronics.
Table 1 illustrates the working speciifications of the Constructor fuel cell stack used in the experiement.

SET UP AND SYSTEMS CONFIGURATION

In Figure 1, the set-up and systems configuration of the experiment is technically detailed. The fuel cell is directly connected to an input filter, followed by the converter, the battery, the output filter and the load.
The purpose of this hybrid configuration set-up is to analyze the power transfer of the fuel cell and battery to the load and to study the response of the system when an increase of load occurs. Table 2 shows the components needed for the experiment. They are critically analysed to provide closer value to what is expected.

EXPERIMENTAL RESULTS AND MEASUREMENTS

The results achieved from this experiment are limited to the resistive variable load only. Due to the low specification of the components used, errors are considerably present on this experiment.

Fuel cell configuration

Figure 2 represents the fuel cell and load configuration. This arrangement specifies the allowable maximum load or the maximum power point of the fuel cell alone. It was completed by directly connecting the fuel cell and adjusting the load gradually to its maximum value. The output load voltage, current and effective power of the fuel cell is varying or unregulated.
It was found out that the measured open voltage of the fuel cell is 8.73 V_{dc}. The variable load was connected and gradually varies from 2 to 40 Ω, which shows the maximum and minimum voltage of 7.43 and 4.72 V_{dc}, respectively. The maximum voltage of 7.43 V_{dc} happened at 40 Ω load and the minimum voltage of 4.72 V_{dc} occurred at 2 Ω load. Furthermore, the maximum and minimum currents are 4.27 and 0.22 A_{dc}, respectively.
The result in this configuration shows the minimum and maximum power capabilities of the fuel cell alone. It was also noted that the measured voltage and current of the fuel cell takes some time to respond when load was gradually changed. By means of the power formula:

$$P_{fc\,(max)} = I_{(max)}^{2} \; x \; R_{Load} \tag{1}$$

Where, $P_{fc\,(max)}$ – maximum power of fuel cell; $I_{(max)}$– maximum current of fuel cell; R_L – load resistance.
The maximum current on this experiment is 4.27 A_{dc} when the load is 2 Ω. Therefore, the maximum power of the fuel cell in this configuration is 36.465 Watts.

Battery configuration

In Figure 3, the specification of the Lead acid battery used in this experiment is a 12 V, 0.8 Ah rechargeable battery. Generally, the formula used in calculating the energy of battery is:

$$P_{bat} = E_{bat} \; x \; (Ampere/hour)_{bat} \tag{2}$$

The computed value of the battery in this experiment is *9.6 W/h*. This value is assumed using the Ideal battery characteristic which means that the battery has no

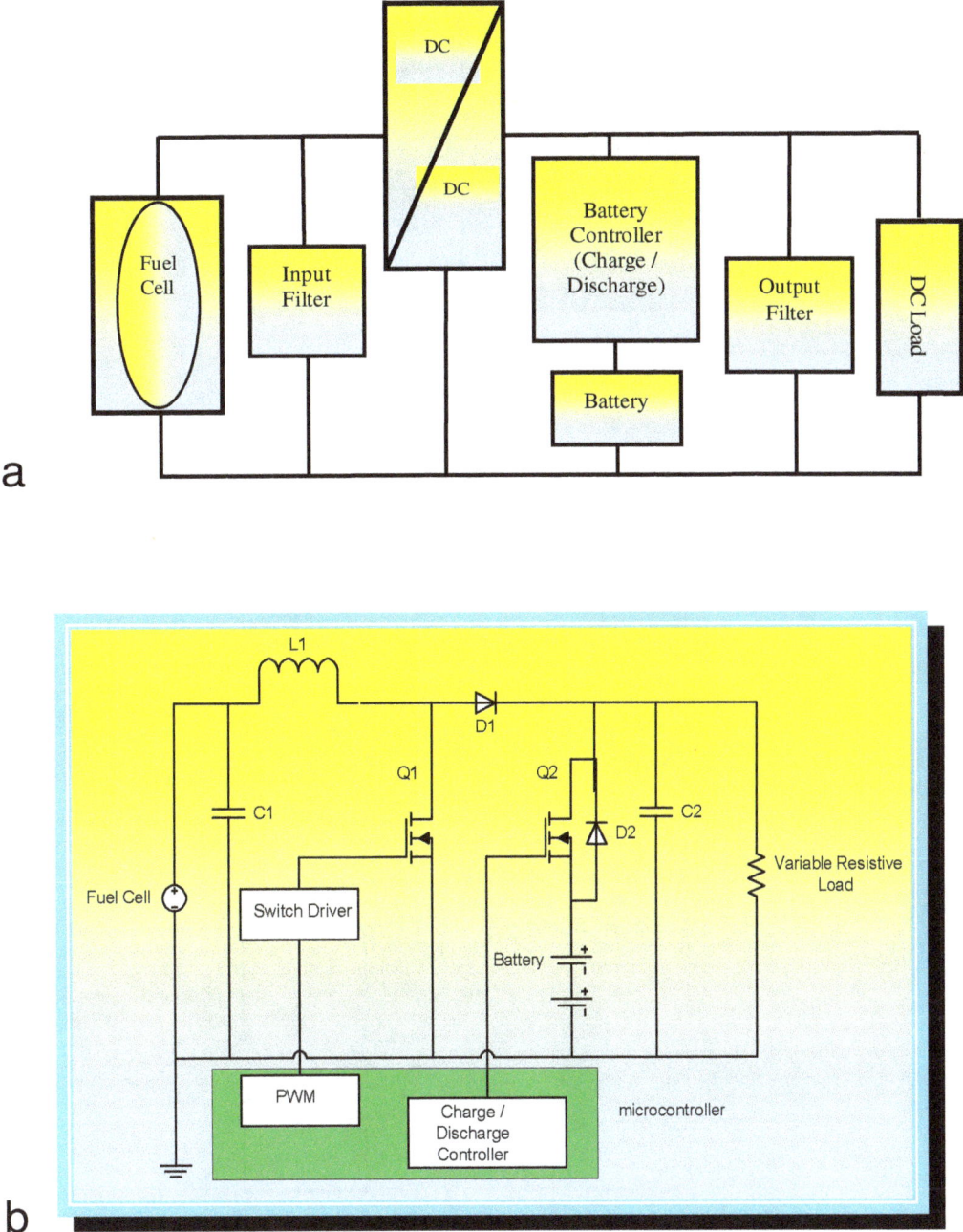

Figure 1. (a) Block diagram; (b) circuit diagram of hybrid fuel cell/battery configuration.

Table 2. Specifications of hybrid power source.

Power Source	Specification
Fuel cell (Constructor)	50 W, PEM fuel cell, 10 cells in series
Battery pack	12 V 0.8 Ah rechargeable, lead acid battery
DC-DC converter	Input voltage 4-8 Vdc, output 12 Vdc (boost)
Microcontroller	PWM generator, Battery controller
Resistive load	Variable load 0-40Ω

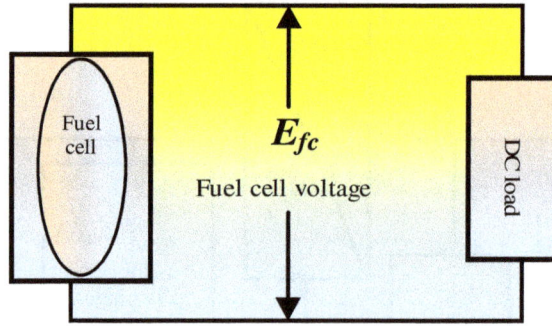

Figure 2. Fuel cell and load.

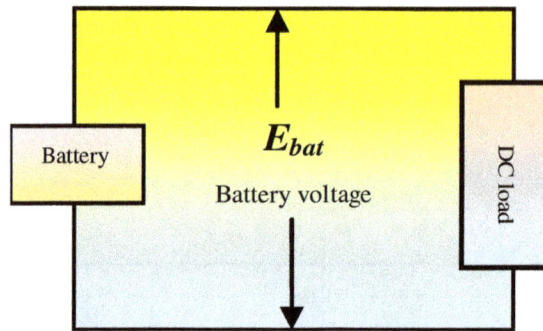

Figure 3. Battery and load.

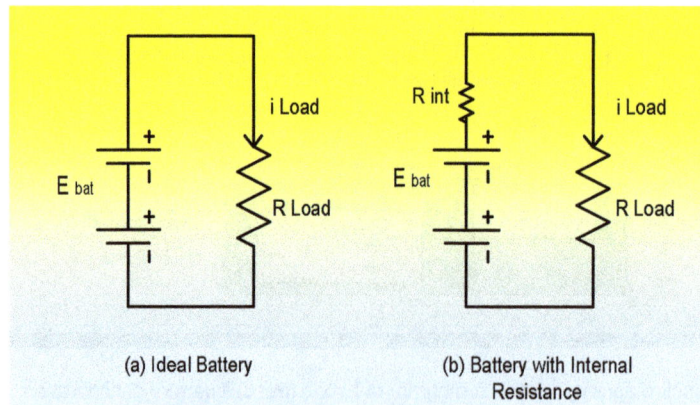

Figure 4. (a) Ideal battery; (b) battery with internal resistance.

internal resistance. The voltage across the battery is exactly the same voltage across the load. This is commonly used in theoretical analysis and computer circuit simulation of battery.

On the other hand, the battery with internal resistance is the actual analysis of an equivalent battery. It has an internal resistance and builds up a voltage drop in a circuit. Therefore, the actual energy that the battery will supply in the system will be less than *9.6 W/h* due to the voltage drop from the internal resistance of the battery.

The following formula illustrates the difference of the ideal battery and the actual battery in Figure 4:

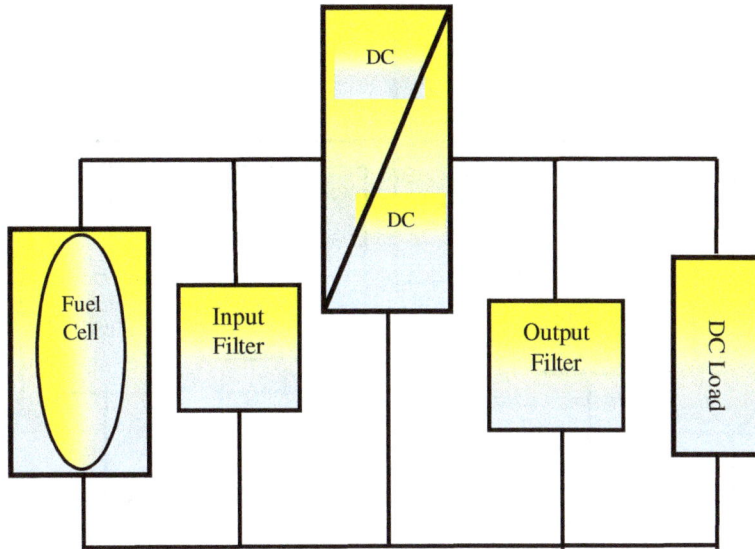

Figure 5. Fuel cell, DC-DC converter and load configuration.

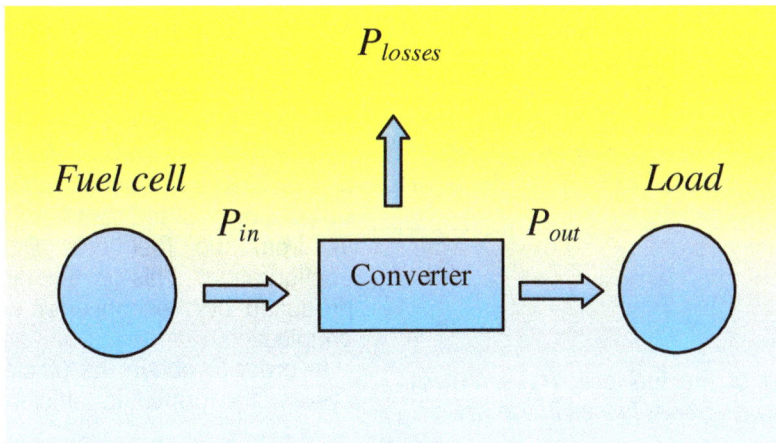

Figure 6. Energy conversion of Figure 5.

$$E_{bat} = i_L \, (R_L) \tag{3}$$

$$E_{bat} = i_L \, (R_{int} + R_L) \tag{4}$$

Fuel cell, DC-DC converter and load configuration

Another type of configuration tested was the fuel cell, DC-DC converter and load arrangement. The experiment was done by connecting the entire component following Figure 5. The load is gradually adjusted from zero to 40 Ω and the value of currents and voltages are recorded.

The output power of the fuel cell serves as the input power to the converter. The boost converter performs the energy conversion process by converting the power of the fuel cell at low voltage to a higher designated usable voltage.

In the process of energy conversion, the three important aspects to consider are the input power, output power and power losses in the system. The converter converts energy and dissipates power, while the conversion process takes place. Figure 6 illustrates the energy conversion of the system fuel cell, converter and load. Power losses in the system are vital. Generally, the

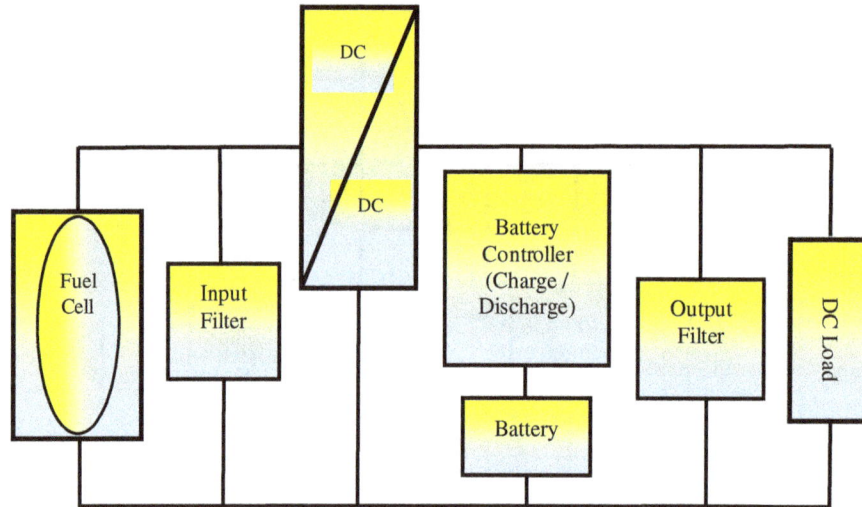

Figure 7. Fuel cell, DC-DC converter, battery and load configuration.

conduction and switching losses of silicon devices are accountable on power losses in a converter. In conduction losses, all silicon devices have resistance associated with it and this causes power loss. Similarly, the switching losses are linked to semiconductor components, like transistors, that continuously switch and dissipate power when in operation. Aside from these, heat is also a factor.

The formula representing Figure 6 in terms of power conversion is:

$$P_{in} = P_{out} + P_{losses} \qquad (5)$$

$$P_{out} = P_{in} - P_{losses} \qquad (6)$$

Where P_{in} – actual power of the fuel cell; P_{out} – output power of the converter; P_{losses} –power losses in the syste .

Fuel cell, DC-DC converter, battery and load configuration

Figure 7 demonstrates the hybrid configuration of the fuel cell/battery as portable power source in a system. Figure 8 illustrates the equivalent block diagram of the fuel cell and battery in terms of power conversion. The power formula of Figure 8 in terms of power conversion is:

$$P_t = P_{out} + P_{bat} \qquad (7)$$

Where: $P_{out} = P_{in} - P_{losses}$

Therefore:

$$P_t = P_{in} - P_{losses} + P_{bat} \qquad (8)$$

Where P_t – total power of the hybrid; P_{in} – actual power of the fuel cell; P_{out} – output power of the converter; P_{bat} – output power of the battery; P_{losses} – power losses in the system .

The actual maximum power generated from the fuel cell is 36.465 watts which is the input power to the converter, P_{in}. The conversion process takes place on the converter and produces the usable power to a higher voltage. The output power of the converter was measured to be 23.835 W from the fuel cell, DC-DC converter and load configuration. This is the maximum allowable power produced by the converter when the battery is not yet coupled.

In order to obtain the efficiency of the converter in the system, the formula for efficiency is used:

$$\% \ efficiency = \frac{P_{out}}{P_{in}} \ x \ 100\% \qquad (9)$$

The percent efficiency of the converter is 65.36%. Higher efficiency of the converter may be achieved by critically analyzing the selection of electronic components to be used in improving the power density and thermal performance of a converter (Moxey, 2008).

The experimentation of the hybrid configuration was done by gradually changing the load demand from zero to 40 Ω. It was found out that the maximum total power of the hybrid system is 26.542 W.

The power of the battery delivered to the load can be calculated using formula (7):

$$P_{bat} = P_t - P_{out}$$

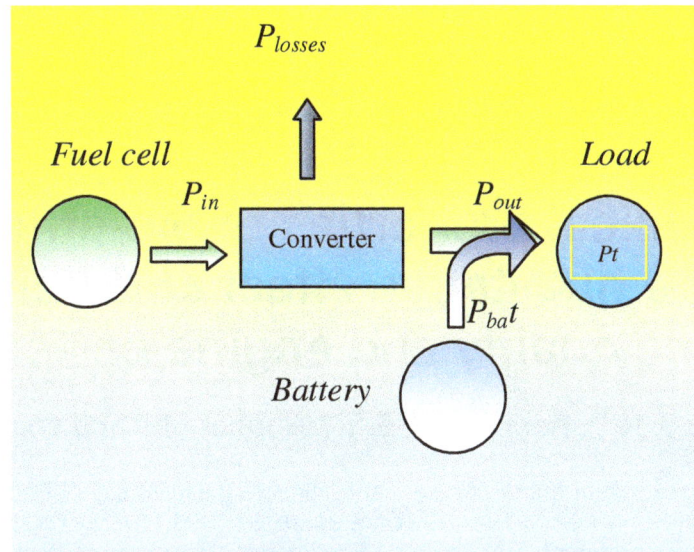

Figure 8. Energy conversion of Figure 7.

The maximum power delivered by the battery to the load is 2.707 W for a period of time. The battery operates to provide power both during load transients and during peak loads of the system. The percent increased of power to the system can be calculated using the formula:

$$\% \ increased = \frac{P_t - P_{out}}{P_{out}} x \ 100 \tag{10}$$

The percent increased by the battery to the system is 11.36%.

Conclusion

The analysis on a hybrid fuel cell/battery is significant compared with other configurations which are used as portable power source generator. It defies the downside of other types of hybrid power resources. It was also evaluated on the experiment that the fuel cell is slow in responding on instantaneous varying load demands. Thus, the addition of battery to the system helps the fuel cell to handle the load where it draws the power from the battery until the fuel cell supports the increased of load demand. The battery provides additional power of 2.707 W for a period of time and a power increased of 11.36% in the system.

The converter is responsible for the energy conversion of the fuel cell and converts the power to a higher voltage application. The size of the converter and its components are essential for energy conversion process. This means fewer components involve, means less power losses in energy conversion stage. Moreover, the addition of microcontrollers and high efficient power converters are considered new and must be incorporated in further researches on hybrid power systems.

REFERENCES

U.S. Energy Information Administration (2010). International Energy Outlook 2010, U.S. Depatment of Energy.

Maskey RK, Mult HC, Nestmann F (2010). Hydro based renewable hybrid power system for rural electrification: A concept paper, [Online] Viewed 2010, 5 November. Available: http://www.mtnforum.org/apmn/hybridconcept.htm.

Dagdougui H, Minciardi R, Ouammi A, Robba M, Sacile R (2010). Modelling and control of a hybrid renewable energy system to supply demand of a green-building, International Environmental Modelling and Software Society *(iEMSs)*, Ontario, Canada, pp. 1-8.

Batteries (2009). Kyocera Solar Electric Products Catalog,pp. 44-49.

Larminie J, Dicks A (2000). Fuel Cell Systems Explained, John Wiley and Sons Ltd., England, pp. 21-24.

Gao L. Jiang Z, Dougal RA (2004). Performance of Power converters in Hybrid Fuel Cell/ Battery Power sources. 35th IEEE Power Electronics Specialists Conference. Columbia, SC., 3: 2018-2022.

Cygan PJ, Atwater TB, Jarvis LP (1998). Hybrid power sources for military applications. 13th Annual Battery Conference on Applications and Advances, pp. 85-90.

Saiju R, Heier S (2008). Performance Analysis of Lead Acid Battery Model for Hybrid Power System. Transmission and Distribution Conference and Exposition. Chicago, IL, pp. 1-6.

Heliocentris Constructor Instruction Manual (2008), Version 0.0, March 2008.

Moxey G (2008). DC/DC converter efficiency revisited, [Online] Viewed 2010, 10 December. Available: http://www2.electronicproducts.com/Dc_dc_converter_efficiency_revisited-article-FAJH_Energy_Jul2008-html.aspx.

20

Physico-chemical properties of bio-ethanol/gasoline blends and the qualitative effect of different blends on gasoline quality and engine performance

Tangka J. K.[1*], Berinyuy J. E.[2], Tekounegnin[3] and Okale A. N.[3]

[1]Renewable Energy laboratory Department of Agricultural Engineering, University of Dschang, P.O. Box 373, Dschang, Cameroon.
[2]Crop Processing laboratory, Department of Agricultural Engineering, University of Dschang , P.O. Box 373, Dschang, Cameroon.
[3]Department of Agricultural Engineering FASA University of Dschang Cameroon.

Physico-chemical and operational properties of various gasoline bio-ethanol blends were evaluated. Bio-ethanol was obtained through distillation from maize (*Zea mays*), sugar cane (*Saccharum* L), raffia (R*affia vinefera*) wine, and palm wine and then purified using a rotavapor. Engine trails involved combinations of various ratios of gasoline/bio-ethanol as fuel in a small unmodified gasoline engine connected to a dynamometer. The vapour pressure, octane number, flash point, specific gravity, and energy density of various compositions of the blends were evaluated. Sugar cane gave the highest yield of alcohol 97.99 g per kg of produce while the lowest amount of alcohol of 10.5 ml per kg of produce was obtained from palm wine. Engine power decreased from 0.400 kW with 100% gasoline as fuel to 0.108 kW with a gasoline ethanol ratio of 1: 10. The octane number increased from 93 at E10 to 106 at E90. The energy density decreased from 33.180 MJ/l at E10 to 23.600 MJ/l at E90. Other physical observations suggest that to successfully run a gasoline engine with bio ethanol/gasoline blends some modifications would have to be done on the engine, including advancing of ignition timing, provision of air tight fuel conduit network, and modification of piston heads to improve pre-combustion fuel homogenisation.

Key words: Ethanol gasoline blends, bio-ethanol, octane number, bio-fuels, renewable energy.

INTRODUCTION

The prices of petroleum products are generally on an increasing trend and consequently affecting the general cost of living. This continuous increase has resulted in the prices of some products tripling in the last decade. In some communities, in the developing nations these products are not only expensive but they are not readily available because of the poor road infrastructure necessary for their distribution. The consumption of petroleum products also has other inconveniencies such as environmental pollution and the emission of green house gasses generally believed to be responsible for global warming. The general trend all over the world now is to reduce the over dependence on petroleum products so as to help reduce the effects of global warming. Other possible advantages of abandoning petroleum products is the fact that alternative sources can be produced from renewable resources that are available almost everywhere that there is life. This avoids the employment of heavy infrastructure for long distance transportation and distribution.

*Corresponding author. E-.mail: tangkajkfr@yahoo.fr.

Abbreviations: M$_m$, mass of the maize (kg); **V$_b$,** Volume of beer (l); **R$_1$,** specific yield (l kg^{-1}); **V$_{al}$,** Volume of alcohol (l); **R$_2$,** Specific yield (%); **R,** juice yield, l kg^{-1}; **V$_j$,** quantity of juice obtained(l); **M$_s$,** mass of the sugar cane used (kg).

The need for alternative sources of energy especially those that can be produced and utilized in enclave areas cannot be over emphasised. Amongst the various alternatives that are attracting attention today, is the use of ethanol as fuel for the motor vehicle engine.

Various fractions of ethanol are already being used in many countries around the world to blend with gasoline and diesel. Proponents of this technology argue that an alcohol fuel provides high quality, low cost fuel for exceptional engine performance (RFA, 2001). Ethanol-blended fuels account for approximately 30% of all automotive fuels sold in the U.S. and ethanol acts as antifreeze in the engine during winter (RFA, 2001). This high quality and high-octane fuel is capable of reducing air pollution and improving automobile performance (Wei-Dong et al 2002; Al-Hassan 2003). One of the major advantages of using ethanol as fuel for the car is that it is a renewable source of energy with possibilities of sustainable production and exploitation.

The problem however is that of other aspects surrounding the use of ethanol blended gasoline. Ethanol has some lower physico-thermal characteristics when compared to gasoline. The effect of blending these two fuels at different ratios on engine performance serviceability and service life has not been published. What are the necessary modifications that have to be made on a modern compression ignition engine if this potential fuel is to successfully replace or complement petroleum fuels? There is a considerable dearth of knowledge on the physico-chemical properties of various compositions of gasoline ethanol blends. Many trails and development exercises have been carried out in many countries around the world but very little data is available for designers.

Earlier work, such as that of Johansen and Schramm (2009), investigated the low-temperature miscibility of ethanol gasoline-water blends in flex fuel applications at -25 and -2°C. It was found out that the blend can be successfully used without phase separations within the tested temperature range. A study of some fuel properties of ethanol blended with diesel was carried out by Ajav and Akingbehin (2002) and it was concluded that ethanol blend percentages of up to 20% had acceptable fuel properties for use fuel in farm machines. This has also been confirmed by RFA (2001) and Amulya et al. (1997). The performance and pollutant emissions of a four stroke SI engine operating on ethanol blends of 0, 5, 10 , 15 and 20% fuel using artificial neural network was investigated by Kiani et al. (2010), Najafi et al. (2009) and Mustafa et al. (2009). They found a decrease in CO and HC emission with the introduction of ethanol into gasoline.

Although there has been a wide spread call for the production of bio-ethanol for engine fuel, the performance of the engine with this fuel seems not to have been mastered by the technological industry. The call for bio-ethanol seems to be politically motivated and is aimed at

fuel autonomy, decreasing dependence on foreign oil, environmental sanitation and less on machine life and fuel performance. Bio-ethanol has unavoidable anti fuel characteristics like moisture, lower calorific value and a higher flash point. What are therefore the effects of the various blends on these characteristics and what is the best suitable mixture?

This study was carried out in order to:

(1) Publish the physico-chemical properties of a wide range of ethanol gasoline blends.
(2) Investigate the amount of bio-ethanol obtainable from some tropical bio products.
(3) Determine the performance of an unmodified gasoline engine using various blends.
(4) Propose necessary modifications that have to be carried out on the modern SI compression engine if this fuel is to partially or fully replace gasoline or diesel.

The study would help energy stake holders in mapping out the choice of strategies towards energy self sufficiency using bio-ethanol. Many developing countries that are not car manufacturers would like to be involved in bio-ethanol production for car use. However, they soon discover that for bio ethanol to be successfully used in blending engine fuels, the car manufacturing industry has to be aware ahead of time to make the necessary modifications on the engine. These modifications depend on the physico-chemical properties of the bio ethanol/gasoline ratio that is chosen. While much work has been done on this area in countries like Brazil and the United States of America, very little data is published on the effects of a wide range of mixtures on the physico-chemical properties of the blends.

METHODOLOGY

Substrates and production of ethanol

Sugar cane

The stems of the cane were cut into pieces of length 25 cm each. This was to make sure that they would go into the hydraulic press used in the extraction of the juice. The pieces were weighed on a balance of sensitivity 0.01 g. Pressure applied on the cane was of the order of 70 MPa until all the mechanically extractible juice was expelled from the cane. The juice was collected and the volume measured in a graduated cylinder. The juice yield of the cane was expressed using Equation 1.

$$R = \frac{V_j}{M} \qquad (1)$$

where R is the juice yield in L kg^{-1}; V$_j$ is the quantity of juice obtained in L ; and M is the mass of the sugar cane used kg.

The juice obtained was diluted which double the volume of water and one litre of the mixture put inside a 1.5 L plastic bottle, sealed air tight and left for fermentation. These bottles were left for 12 days under ambient temperatures between 18 to 22°C with distillation trials being carried out each day until it was clear that there would

Table. 1. Composition of bio-ethanol/gasoline blended samples used for analysis.

Sample code	% Ethanol	% Gasoline
E10	10	90
E20	20	80
E30	30	70
E40	40	60
E50	50	50
E60	60	40
E70	70	30
E80	80	20
E90	90	10

Figure 1. Dynamometer and engine assembly used to test the performance of bio-ethanol/gasoline blended fuel.

be no further increase in the yield of alcohol produced.

Palm wine

Wines obtained using the traditional method from *Elaeis guineensis* and from *Raphia vinifera*, were tested. The traditional method involves cutting the shoot of a central stem in a mature cluster of palms and attaching a container for the collection of the sap. The sap or wine that comes out is usually sweat and becomes very alcoholic as days go by. Different samples obtained from palm bushes in the western region of Cameroon were used. One litre of each wine was put in 1.5 L plastic bottle and sealed airtight. Fermentation was done at 18 to 22°C for 14 days with distillation trials on samples every day until alcohol production became maximum.

Maize

The grains were weighed on a sensitive balance and then soaked in water for 24 h and washed thoroughly. They were then aerated

on a platform in a shade during which the seeds were watered every morning and evening. This was to provoke the germination of the seeds. The germinated seeds were then sun dried to moisture content of 7% and then milled to obtain corn flour. The milling process was made to provide finely ground grain so that the starch would easily absorb water. The flour was then soaked in water for 30 min, stirred and left for about 2 h for the starch to settle. After this excess water was drained from it and the paste was boiled slightly and the drained water was re-added to it and was left for 12 h. The substrate was sieved from the water and the remaining water boiled again for two hours. The liquid was then filtered and fermented to obtain malt. The specific yield R_1 of alcohol in litres per kg of maize was estimated using Equation 2 and the percentage specific yield was estimated using Equation 3.

$$R_1 = Q_b/M_m \tag{2}$$

where M_m is the mass of the maize kg; Q_b is the quantity of malt in L; and R_1 is the specific yield L kg^{-1}

$$R_2 = \frac{Q_{al.} \times 100}{Q_b} \tag{3}$$

where Q_{al} is the quantity of alcohol L; Q_b is the quantity of malt, and R_2 is the specific yield %.

Distillation step

The prepared substrates were distilled in a Liebig condenser while heating the liquid in a 3 L round bottom flask and the temperature of the water in the heating jacket was set at 90°C. The temperature was first adjusted until the boiling point of methanol, 53°C in order to separate possible methanol from the final product. The distillate made of water and ethanol was collected at a temperature of 90 to 91°C and this temperature was maintained until there was no more alcohol condensing in the system. The distillate, a mixture of water and ethanol also known as azeotrope was collected in a graduated cylinder and then purified in a rotavapor to obtain 95% pure ethanol.

Nine blends of ethanol gasoline mixtures were used. The blends are usually referred to as EX, where E represented bio-ethanol and X represented the percentage of the bio ethanol in the blend. For example, E10 means a blend composition in which bio ethanol is 10% and gasoline is 90%. The various compositions are shown in Table 1.

Dynamometer/engine calibration

During the engine trails, various ethanol gasoline blends were used as fuel to run a small gasoline engine attached to a dynamometer and the break power developed by the engine was computed using the torque output and the maximum engine speed developed. Before this was done it was necessary to calibrate the equipment using gasoline only as fuel.

The equipment used was a 4 hp engine attached to a dynamometer of the type Power Lab PL 100D. The engine dynamometer assembly is shown in Figure 1. This dynamometer consisted of a fined rotor mounted in cradle housing. The engine being tested was attached to drive the fine rotor section as described in the PL-D dynamometer laboratory and technical manual. During tests water was introduced into the dynamometer under pressure into the water absorption unit mounted in an

Table 2. Ethanol production from maize grain.

Trial	Quantity of malt (L)	Quantity of ethanol (cl)	Ethanol yield (cl/L)
1	4.55	12.50	2.75
2	5.10	15.00	2.97
3	5.00	16.50	3.30
Mean	4.87		3.01

adjustable bracket system. As the engine being tested had to pump the water, this placed a load on the engine and the load was varied by decreasing or increasing the water in the system. The outward movement of the water caused the housing to rotate in its cradle and a scale attached to the housing enabled the torque to be measured using a load cell and a bourdon tube gauge. The said system was designed to measure brake horsepower and torque at various data points throughout the speed range of the engine being tested. The mechanical energy of the engine was converted into heat energy in the form of hot water. As the hot water left the dynamometer, the system was cooled and therefore no auxiliary cooling was required.

The values obtained during calibration were taken, as standard values so that the readings obtained using different fuel combinations would be compared to these.

Fuel combustion rate

The fuel consumption characteristics of the engine were studied using various combinations of ethanol gasoline/ethanol blends as shown in Table 1. This was necessary in order to determine the effect of the various blends on mileage. The engine was maintained at constant acceleration. The rate of fuel consumption was determined using the following equation:

$$F_r = \frac{V_f}{T} \qquad (4)$$

where F_r is the average fuel consumption rate of the engine (litres / hour); V_f is the volume of fuel in litres; and T is the time used in consuming the fuel in hours.

Measurement of physico-chemical properties

The flash points of the blends were measured using the Pensky-Martens closed cup method as detailed in ASTM D93 as described by Ajav and Akingbehin (2002). The octane number of the various blends was measured using a portable cetane/octane meter adapted to ASTM D 2699-86 and D 2700-86 methods while the energy densities of the different gasoline ethanol blends were determined by burning a known amount of blend in a Gallenkemp ballistic bomb calorimeter. The specific gravity of the various bio-ethanol/gasoline blends was determined using the specific gravity bottle as described by Ajav and Akingbehin (2004). The vapor pressure of the blends was determined using the ASTM test method D4953-99a standard test method for vapor pressure of gasoline and gasoline oxygenated blends (dry method). The experimental set up was similar to the one described by Kar et al. (2009). The auto ignition point was measured according to the procedure described in ASTM E659.

RESULTS AND DISCUSSION

Results of specific bio ethanol yield from each crop are shown in Tables 2, 3, 4 and 5. Analysis of the ethanol production from maize (Table 2) grain shows that a kilogram of this maize variety would produce about 10.58 cl of the ethanol 95°or about 1.06 L from 1 per kg of maize. Grain ethanol yield from maize literature shows a lot of variation. Grain ethanol yields of 64 L of ethanol from 100 kg of maize of grain or about 0.64 L per kg (Zeinlin'ska et al., 2009). Research at the University of Arkansas reports more than 3.90 L of ethanol from per kg of maize. Therefore the ethanol yield obtained in this investigation was lower. This could be due to the variation in species and agronomic practice. This is confirmed by Persson et al. (2009, 2010) who stated that the production potential of ethanol from maize varies with weather and climatic conditions as well as crop management practices The maize used for the experiments was locally grown by the villagers with probably very little scientific input. No literature was available for comparison of alcohol yield from palm wines and raffia wine shown on Tables 4 and 5. From Figures 2 and 3, it can be seen that the yield of ethanol is the maximum after three days of fermentation. The maximum for palm wine is about 3.2 cl/L, while the maximum yield for raffia wine is 1.2 cl/L. The maximum yield of alcohol for sugar cane juice is about 100 cl/L and for corn it was about 3.33 cl/L of malt. After three days of fermentation, the alcohol yield drops considerably for all the four substrates. This can be explained by the fact that after three days of fermentation, oxidation reactions proceed in the formation of acetic acid and ethanoic acid, thereby decreasing the alcohol yield. Alcoholic fermentation is catalysed by yeasts. The drop in alcohol production is due to the action of vinegar production bacteria. They are as abundant as yeast in nature and they feed themselves on the alcohol and produce acetic acid. Table 6 shows the results of the calibration experiment. As the loading increased, the speed of the engine decreased and the break power developed increased. The maximum power developed was about 405.0 W corresponding to an engine speed of 1900 rpm. The engine developed the lowest power of 136.3 W at a maximum engine speed of 2400 rpm. These values were used to standardise the engine/dynamometer assembly since the unmodified engine was designed to run with E0 or 100% gasoline.

Table 3. Production of ethanol from sugar cane juice.

Trial	Mass of sugar cane (kg)	Quantity of juice (L)	Quantity of water added (L)	Quantity of ethanol (cl)	Ethanol yield (cl /kg) of sugar cane	Ethanol yield cl/L of sugar cane juice
1	4.234	1.90	4.0	190	44.87	100.00
2	4.574	2.19	4.0	208	45.47	94.98
3	4.312	1.94	4.0	192	44.52	98.99
Mean	4.473	2.01	4.0	192	44.08	97.99
SD					0.48	2.66

Table 4. Ethanol production from fresh palm wine.

Trial	Quantity of palm wine (L)	Quantity of ethanol (cl)	Ethanol yield (cl/L)
1	3.00	13.50	3.50
2	4.00	13.80	3.45
3	5.00	17.50	3.50
Mean	4.87		3.48
SD			0.03

Table 5. Ethanol production from fresh raffia wine.

Trial	Quantity of raffia wine (L)	Quantity of ethanol (cl)	Ethanol yield (cl/L)
1	3.00	3.6	1.20
2	4.00	4.9	1.23
3	5.00	5.8	1.16
Mean	4.87		1.20
SD			0.04

Table 6. System calibration using gasoline as fuel.

Load (turns)	Torque (Nm)	Speed of rotation (rpm)	Break power (Watt)
0	0.0553	2400	136.30
1	0.0553	2300	130.57
2	0.0691	2200	155.85
3	0.1659	2100	357.79
4	0.1936	2000	397.46
5	0.2074	1900	404.62

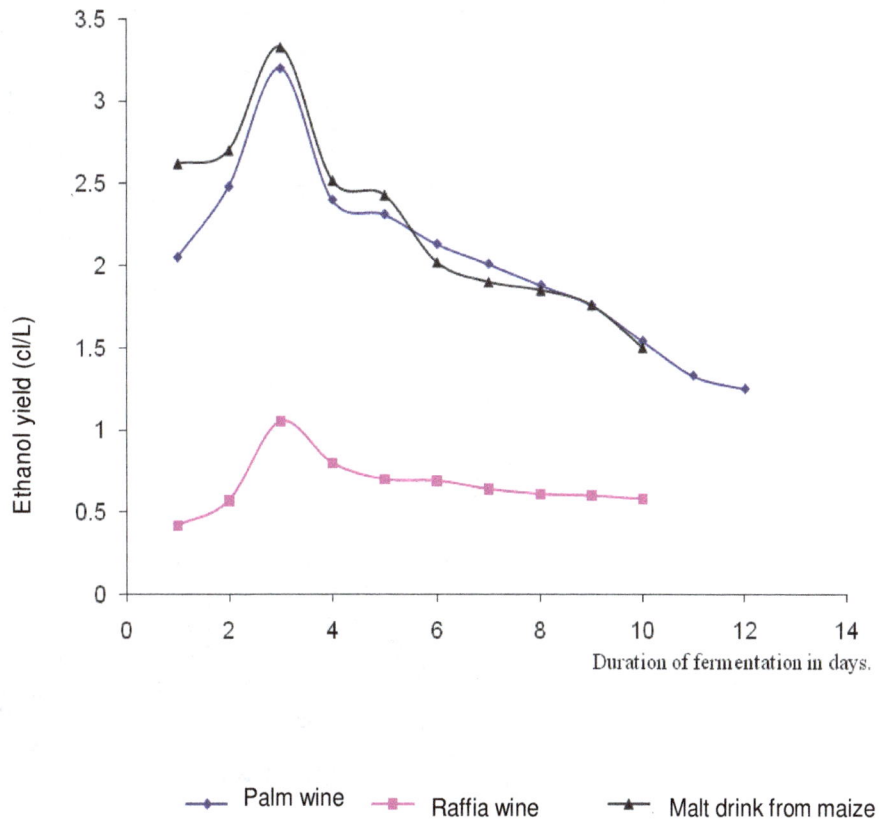

Figure 2. Variation of ethanol yield with duration of fermentation of three bio-materials.

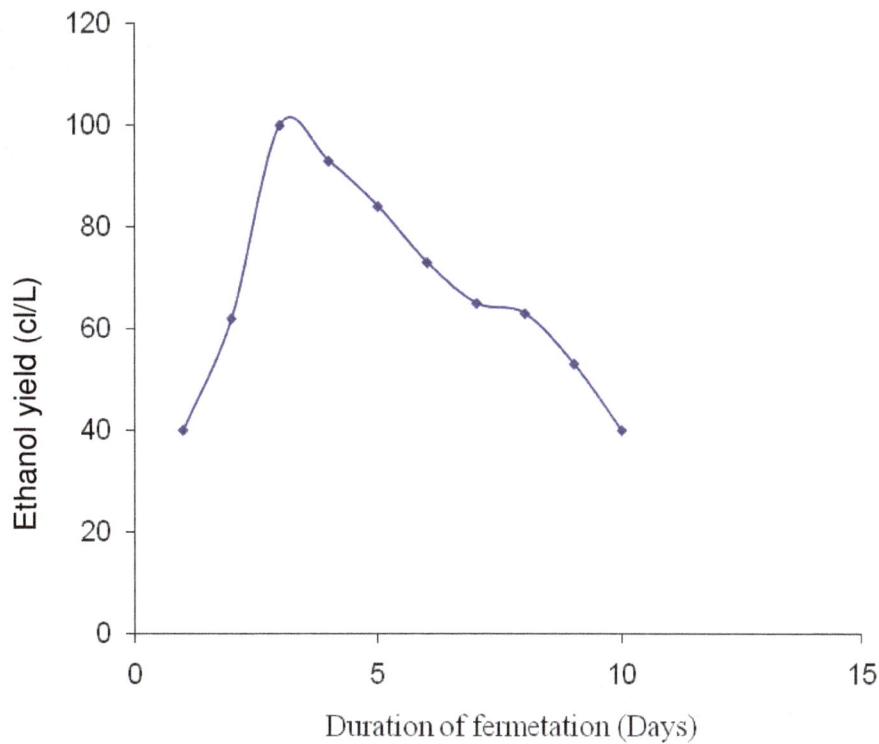

Figure 3. Variation of ethanol yield with duration of fermetation of sugar cane juice.

Table 7. Properties of gasoline fuel blended with various percentages of ethanol (Average values).

Sample code	% Ethanol	% Gasoline	Flash point (°C)	Auto ignition temperature (°C)	Vapour pressure (Kpa at 37.8°C)	Energy density (MJ/L)	Octane number	Specific gravity
E0	00	100	-65	246	36	34.2	91	0.7474
E10	10	90	-40	260	38.9	33.182	93	0.7508
E20	20	80	-20	279	39	32	94	0.7605
E30	30	70	-15	281	38	31.5	95	0.7782
E40	40	60	-13.5	294	35.6	30	97	0.7792
E50	50	50	-5	320	34	29	99	0.7805
E60	60	40	-1	345	31	28	100	0.7812
E70	70	30	0.00	350	28	27	103	0.7823
E80	80	20	5	362	24	26.5	104	0.7834
E90	90	10	8.5	360	18	23.6	106	0.7840
E100	100	00	12.5	365	9	23.5	129	0.7890

The measured properties of the various blends are shown in Table 7. Energy density decreases as ethanol concentration increases therefore more fuel should be consumed per mile as the ethanol fraction in the blend increases. The specific gravity of blends increased from 0.7508 for E10 to 0.7890 with E100. This was expected especially as bio ethanol is heavier than gasoline. This explains why when a mixture of bio ethanol gasoline is viewed inside a transparent container; ethanol is seen to settle at the middle sandwiching gasoline on top and water at the bottom. The immiscibility of the two liquids especially and higher ethanol fractions should be responsible for the unsteady engine production noticed during the tests. Amongst other modifications to be done in an SI engine should include the reduction of the carburettor jet sizes and modification of the piston head configuration to improve on the atomisation and homogenisation of the compound fuel.

The flash point increases from -65 at E0 to 12.5 °C with E100. The flash point indicates the temperature at which the fuel can vaporise to produce an ignitable mixture with air. At the flash point an applied flame gives momentary flash instead of some steady combustion. Therefore the flash point gives some indication on the flammability of the liquid. This shows that with the addition of ethanol fractions to gasoline the burning characteristics of the mixture reduces. It is important to note that the higher the flash point of a fuel the more difficult it is to start the car. For example, the flash point of E100 was found to be 12.5°C meaning that starting will be very difficult at temperatures of about 9°C. Therefore a gasoline engine converted to an E 100 engine would need to use pre heater plugs in winter.

The specific gravity increases from 0.7474 at E0 to 0.7890 at E100. This shows that addition of ethanol to gasoline produces a fuel blend denser than gasoline. The specific gravity of a fuel is the weight of the fuel compared to the weight of the same volume of distilled water at a given temperature. It gives an indication of the purity of a fuel. When fuel is contaminated with another liquid, the specific gravity will either increase or decrease depending on the specific gravity of the contaminant.

There is only one air/fuel ratio that gives the best performance of each combustion engine.

The ratio is determined by the weight of the air/fuel mixture. The fuel supply system of all engines such as carburetors and injection systems measure fuel by volume and because of this the injection jets will have to be modified every time the fuel specific gravity changes. The volume will have to be adjusted to maintain the weight of the fuel going into the engine for combustion.

Figure 4 shows the variation of research octane number with percentage of bio-ethanol. The Octane number increases from 91 at E0 (100% gasoline) to 106 with E90. Octane number of a fuel indicates its ability to resist pre- ignition and burn evenly. The Octane number of pure bio-ethanol is 129. This shows that the addition of bio-ethanol to gasoline improves considerably, the Octane number consequently increasing the activation energy of the fuel or the energy necessary to start a reaction. The research octane number RON or the measure of a blends' resistance to auto-ignition in the engine is defined

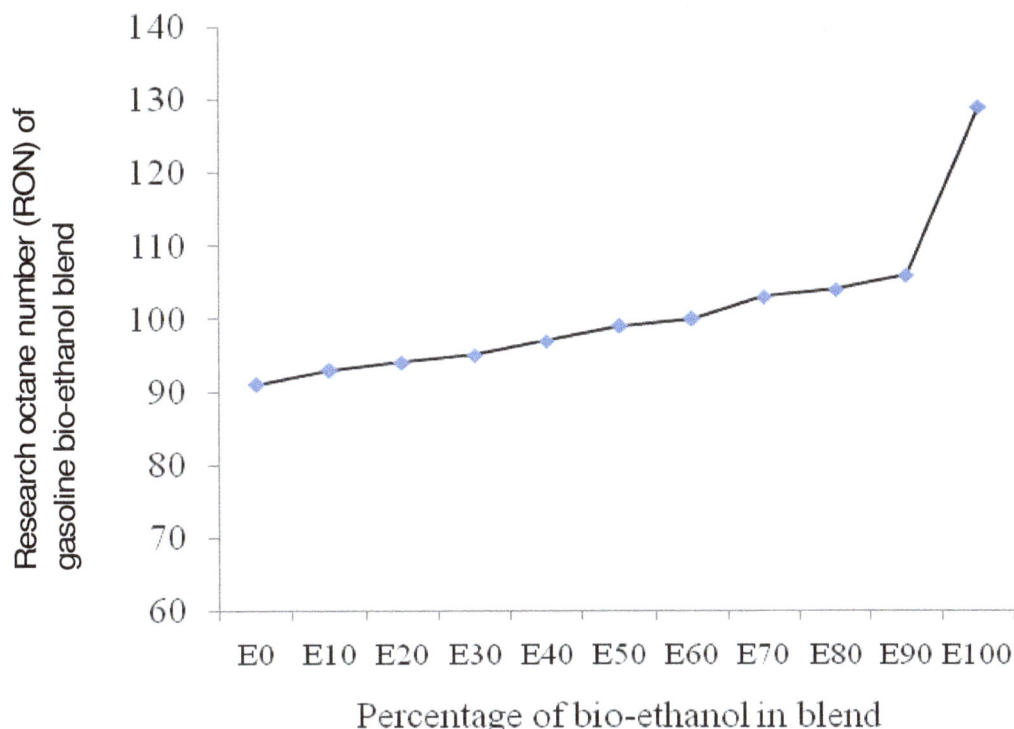

Figure 4. Variation of octane number with bio-ethanol percentage on blended gasoline.

by comparison with the mixture of iso-octane and heptane which would have the same anti-knocking capacity as the fuel under test. It is a measure of the fuel's tendency to burn in a controlled manner or a measure of its tendency to knock in an SI engine. As far as ignition only is concerned, the best blend would be E60 that gives exactly the same performance like iso-octane, that is, with an octane number of 100. Investigations on the auto ignition characteristics of ethanol in the 667 to 743K temperature range at one atmosphere were made by Bolentin and Wilk (1996), and conclusions were drawn that the ethanol concentrations proved to have a greater effect on the ignition delay times than did oxygen concentrations.

The vapor pressure of the blends decreases from E10 to E100. The vapor pressure of a liquid is very important because it affects the starting and warm up of spark ignition engines (Davis et al., 2002). At high altitudes and during high operational temperatures the cause of vapor lock in fuel pumps depends on the vapor pressure of the fuels. Therefore blending ethanol with gasoline negatively affects the ease of starting the engine.

Engine performance using various gasoline ethanol mixtures

The observations made using all compound fuels (various ratios of gasoline/bio ethanol mixture) were similar. We noted a very difficult starting of the engine and at times vibrations due probably to the water factions in the fuel and the fact that the improved octane rating produced uncontrolled ignition. Uncontrolled ignition in an internal combustion engine is very undesirable because it leads to reduced power no matter the energy density of the fuel. Therefore for an increase in the octane number to produce an advantage, the engine has to be modified to advance the timing and also increase the compression ratio. Other possible modifications could be; increasing the carburettor jet size, modifying the intake manifold for better fuel atomisation, modifying the piston head configuration. The rattling of the engine especially experienced with higher percentages of bio-ethanol was due to the moisture content of ethanol. In spite of the fact that the bio-ethanol obtained from distillation was purified in a rotavapor only 95% purity could be obtained and this is what is practically possible because of the hygroscopic nature of bio-ethanol. The remaining 5% is water residue from the distillation process and this contributes to some anti-fuel properties. The vibrations rattling and knock are due to the passage of fractions of ethanol containing water into carburettor. Bio-ethanol should be corrosive due to this moisture content and if higher percentages are used in compound fuels, degradation of some materials in the engine and the fuel system should be expected. Fuel tanks for blended fuels might therefore need to have a water trap that can be drained as soon as the fuel is put into the engine. However since the fuel

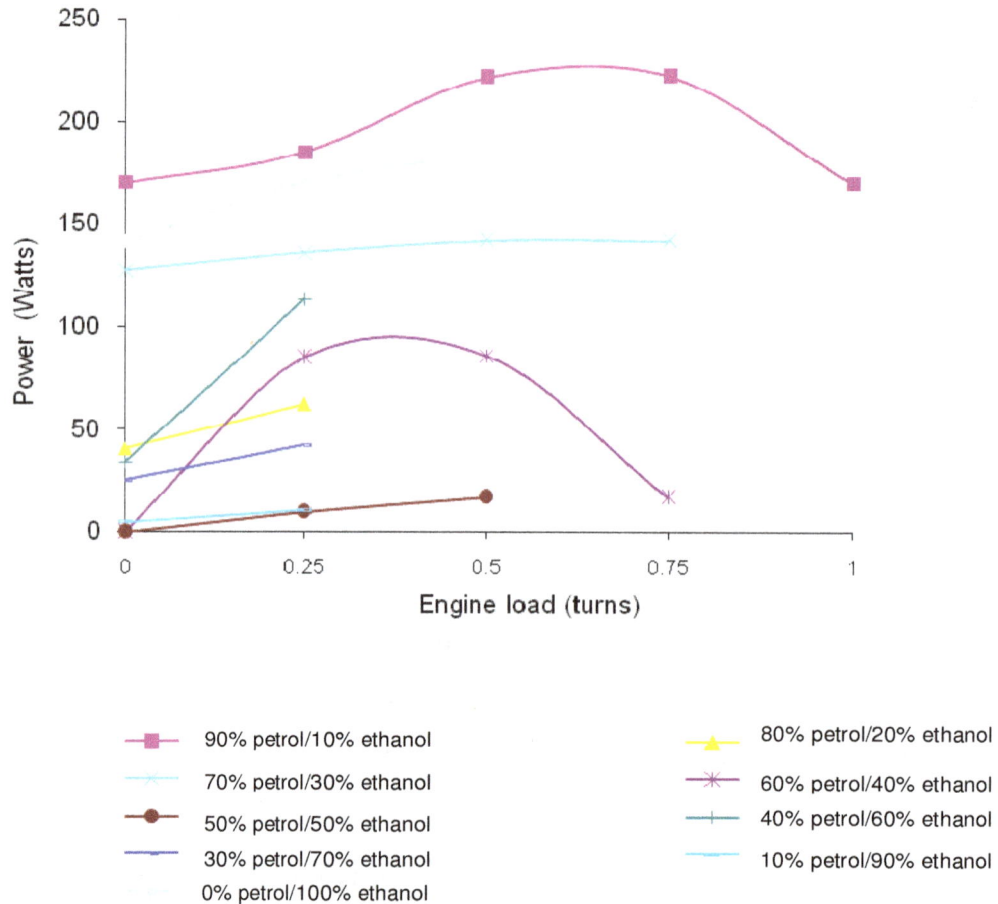

Figure 5. Variation of engine power with various loads for different qualities of ethanol gasoline blended fuel.

system is not airtight, more moisture should be expected in the system due to the activities of hygroscopic ethanol. Earlier researchers Gibson (1949) and Fubing et al. (2007), proposed the use of reformed ethanol in the engine to avoid some of these problems.

The maximum speed of the engine was about 2400 rpm for all mixtures including E0. Any attempt to load the engine using the load control valve resulted in turbulent combustion. This turbulence was due to the immiscibility of the two fractions of the fuels. Figure 5 shows the behaviour of the engine with different ethanol/gasoline ratios as fuel. We note that the engine develops the fastest speed with a fuel composition of 100% gasoline and 0% ethanol. The maximum engine power using the compound fuel was obtained with E10. E20 also gives some good results but the maximum is E30 after which the engine performance is no longer satisfactory. With E10, the engine does not tolerate any loading. However E10 gave the best engine increasing power without vibrations. As seen on Figure 3 the curve for E10 is smooth tolerating a wide range of load variations. E10 has also been recommended by Ceviz and Yüksel

(2005). The maximum loading obtainable is ¼ turns. The engine develops the slowest speed when using fuel of composition, 30% gasoline and 70% ethanol.

Necessary engine modifications

From the results obtained in this investigation some recommendations can be made for modification of the standard SI gasoline engine to run successfully on gasoline/bio ethanol blends. These include advancing the engine timing, and increasing the compression ratio. Bio ethanol is lower energy fuel than gasoline. Gasoline has a negligible solubility in water while the blended fraction is very soluble. Phase separation occurs in the tank leading to lower octane component in the petroleum phase. It is therefore necessary to improve on pre and post induction mixing to neutralise the phenomenon of phase separation. This can be achieved amongst other possibilities by roughening the piston head to improve on atomisation.

Because bio ethanol is a very good solvent extra filtration is necessary between the tank and the engine to remove rust, and other dirt. A 10 micron filter can be good for such filtration. Higher percentages of ethanol may damage seals diaphragm, aluminium, and zinc plated components.

Conclusions

Bio-ethanol production from the chosen crops was feasible. Bio-ethanol production is highest after three days of fermentation using natural methods. The highest bio-ethanol production came from maize although the values got were lower than those obtained in the United States. Addition of ethanol to gasoline reduces the energy density of the volume. Engine speed increases and engine torque decreases. However overall break power increases. There is loss of fuel economy and consequently mileage with blended rations of bio-ethanol gasoline because the alcohol has a lower energy value than gasoline. Blended fuel is not yet a good option for less developed countries that are not manufacturing resistant parts. Anti-fuel properties of bio-ethanol blended fuels calls for serious engine modifications to avoid knock and unwanted emissions. These modifications include the advancing of engine timing, provision of airtight conduit lines, increasing the compression ratio and roughening of the piston head.

REFERENCES

Al-Hasan M (2003). Effect of ethanol unleaded gasoline blends on engine performance and exhaust emissions. Energy Conversion Manage., 44: 1547-1561.

Amulya KNR, Robert HW, Thomas BJ (1997). L'Energie après Rio. Perspectives et défis. Synthèse d'une étude thématique. Programme des Nations Unies pour le développement. Energy after Rio. Perspectives and challenges. Synthesis of a study. United Nations Development Program.

Ajav EA, Akingbehin OA (2002). A study of some fuel properties of ethanol blended with diesel. Agric. Eng. Int. CIGR J. Sci. Res. Dev. Manuscript EE 01 003 Vol. (4).

Bolentin JW, Wilk RD (1996). Auto ignition Characteristics of Ethanol. Paper presented at the International Fuels and Lubricants Meeting and exposition May 1996, Dearborn, MI, USA, Session: International Fuels and Lubricants Meeting and Exposition.

Ceviz MA, Yüksel F (2005). Effects of ethanol–unleaded gasoline blends on cyclic variability and emissions in an SI engine. Appl. Therm. Eng., 25(5&6): 917-925.

Davis GW, Heil, Rust ET, Kettering UR, Flint MI (2002). Ethanol vehicle cold start improvement when using a hydrogen supplemented E85 fuel, 1: 303-308.

Fubing Y, Giseng L, Xiaohong G (2007). Proceedings of ISES World Congress 2007. Edited by D. Yogi Goswami and Yuwen Zhao. Solar Energy Hum. Settlement, 1&5: 2418-2421.

Gibson HJ, Factors G, Gary HE, Langston J (1919). Production and Use of Ethanol. Comprehensive Affecting Octane-Number requirement. SAE J. (Abstract) Washington, D.C., USA.

Johansen TJ, Schramm L (2009). Low-Temperature Miscibility of Ethanol-Gasoline-Water Blends in Flex Fuel Applications Energy Sources. Part A: Recovery, Util. Environ. Effects, 31(18): 1556-7230, 1634–1645. DOI: 10.1080/15567030903021897.

Kar K, Tristan L, Haywood C, Raine R (2009). Measurement of Vapor

Pressures and Enthalpies of Vaporization of Gasoline and Ethanol Blends and Their Effects on Mixture Preparation in an SI Engine SAE. Int. J. Fuels Lubricants., 1(1): 132-144.

Kiani D, Kiani M, Ghobadian B, Tavakol T, Nikbakht AM, Najfi G (2010). Application of artificial neural networks for the prediction of performance and exhaust emissions in SI engine using ethanol-gasoline blends. Energy, 35(1): 65-69.

Mustafa K, Yakup S, Tolga T, Hüseyin SY (2009). The effects of ethanol–unleaded gasoline blends on engine performance and exhaust emissions in a spark-ignition engine. Renew. Energy, 34(10): 2101-2106.

Najafi G (2010). Application of artificial neural networks for the prediction of performance and exhaust emissions in SI engine using ethanol-gasoline blends. Energy, 35(1): 65-69. doi: 10.1016/j.energy.2009.08.034.

Najafi G, Ghobadian B, Tavakoli T, Buttsworth DR, Yusaf TF, Faizollahenejed M (2009). Performance and exhaust emissions of a gasoline engine with ethanol gasoline fuels using artificial neural network. Appl. Energy, 86(2009): 630-639. Elsevier Science Publishers.

Persson T, Garcia Y, Garcia A, Paz J, Jones J, Hoogenboom G (2009). Maize ethanol feedstock production and net energy value as affected by climate variability and crop management practices. Agric. Syst., 100(1-3): 11-21.

Persson T, Garcia Y, Garcia A, Paz JO, Ortiz BV, Gerrit (2010). Simulating the production potential and net energy yield of maize-ethanol in the southeastern USA. Eur. J. Agron., (Article in press as corrected proof) doi:10.1016/j.eja.2010.01.004.

Wei-Dong H, Rong-Hong C, Tsung-Lin W, Ta-Hui L (2002). Engine performance and pollutant emission of an SI engine using ethanol–gasoline blended fuels. Atmos. Environ., 36(3): 403-410.

RFA (2001). Renewable Fuels Association. http://www.ethanolrfa.org/exchange/2 consulted January 2001.

Zeinlin'ska KJ, Miecznikowski AH, Stecka KM, Suterska AM, Bartosiak EE (2003). The silage of maize grain with the addition of biopreparations as raw material to production of ethanol. Instytut Biotechnologii Przemyslu Rolno-Spozywczego, Poland.

21

Characterization and evaluation of African bush mango Nut (Dika nut) (*Irvingia gabonensis*) oil biodiesel as alternative fuel for diesel engines

E. I. Bello*, A. O. Fade-Aluko, S. A. Anjorin and T. S. Mogaji

Department of Mechanical Engineering, The Federal University of Technology Akure, Ondo State, Nigeria.

The African bush mangos (*Irvingia gabonensis*) were collected from the wild, dried and the kernels were split to release the nuts. The oil from the nuts was extracted in a soxhlet extractor that was operated at 60°C using normal hexane as solvent. It gave an oil content of 60% which was mainly triglyceride and was solid at room temperature. It was heated to 40°C to liquefy it and a methanol molar ratio of 6 to 1 was used along with 4 g/L of sodium hydroxide as catalyst during transesterification to biodiesel. Chromatography analysis of the biodiesel gave 59% linolenic fatty acid and tests show that the biodiesel has properties that are within the American society for tests and materials limits for biodiesel. Of particular significance is the cloud point of -14°C which allows it to be used in cold conditions and a flash point of 97°C that makes it a safe fuel during storage. The performance characteristic is similar to that of diesel fuel and the specific fuel consumption is 8% higher than diesel fuel which is consistent with the difference in their heating values. The oil will thus be a potential source of alternative fuel for diesel engine.

Key words: African bush mango (*Irvingia gabonensis*), fuel properties, biodiesel, transesterification.

INTRODUCTION

The diesel engine is predominantly used to power trucks, farm tractors, electrical power generating sets and earth moving equipment because of better fuel economy, longer engine life and higher maximum power output than petrol engines. It has however, created proportional increase in pollution from diesel engine that is having adverse effect on the environment hence the search for alternative fuels for diesel engines. The search has been focused on vegetable oils because of the similarity of their molecular structure to diesel fuel and its renewable source from agriculture and animal fats.

However, the relatively high viscosity of vegetable oils, the consequential poor fuel atomization characteristic and the associated fuel injector blockage and cold starting problems, have made vegetable oils unsuitable for such use in neat form thus creating the need for a method of conversion to vegetable oil esters which is commonly known as biodiesel (Mangech, 1999; Neihaus et al.,

1985; Schlick et al., 1988; Peterson et al., 1981; Peterson, 1986; Chowhury, 1942; Pryor, 1982).

Biodiesel has been defined as a mono-alkyl ester of vegetable oils or animal fats (NBB, 1997; Walton, 1938; Knothe et al., 2005; Mittelbach and Remschmidt, 2004). It has very similar physical properties to diesel fuel which allows it to be used as substitute fuel in diesel engines without any modification. Since it is produced from renewable, domestically grown and sometimes under-utilized feed-stocks, it can reduce the demand for petroleum based fuels and possibly lower the overall cost of diesel fuel in the long term. It also contains no or very little sulfur and thus offers promise to reduce particulate and toxic emissions, which is one of the primary objections to diesel fuel.

Africa has a variety of trees and plants that grows in the wild and produce oil yielding seeds. One of such seeds is the African bush mango or Dika nut (*Irvingia gabonensis*), the yielding tree has a conical shape and grows mostly in the tropical rain forest. It can rise to a height 40 m and has a peculiar dense dark green foliage.

---

*Corresponding author. E-mail: eibello2005@yahoo.com.

The mangos are usually ripe for harvest in the month of December and they drop on their own to the ground when ripe except when plucked or disturbed by birds or bats. The African bush mango has great economic value and is traded widely. The fruit is usually collected from the wild, dried and the kernel is split to release the nut. The kernel is usually processed into press cake and added to soup in mashed form to thicken it, confer flavor, aroma and drawability. It contains mainly triglyceride and is thus a potential feedstock for biodiesel production. The aim of this work is to extract the oil, transesterify, characterize and test it as possible alternative fuel for use in diesel engines.

MATERIALS AND METHODS

The nuts were harvested from the wild, dried in the sun and the kernel remove by splitting the nut into two halves using a nut splitting tool. The kernels were dried in the sun to reduce moisture content. The oil was leached out using normal hexane solvent in a soxhlet extractor operated at 60 ℃. The oil extracted turned to solid at room temperature. The normal hexane and sodium hydroxide used were to analytical grade and were purchased from Finlab company in Akure, Nigeria. The experiments were carries out in the chemistry and engine laboratories of the Federal University of Technology, Akure, Nigeria.

Transesterification process

The extracted oil which was solid at room temperature was heated to 120 ℃ for 1 h to remove any water content and then allowed to cool to 60 ℃. 4 g of sodium hydroxide was added to 1 L of methanol in a mixer and stirred at 500 rpm until it is completely dissolved to form sodium methoxide. The sodium methoxide was next mixed with the neat oil at different molar ratios in the reactor and stirred at 1000 rpm (Doranto et al., 2002) for 3 h at a constant temperature of 60 ℃. The mixture was then allowed to settle for 8 h to drive the reaction to completion and allow the mixture to separate into two layers of biodiesel and glycerol which is denser, at the bottom. Finally, the mixture was separated using a separating funnel.

Optimum condition for biodiesel yield

The amount of catalyst and molar ratio for optimum biodiesel yield by transesterification were not available in the literature and hence determined by an iterative experimental design. The catalyst amount was varied from 2.5 to 5 g/L in a step of 0.5 g. For each value, molar ratios of 1:1, 4:1, 6:1 were used which gave 18 test samples that were transesterified and the proportion of biodiesel in each sample was determined.

Purification process

The biodiesel was washed with distilled water to remove impurities such as diglycerine, monoglycerine, catalyst, soap and excess methanol. The washing was done by mixing with 20 vol. (%) distilled water and stirred gently for 30 min. It was allowed to settle and gave a two phase mixture from which the biodiesel was separated. The procedure was repeated three times (Doranto et al., 2002) to obtain a clear biodiesel. It was finally heated to 110 ℃ to remove any water vapour still present.

Measurement of properties and fatty acid profile

The main properties of the oil and its biodiesel were measured using mainly the American society for testing and materials (ASTM) protocols for biodiesel fuels. The weight proportions of the composition of the saturated and unsaturated fatty acids in the oil were analyzed by a gas chromatography (GC) analyzer (GC-17A model Shimadzu inc., Japan) using HP 5/Bp 5-capillary column 30 m long, 0.25 mm diameter, 0.25 mm film and a spit/splitless injection port. The injector was maintained at 230 ℃ and the detector temperature at 240 ℃. Nitrogen was the carrier gas at a flow rate of 45.0 ml/min and pressure of 2.0 psi. The oven temperature was set at 40 ℃, held for 1 min, programmed to 120 ℃ at 20 ℃/min for 25 min and then to 230 ℃ at 35 ℃/ min for 35 min. The total run time was 68.14 min. The fatty acid composition was expressed as the peak area of each acid divided by the total peak area of all the acids.

Engine testing

The test engine is a Lister 8/1 VA low speed single cylinder diesel engine that is connected to an electrical dynamometer.
The specification of the engine is:

Nominal power output: 6 kW @ 850 rev/min and 4.5 kW @ 650 rev /min.
Bore: 114.3 mm
Strok: 139.7 mm
Swept volume: 1.433 L.

The field is excited separately by current from the mains and adjusted by a rheostat. The engine is loaded by switching in the 8 electrical elements that are capable of absorbing the full output of the dynamometer and are selected by means of a selector switch. The generated current is fed to the load resistance in the generator circuit. The engine was operated at full throttle opening and the load was increased gradually by switching on the load elements one at a time until the speed was reduced to the minimum value at which the engine will just run smoothly. The plint fuel guage used to measure fuel consumption consists of a glass tube containing four knife-edged spacers, which are positioned to accurately calibrate the volume of fuel between them. The spring balance reading, the engine speed, revolution counter, quantity of fuel consumed and time were all recorded during each test run and used to compute the engine performance parameters. The test was done for B100 and diesel fuel for comparison.

RESULTS AND DISCUSSION

The oil extracted was solid at room temperature. The optimum catalyst amount and molar ratio required were determined by iterative experimental design and the results are shown in Figure 1. The best molar ratio from the economic point of view is 4 to 1.
The free fatty acid composition obtained from chromatography analysis is as shown in Table 1 and consists essentially of 59% linolenic fatty acid which is a polyunsaturated acid with three bonds. It has a total unsaturation of 95.28% which is similar to those of castor and safflower oils biodiesel and higher than those of jatropha and soya bean oils.
In view of the fact that the oil was solid at room temperature and had to be preheated to liquefy it and the

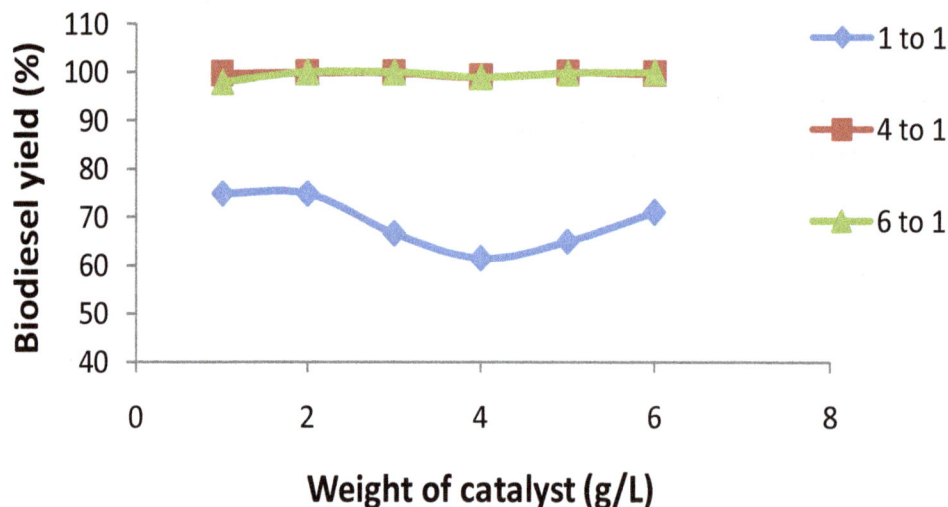

Figure 1. Variation of Dika oil biodiesel yield with catalyst amount and molar ratio.

Table 1. Composition of African bush mango nut and other oil biodiesels.

Fatty acid	Class	African bush mango	Jatropha[a]	Soyabean[b]	Safflower[b]	Castor oil[c]
Dihydroxystearic acid	C18:1	33.9	-	-	-	0.7
Eicosanoic acid	C18:0	0.31	-	-	-	0.3
Linolenic acid	C18:3	59.0	-	8.3	0.1	0.3
Linoleic acid	C18:2	2.2	48.18	54.1	75.5	4.2
Myristic	C14:0	-	-	0.1	-	-
Oleic acid	C18:1	-	28.46	22.5	14.4	3.0
Palmtic acid	C16:0	0.18	18.22	10.3	-	1.0
Stearic acid	C18:0	4.2	5.14	4.7	3.3	1.0
Ricinoleic acid	C18:1	-	-	-	-	89.5
Total saturated	-	4.50	23.36	15.10	10.00	2.00
Total unsaturated	-	95.28	76.64	84.90	90.00	98.00

a: (El-Diwani et al., 2009); b: (Allen et al., 1999); c: (Conceicao et al., 2007).

high unsaturation, a molar ratio of 6 to 1 and catalyst amount of 4 g/L of oil were used and it gave a biodiesel yield of 98%.

The properties of the oil, biodiesel and diesel fuel were measured and the results are shown in Table 2. The crude oil has a relative density of 0.93 at 60 °C and reduced to 0.91 after transesterification that is still higher than 0.90 maximum for biodiesel but can be reduced further by blending with diesel fuel.

The cloud point is comparable to diesel fuel while the pour point of -6 °C is higher than the -20 °C for diesel fuel which limits its cold temperature and aviation applications. With a flash point of 125 °C, the biodiesel can be classified as a non hazardous fuel. Safe for storage and handling.

The kinematic viscosity reduced from a high 250 mm^2/s at 40 °C for the crude oil to 5.5 mm^2/s for the biodiesel

which is within the ASTM limits for biodiesel. It can be seen from the chemical composition of the oil in Table 1 that it contains 59% linolenic fatty acid which has a molecular weight of 278.

Figure 2 shows the variation of torque developed with engine speeds. Although the torque developed is less than that when using diesel fuel, it has flat top like diesel fuel and most suitable for an engine working between 1200 and 1800 rpm. The torque spectrum is wider and is capable of keeping the engine running to a much lower speed. This means that the engine can operate at a much lower speed which is a good characteristic for an engine working in hilly market. The reason for the wide spectrum is the more complete combustion due, largely, to the oxygen content of the fuel.

Figure 3 shows the specific fuel consumption over a range of speeds. The specific fuel consumption for dika

Table 2. Properties of African bush mango nut, its biodiesel and diesel fuel.

Property	ASTM code	ASTM limits	African bush mango nut oil	African bush mango nut oil biodiesel	Diesel fuel
Relative density	D1298	-	0.93	0.91	0.85
Pour point (℃)	D2500	-	28	-6	-20
Cloud point (℃)	D2500	-	23	-14	-12
Flash point (℃)	D93	130 min	300	140	55
Kinematic viscosity (mm^2/s at 60℃)	D445	1.99 to 6	45	3.2	2.8
Heating value (MJ/kg)	D240	-	28	39	42.7
Carbon residue (%)	D4530	0.05	0.10	0.05	0.05
Acid number (mg KOH/g oil)	-	0.80 max	1.2	.01	-

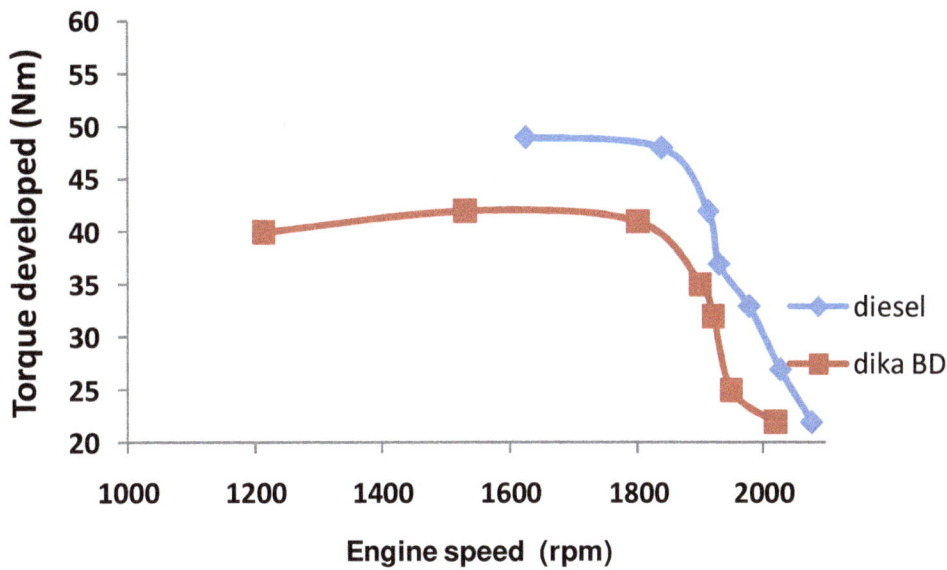

Figure 2. Variation of torque with engine speed of African bush mango nut oil biodiesel.

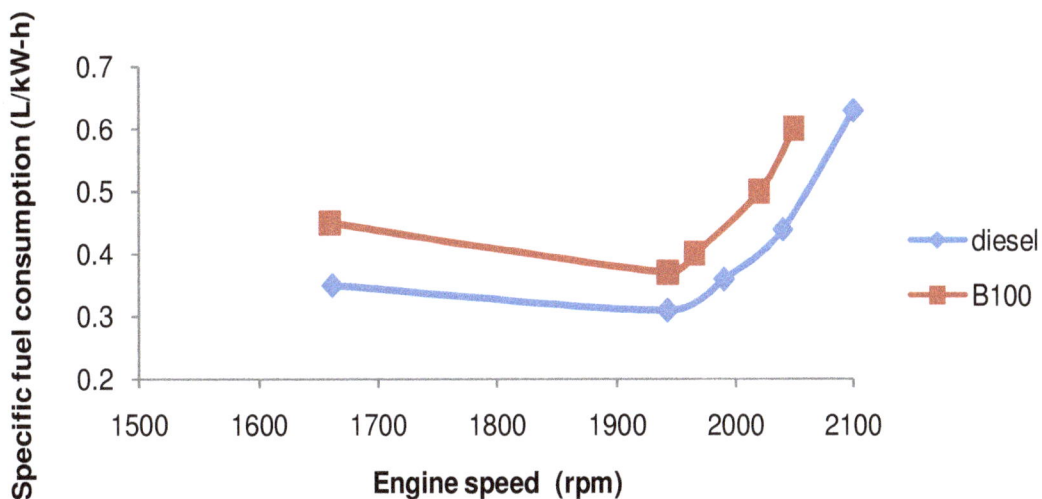

Figure 3. Variation of specific fuel consumption with engine speed of African bush mango nut oil biodiesel.

oil biodiesel is about 8% lower than that of diesel fuel and consistent with the difference in their heating values. The minimum specific fuel consumption occurred at 1950 rpm

Conclusion

African mango nut oil biodiesel has similar properties to diesel fuel and superior cold flow properties.

The flash point is also much higher than that of diesel fuel, which makes it a suitable alternative fuel for diesel engines. Dika oil is solid at ambient temperature of 25°C. Its wild source, high oil content and high conversion rate would make it a competitive fuel to diesel and other biodiesel fuels.

REFERENCES

Allen CAW, Watts KC, Ackman RG, Pegg MJ (1999). Predicting the viscosity of biodiesel fuels from their fatty acid ester composition. Fuel, 78: 1319-1326.

ASTM. American Society of Testing and Materials. ASTM D6751-02 Requirements for biodiesel ASTM, Easton, Maryland, USA.

Conceicao MM, Candeia RA, Silva FC, Bezerra AF, Fernandes VJ, Souza AG (2007). Thermochemical Characterization of Castor Oil Biodiesel Renewable and Sustainable Energy Reviews, 11(5): 964-975. Chowhury DH (1942). Indian Vegetable Fuel Oils for Diesel Engines. Gas and Oil Power, 1(37): 80.

Dorado MP, Arnal JM, Gomex J, Gill A, Lopez FJ (2002). The Effects of a Waste Vegetable Oil Blend with Diesel Fuel on Engine Performance. Transactions of ASAE, 45(3): 519-523.

El-Diwani G, Attia NK, Hawash SI (2009). Development and evaluation of biodiesel fuel and by-products from jatropha oil. Int. J. Environ. Sci. Technol., 6: 219-224.

Mangech AN (1999). Biomass Energy Development, Vegetable Oils as Fuels. Energy Research Group. Wiley Eastern Publications.

Mittelbach MP, Remschmidt C (2004). Biodiesel Comprehensive Handbook, Published by M. Mittelbach, Karl-Franzens-Universitst Graz, Graz, Austria.

(NBB) National Biodiesel Board (1997). Jefferson City, MO. USA.

Niehaus RA, Goering CE, Savage LD, Sorenson SC (1985). Cracked Soybean Oil as a Fuel for a Diesel Engine. ASAE Paper No. 85-1560. ASAE, St. Joseph, MI.

Pryor RW, Hanna MA, Schinstock JL, Bashford LL (1982). Soybean Oil Fuel in a Small Diesel Engine. Trans. ASAE, 26(2): 333-337, 342.

Peterson TL, Auld DL, Thomas VM, Withers RV, Smith SM, Bettis BL (1981). Vegetable Oil as an Agricultural Fuel for the Pacific Northwest. University of Idaho Agricultural Experiment Station Bulletin No. 598.

Peterson CL (1986). Vegetable Oils as a Diesel Fuel: Status and Research Priorities. Trans. ASAE, 29(5): 1413-1421.

Schlick ML, Hanna MA, Stinstock JK (1988). Soyabean and Sunflower Performance in a Diesel Engine. Trans. ASAE, 31(5): 1345-1349.

Walton J (1938). The Fuel Possibilities of vegetable Oil. Gas and Oil Power, 33: 67-168.

Knothe G, Dun RO, Bagby MO (1997). Biodiesel: The Use of Vegetable Oils and Their Derivatives as Alternative Diesel Fuels, in ACS Symp. Ser. 666 (Fuels and Chemicals from Biomass), American Chemical Society, Washington, DC, pp. 172-208.

Knothe GL, Van Gerpen J, Krahl J (2005). The Biodiesel Handbook, AOCS Press, Champaign. Illinois. USA.

22

Properties of fossil diesel oil blends with ethanol and FAME as fuels for diesel engines

Andrea Kleinová* and Ján Cvengroš

Faculty of Chemical and Food Technology, Slovak University of Technology (STU), Bratislava, Slovak Republic.

Ethanol is an attractive oxygenate not only for gasolines but also for diesel fuels. The paper presents the results of the study of stability of blends fossil diesel oil-ethanol, at the temperature of -10°C. Low-temperature phase stability was significantly influenced by the presence of water in the blend. The stability of blend was increased by the addition of FAME. The ternary mixture fossil diesel oil – FAME – ethanol was stable at -10°C, if the ratio FAME: ethanol (0.70% w/w of water) was higher than 2. Selected ternary blends with the content of bio-components 7 and 10% v/v were used in performance and emission tests with two types of testing engines. The presence of oxygenates had no negative influence on performance parameters of the engines and they positively influenced the emissions, especially opacity. Except of the flash point, the fuel parameters of blended fuels were not impaired and excellent lubricity was achieved.

Key words: Ethanol – diesel oil blends, diesel oil – FAME – ethanol blends, oxygenates, performance and emission tests, opacity.

INTRODUCTION

Interest in bioethanol (EtOH) as attractive extender of EtOH and fossil diesel oil blends utilizable as a fuel for diesel engines started in the eighties during the petroleum crisis. It was shown that EtOH as typical component of liquid fuels for ignition engines can be used also in a blend with fossil diesel oil (FDO) in standard diesel engines. The utilization of EtOH as a fuel component for diesel engines is discussed in detail in comprehensive study (Hansen et al., 2005). Due to high price of EtOH, this type of fuel was considered as an alternative only for crises periods. Recently, the economic situation has changed in favor of EtOH, which is actually able to compete with the standard diesel oil. Despite the fact that after EtOH addition to the FDO the characteristics of blended fuel are generally less

favorable compared to FDO alone, there exists serious economic pressure to use EtOH also as a component of diesel fuel. This applies especially for the European region with preferential interest in diesel fuels and lesser interest in gasoline fuels. EtOH is attractive as a fuel from renewable sources as well as oxygenate with positive influence on emissions, especially on particular matters.

Suitability of particular fuel for diesel engines is influenced by several aspects. Addition of EtOH to FDO influences properties of the resulting fuel blend as are the phase stability, viscosity, lubricity, power content and cetane number (CN). Corrosiveness and safety aspects characterized by flash point and flammability are also important.

Fuel viscosity and its lubricity play important role in lubrication of high-pressure fuel injection pump. Addition of EtOH to the FDO decreases viscosity (Barabas et al., 2010) and lubricity (Lapuerta et al., 2010a) of fuel blend. The blend of 10% of dry EtOH in FDO has viscosity approaching the minimum value requested by the standard (Wrage and Goering, 1980). Lubricity tests in the system FDO – EtOH using HFRR method according to EN 12156-1 are scarce (Lapuerta et al., 2010a, b). Reported higher corrosiveness of blended fuel FDO – EtOH should be connected with highly hygroscopic

---

*Corresponding author. E-mail: andrea.kleinova@stuba.sk.

Abbreviation: BSFC, Brake specific fuel consumption; **CFPP,** cold filter plugging point; **CN,** cetane number; **CO,** carbon monoxide; **EtOH,** ethanol; **FAME,** fatty acid methyl esters; **FDO,** fossil diesel oil; **HC,** hydrocarbons; **NOx,** nitrogen oxides; **PM,** particular matter.

behavior of EtOH.

Power content of the FDO – EtOH blend decreases by 3% for each 5% v/v of added EtOH and does not represent a key problem (Wrage and Goering, 1980). Decreased power output can be caused also by pump leakage due to lower viscosity of the fuel. Typical CN of FDO ranges from 45 to 50. The CN of EtOH is estimated to be in the range between 5 and 15. It is therefore recommended to apply CN improvers so that CN of fuel blend reaches the prescribed range. Possible additive is triethylene glycol dinitrate (Meiring et al., 1983).

Low flash point of the blended fuel FDO–EtOH represents a serious problem. Blends of EtOH and FDO have the flammability practically identical to that of EtOH regardless of the EtOH content. Flash point of 12°C classifies FDO – EtOH blends as flammable liquids of Class I, while FDO itself belongs to Class III (flash point over 56°C). This applies more stringent requirements on storage and increased safety distances from storage tanks.

During emission tests, substantial decrease of the content of particular matter (PM) by 20 to 27% was monitored at 10% of EtOH addition to the FDO (Spreen, 1999). Other emission components such as CO, HC, NOx were reported by some authors to achieve substantial decrease (Kass et al., 2001), while other authors reported the values ranging in both directions comparing to standard FDO (Schauss et al., 2000; Rakopoulos et al., 2008), depending on speed and load of the engine but still well below the prescribed emissions limit.

EtOH solubility in FDO depends on temperature and water content in the blend (Fernando and Hanna, 2004). Dry EtOH with water content less than 0.20% w/w is at higher temperatures completely miscible with FDO. At temperatures below 10°C, there is a separation of components and formation of two phases at a rate dependent on water content in the system (Lapuerta et al., 2007). Separation can be avoided by addition of an emulsifying agent to keep the emulsion of EtOH in FDO stable or by addition of a co-solvent as a bridging agent for homogeneous blending (Hansen et al., 2005). Approximately, 2% of a surfactant was required for every 5% of aqueous EtOH (5% of water) added to FDO to keep the micro-emulsion stable (Moses et al., 1980). Tetrahydrofurane or ethyl acetate was found as effective co-solvents to prevent phase separation (Lechter, 1983). The consumption of co-solvent increases with increasing water content and decreasing temperature. At 0°C, full miscibility ratio for ethyl acetate to dry EtOH was 1:2. Presence of aromatics in FDO increases EtOH solubility as well as the effects of co-solvents and emulsifying agents (Gerdes and Suppes, 2001).

Fatty acids methyl esters (FAME) can also be used as possible co-solvents preventing separation of FDO – EtOH blends (McCormick and Parish, 2001). The combination of these two oxygenates may help to balance their less favorable properties. EtOH may be expected to improve the low temperature filterability, while high CN of FAME can compensate its decrease in FDO – EtOH blend. It has been proved that engine performance with such fuel blends did not differ significantly from engine performance with FDO (Ali et al., 1995a, b). The blends containing 10, 20 and 30% w/w of EtOH and 5, 10 and 15% w/w FAME, respectively, were studied by Chen et al. (2008) without providing any information on their phase stability. Significant reduction of smoke and PM was recorded but the used high percentage of oxygenates are hardly applicable in praxis from the economical and capacity reasons. In other study, a blend of 20% v/v of EtOH in FAME was added to FDO as an oxygenated additive at the levels of 15 and 20% v/v, respectively (Shi, 2005). PM, CO and HC decreased with increasing oxygenate content but NOx emissions increased.

In the current study, we dealt with the impact of EtOH addition to the FDO with the aim to find real and practically applicable proportion with suitable fuel properties. The influence of FAME as a co-solvent for increasing the phase stability at low temperatures and also as a lubricity enhancer is presented. The role of the influence of water content in binary and ternary blends of FDO, EtOH and FAME on the phase stability is also searched. We discuss the results from low-temperature phase stability tests as well as some performance and emissions characteristics of FDO – FAME – EtOH blended fuels.

EXPERIMENTAL

Materials

FAME was prepared by standard two-step alkali transesterification (Cvengroš and Považanec, 1996) of refined rapeseed oil with the acid value of 0.2 mg KOH/g. Crude FAME were finally treated by washing (1% w/w of water), centrifugation and filtration. EtOH with water content 0.70% w/w and 0.20% w/w containing denaturizing agents were obtained from Slovnaft VURUP, Bratislava, Slovak Republic. Winter, additive- and sulphur-free diesel oil was delivered by the company Slovnaft VÚRUP Bratislava, Slovak Republic. Low-temperature properties of the used FDO were modified by adding commercial additives, detergent (200 ppm) and depressant (300 ppm). Parameters of the used diesel oil without additives are summarized in Table 1. Alkanolamide-based additive for adjustment of anticorrosion properties LUBOL 210D, exhibiting also detergent effect and ability to bind water was supplied by Lubocons Chemicals, Stupava, Slovak Republic.

Preparation of fuel blends

FDO, EtOH and FAME, respectively, were simply blended to form mixture with defined ratio of individual components. EtOH content ranged from 2 to 6% w/w and FAME content from 6 to 10% w/w. Two sets of measurements were performed utilizing 1% w/w of LUBOL 210D.

Table 1. Characteristics of fossil diesel oil (FDO).

Characteristic	Unit	Value	Method
Density at 15 °C	kg m^{-3}	832.6	EN 12185
Viscosity at 40 °C	mm^2 s^{-1}	2.376	EN 3104
Cetane index		50.7	EN 4264
Cetane number		50.6	EN 5165
Water content	mg kg^{-1}	26.2	EN 12937
Flash point	°C	70	EN 2719
CFPP	°C	-30	EN 116
Lubricity	µm	605	EN 12156-1

CFPP – Cold Filter Plugging Point.

Table 2. Characteristics of used engines.

Type of engine	2.5 UI	MD UR IV
Cylinder number	5	4
Bore (mm)×Stroke (mm)	81×95.5	110×128
Volume (L)	2.5	4.8
Compression ratio	19.5:1	18.8:1
Maximal power output (kW/rev)	128/3500	80.8/2200
Maximal torque (N m/rev)	400/2000	407/1480
Injection pressure (bar)	2050	900

Laboratory tests

The model blends of FDO containing EtOH (0.7% w/w of water) and FAME were tested according to the testing methods prescribed by the standard EN 590. For testing the phase stability at low temperatures, the samples were cooled down to -10°C in liquid bath and kept at this temperature for 24 h. After this period of time, their appearance at this temperature was visually evaluated - cloudiness and the presence of new liquid phase, the presence of the new liquid phase being of key importance. This methodology was considered as sufficient for present purposes. In addition, we do not know other options for evaluation of phase stability at low temperatures. The temperature of -10°C was selected with respect to CFPP -10°C. According to EN 590 this CFPP value is actual during three quarters of a year, while -20°C only 3 winter months.

Testing engines

To test blended fuels with bio-component content of 7 and 10% v/v, a passenger car VW Touareg R5 2.5 unit injection (UI) System, engine code AXD, year of production 2007, was used. Basic engine characteristics of the used all-wheel drive car are given in Table 2. The engine is four-stroke inline five cylinder (R5) Turbocharged direct injection (TDI) turbodiesel, gear-driven single overhead camshaft (SOHC), swirl-inducing intake ports, high pressure fuel injection (UI). The engine is liquid cooled and supercharged, with intercooler of compressed air. It has 10 valves total, hydraulic bucket tappets with automatic valve clearance compensation, 5-hole injection nozzle, the hole diameter 0.35 mm.

The measurements were carried out also on a 4-cylindrical engine MD UR IV 8004.000, whose parameters are shown also in Table 2. Diesel engine UR IV is four-stroke, four-cylinder-in-line

OHV, with direct fuel injection. The engine is liquid cooled and supercharged, without intercooler of compressed air. 5-hole injection nozzle, the hole diameter 0.35 mm. The engine is used as a power unit for agricultural and forest tractors, building and ground machinery, electricity generating stations, water pumps and other industrial devices.

Performance and emission tests, fuel consumption measurements

Performance measurements on testing engine 2.5 UI were carried out using the chassis dynamometer MAHA LPS 2000 (MBH Haldenwang/Allgäu, Germany). Emission measurements were performed with an exhaust gases analyzer MAHA MGT 5 by means of the emission determination at steady-state mode during idle running and the constant speeds of 60, 90 or 120 km/h. The analyzer operated on the principle of infrared spectroscopy. The contents of CO, CO_2, and HC were determined based on selective absorption of each gas in the proper range of IR radiation. O2 and NOx measurements were carried out electrochemically using the appropriate sensor. Diesel engine opacity determination was performed by the method of free acceleration with an opacitometer AVL DiSmoke 435. The test is based on the evaluation of the light beam absorption and scattering by exhaust gas in comparison to the standard. Opacity is expressed as 1 to 100% HSU (Hartridge Smoke Unit). Aldehydes were determined by sucking the volume of combustion products measured by gasometer through detection tube, and determined as described in (Potter and Karst, 1996; EPA, 2006). Fuel consumption measurements were done by the weighting of tested fuel samples at constant speed regime of 60, 90 or 120 km/h and time intervals of 60, 40 and 30 s. The obtained results in g/min were then converted to g/kW h on the basis of

Table 3. Phase stability of the FDO – EtOH blends at -10 ℃.

FDO (% w/w)	EtOH (% w/w)		LUBOL 210D (% w/w)	phase stability at -10℃	
	0.7% H_2O	0.2% H_2O		EtOH (0.7% H_2O)	EtOH (0.2% H_2O)
98	2	2	-	Unstable	Stable
96	4	4	-	Unstable	Unstable
94	6	6	-	Unstable	Unstable
97	2	2	1	Unstable	Stable
95	4	4	1	Unstable	Stable

measured power at given constant speeds.

During the measurement of the performance and emission characteristics with the MD UR IV engine, the regulated emissions such as CO, HC were measured continuously using a certificated measuring technique according to EEC requirements. Sampling of gas from the exhaustion pipe of the engine was carried out according to EEC 49 recommendations for regulated and non-regulated emissions. The samples for determination opacity were taken separately with a different probe. The measurements were carried out in a steady state mode after stabilization of the temperature of exhaust gases. The selection of modes was carried out on a basis of the EEC modes, where the most relevant is the mode with the maximum output and the maximum torque. In order to better demonstrate the influence of alternative components of the fuel on the content of aldehydes, the mode of maximum engine revolutions at half engine load was selected. The engine is of older conception and according to the former regulations it was adjusted to opacity value max. 60% HSU (Hartrige Smoke Units). The engine was decelerated with the use of a dynamometer MEZ Vsetín DS 1002-4/V.

The results of individual tests are summarized in the corresponding Tables. The values of the parameters given in the Tables represent the average of five measured values. Other determinations of respective parameters were carried out by standardized procedures.

RESULTS AND DISCUSSION

Phase stability at low temperatures

Phase stability of the FDO – EtOH mixture at -10℃

EtOH with water content 0.2 and 0.7% w/w was used in phase stability tests in a two-component system FDO – EtOH. The criterion for stability evaluation was the presence or absence of cloudiness and formation of a new liquid phase. The cloudiness may be related to achievement of the cloud point and does not necessarily represent phase instability. After temperature rising, cloudy sample clears out without any consequences. The decisive indication of the phase instability is then the formation of a new liquid phase. The new liquid phase has lower density and is placed over the non-polar phase. Similar long-term visual assessment of phase stability was performed by Jimenez et al. (2010) in the temperature range of -18 to 30℃. This methodology was considered sufficient for these purposes. Lapuerta et al.

(2010a) studied the stability of alcohols – FDO blends in the temperature range from -10 to 40℃ using optical equipment Turbiscan, but without any detailed information about the measurement methodology.

The results of the tests are summarized in Table 3. It is shown that in the system FDO – EtOH with the water content 0.70% w/w there exists no suitable FDO – EtOH ratio, which would exhibit long term phase stability at -10℃. Close to this goal comes the system containing also 1% w/w of the LUBOL 210D additive at the FDO – EtOH ratio 97:2. If EtOH with lower water content 0.20% w/w is used, the phase stability of the two-component blend will be increased. The FDO blend with the addition of 2% w/w of such EtOH is stable at the studied temperature of -10℃. The system containing 4% w/w of EtOH with the water content 0.20% w/w is already phase unstable. The effect of the LUBOL 210D is positive; it contributes to phase stability in the case of the blend FDO – EtOH – LUBOL 95:4:1 (water content in EtOH 0.20% w/w).

Phase stability of FDO – FAME – EtOH blend at -10℃

In the study of phase stability of the system FDO – FAME – EtOH, with FAME as a co-solvent, EtOH with higher water content (0.70% w/w) was used, as at lower EtOH contents (0.20% w/w of water) the systems were stable without co-solvent. The results are summarized in the Table 4, and show that the FAME addition increases the stability of blended fuels also in the case when increased amount of water is present in the blends. The mixtures with the EtOH portion 2, 4 and 6% w/w and with the FAME portion 6, 8 and 10% w/w were used for the measurements. The systems are stable at the EtOH content up to 4% w/w and simultaneously at the FAME: EtOH ratio higher than 2. If the ratio is less than 2, the phase stability at -10℃ is impaired.

Physico-chemical properties of blended fuels

The attention has been paid to selected properties of blended fuels, where there the strongest influence of the

Table 4. Some physico-chemical properties of FDO – FAME – EtOH blends, water content in EtOH 0.7% w/w.

FDO (% w/w)	FAME (% w/w)	EtOH (% w/w)	CFPP (°C)	nu ((40°C) (mm²s⁻¹)	ρ (15°C) (kg m⁻³)	Phase stability at -10°C	Lubricity (μm)
92	6	2	-27	2.319	832.9	Stable	-
90	8	2	-26	2.349	833.9	Stable	-
88	10	2	-24.5	2.365	834.6	Stable	-
90	6	4	-28.5	2.237	831.9	Unstable	-
88	8	4	-27	2.256	832.7	Stable	-
86	10	4	-25.5	2.303	833.6	Stable	-
88	6	6	-25	2.170	830.2	Stable	229
86	8	6	-23.5	2.218	831.6	Stable	-
84	10	6	-23	2.267	832.3	Unstable	245
94	6	-	-11.3	2.472	834.6	Stable	-
100	-	-	-21.8	2.391	832	Stable	605

presence of oxygenates could be presumed. Some of these properties are shown in Table 4. Density of blends increases with the content of FAME, and decreases linearly with the content of EtOH. Identical trends are observed also for viscosities of studied blends. The densities and viscosities of the blends FDO – FAME – EtOH with relatively low content of biocomponents meet the requirements of the EN 590 with the density in the range 820 to 845 kg m⁻³ at 15°C and viscosity between 2.0 and 4.5 mm² s⁻¹ at 40°C. The presence of FAME alone in the FDO – FAME blend results in CFPP increase, however in ternary mixtures FDO – FAME – EtOH the CFPP is changed negligibly in comparison to FDO. However, the information about CFPP is misleading due to phase instability at low temperature and separation of the system into two liquid phases. The presence of two phases does not, in principle, influence the filterability at low temperature, but the CFPP value is therefore illusory. Despite poor lubricity of input FDO, the lubricity of blended fuels with EtOH and FAME is excellent also at high EtOH proportion provided the FDO – FAME – EtOH phase stability. FAME act as an efficient lubricant.

Engine performance tests

When selecting modes of testing engines, the considerations were as follows; when measured with a certain fuel, the emissions at maximum performance were set. In the next set of measurements, emissions at maximum torque were determined. In the last part of the emissions measurements, emissions at half load and maximum engine speed were stated. By these conditions, the increase production of aldehydes is expected (Henein, 1973, 1973) which is related to lowering the combustion temperature at lower loads, at higher speeds and shorter time of the combustion process. The aim of these measurements was to show that the tested materials have useful properties comparable to that of fossil fuel.

Table 5 summarizes the results of engine performance tests, acquired with prepared blended fuels with the use of the testing engine MD UR IV. As expected, the results in Table 5 show the highest value for maximum torque and maximum power output in the case of FDO, which applies for all tested regimes. Brake specific fuel consumption (BSFC) shows identical trend. This is related to lower energetic content of oxygenate and roughly corresponds to their proportion in the blend. The differences in output parameters are therefore not large and do not exceed 5%.

The same set of blended fuels was tested also with another testing engine 2.5 UI. The results are shown in Table 6. Maximum values for torque and output belong naturally also here to FDO as the fuel with the highest energy content, but the differences are even smaller than in the case of the testing engine MD UR IV. Specific consumption is lowest for FDO at all performances at given speeds. At the 2.5 UI testing engine acceleration was also evaluated, its values are shown in Table 7. The changes of

Table 5. Basic parameters of the MD UR IV engine. Regulated emissions at maximum power output, maximum torque and half load at maximum engine revolutions.

Fuel	Rev. (min^{-1})	Max.torque (N m)	Max. power (kW)	BSFC (g/kWh)	CO (vol. %)	NO$_X$ (ppm)	CH$_x$ (FID) (ppm)	Opacity HSU (%)
A	2200	282	64.9	273.0	0.134	441.8	18	49.9
	1640	392	67.3	244.3	0.246	500.0	20	43.9
	2233	222	51.9	292.6	0.112	298.4	18	33.1
B	2210	279	64.6	274.4	0.160	445.8	17	46.0
	1600	386	64.6	249.3	0.192	601.8	16	41.4
	2223	223	45.2	289.0	0.060	313.0	14	20.9
C	2210	280	64.8	278.7	0.178	461.0	17	38.7
	1660	376	65.3	252.2	0.236	652.0	24	33.2
	2224	194	45.1	287.6	0.070	380.4	19	18.2

A – Fossil diesel oil. B – 93% FDO + 2% EtOH + 5% FAME. C – 90% FDO + 3% EtOH + 7% FAME. Water content in EtOH 0.7% w/w.

Table 6. Performance characteristics, break specific fuel consumption and opacity measured on 2.5 UI engine.

Fuel	Max. output (kW)	Max. torque (N m)	BSFC (g/kWh) 3.3 kW	7.9 kw	19.8 kW	Opacity HSU (%)
A	127	476	705	524	361	17
B	125	466	760	547	385	14.3
C	126	464	767	545	383	12.6

Abbreviations – see Table 5.

Table 7. Acceleration times in seconds of tested fuels measured on 2.5 UI engine.

Fuel	40→80 km/h (2nd gear)	60→100 km/h (3rd gear)	80→120 km/h (4th gear)
A	5.3	10.2	21.7
B	5.4	10.2	21.6
C	5.2	9.9	20.2

Abbreviations – see Table 5.

acceleration are minimal, oxygenate components do not impair the vehicle dynamics.

The presence of oxygenates up to 10 vol. % does not impair the fuel parameters of blends, with the exception of the flash point. Oxygenates

of ester type (FAME) ensure excellent lubricity of the blend. Performance characteristics of testing engines fuelled with blends of oxygenate with FDO are not significantly different from the characteristics for the FDO itself. This finding is

also consistent with observations of Torrez-Jimenéz (2011) where no significant changes were observed at ethanol content up to 15%. Slightly lower power output and higher BSFC of blended fuels are the result of lower energetic

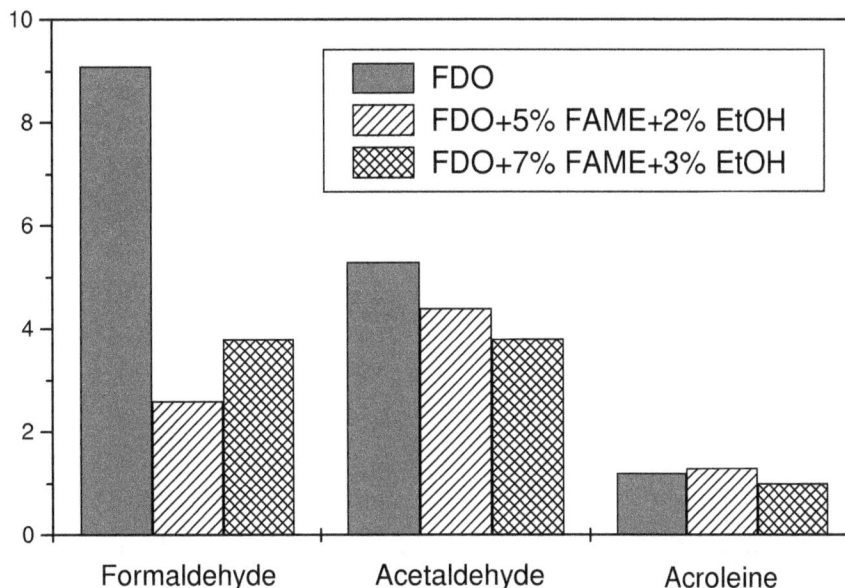

Figure 1. Concentration of formaldehyde, acetaldehyde a acroleine in µg/l in exhaust gases at maximum engine revolutions and at half load of the engine measured on MD UR IV engine. Water content in EtOH 0.7% w/w.

content of these fuels.

Emission tests

Regulated CO, NOx and HC emissions of tested blends measured during the operation of the engine MD UR IV and summarized in Table 5 depend on the engine regime. The values at the level of about one half in comparison to FDO were achieved at maximum engine revolutions and half load for both tested oxygenated fuels. For the engine with direct injection, the regime of maximum torque is strongly influenced by adjustment of fuel injection. In non-supercharged engines the addition of EtOH and FAME can lead to increase of NOx, in supercharged engines to decrease of NOx. The system is very sensitive to excess of air in fuel beam (Henein, 1973).

The opacity of oxygenated blends is lower than that of FDO, which is in accordance with published data. Marked reduction of opacity was observed at maximum engine revolutions at half load for both tested oxygenated blends. Figure 1 shows the concentration of formaldehyde, acetaldehyde and propenale (acroleine) in exhaust gases at maximum engine revolutions and half engine load. The presence of oxygenate has not marked influence on the content of aldehydes in exhaust gases in this working regime, as well as other tested regimes and the contents of non-regulated emissions are relatively low. Aldehydes are formed especially in the engine regimes with high engine revolutions and low loads and imperfect combustion of lean blends at the outer edge of

the fuel beam – quenching zone with short reaction time (high engine revolutions) at low temperatures (low engine load) (Henein, 1973). At the same volume of fuel batch, different chemical composition of the fuel is reflected in the change of the theoretical air consumption. Diesel engine then works with increased excess of air in comparison to FDO, because the alternative oxygen-containing fuels deplete the blend. In case of EtOH the vapor pressure of the fuel increases, this influences the volume of quenching zone, i.e. the zone of aldehyde formation (Henein, 1973). By shortening the ignition delay, the thickness of quenching zone decreases, which decreases the cetane number of the fuel.

Emissions measured during the tests of studied blends with the engine 2.5 UI are shown in Table 8 at steady-state regime during idle running at the constant speeds of 60, 90 or 120 km/h. The CO emissions are at zero level for all fuels at all speed regimes including idling. Zero content of CO in the emissions is associated with the supercharged engine used in the tests. With a high excess of air, the fuel has a chance to completely burn out. The blends of fuels with oxygenate component have usually lower HC emissions, similar tendency is observed for NOx. The comparison of the emission levels from both tested engines is problematic due to different conditions of measurement and also a different conception of the engines. However, the trends are maintained, the blended fuels exhibit comparable emission characteristics for both testing engines. Blended fuels with the content of oxygenate biocomponent exhibit markedly lower opacity in comparison to FDO. Non-regulated emissions (aldehydes) vary at the level of tenths to units of µg/l

Table 8. CO, HC and NOx emissions of tested fuels measured on 2.5 UI engine.

		A	B	C
CO (vol. %)	Idle	0	0	0
	60 km/h (3.3 kW)	0	0	0
	90 km/h (7.9 kW)	0	0	0
	120 km/h (19.8 kW)	0	0	0
HC (ppm)	Idle	0	0.4	1.6
	60 km/h (3.3 kW)	19.6	15.8	13.8
	90 km/h (7.9 kW)	31	25.2	25.8
	120 km/h (19.8 kW)	20	20.4	19.6
NOx (ppm)	Idle	44	33	9.4
	60 km/h (3.3 kW)	87.6	80.8	83.6
	90 km/h (7.9 kW)	294.6	277.6	288
	120 km/h (19.8 kW)	476.2	459.2	462.6

Abbreviations – see Table 5.

without significant influence of the engine regime.

It is difficult to explain the impact parameter changes of the fuel and engine regime on the unregulated emissions, because the mechanism of their formation and conditions in the combustion chambers are not well known (Meyer, 2010).

Conclusions

The performed study showed that EtOH is an efficient oxygenate of diesel fuel blends decreasing the PM in emissions. Low temperature phase stability of two-component blended fuels FDO – EtOH down to -10°C, is strongly dependent on the water content in the blend. FDO and EtOH form phase stable systems at temperatures down to -10°C at the portion of 4% w/w of dry EtOH with water content up to 0.20% w/w. The water content above 0.70% w/w in EtOH does not guarantee the system stability. Alkylamides-based additive (emulgator) at the content up to 1% w/w increases the low-temperature stability and acts as an anticorrosive agent. In ternary blends FDO – FAME – EtOH, the FAME act as a co-solvent and increase the solubility of EtOH in FDO, also at increased water content in the blend. EtOH with 0.70% w/w of water at the 4% w/w portion forms a stable system with FDO at -10°C, if the portion of FAME is at least twice the amount of EtOH. The presence of oxygenates does not impair the fuel parameters of blends, with the exception of the flash point. The presence of oxygenates of ester type (FAME) ensures excellent lubricity of the blend. Performance characteristics of testing engines with oxygenate – FDO blends are not significantly different from the characteristics for the FDO itself. Slightly lower power output and higher BSFC of blended fuels are the result of

lower energetic content of these fuels. Opacity of blended fuels is significantly lower in comparison to standard FDO. Regulated emissions (CO, HC, and NOx) depend on the engine regime. Non-regulated emissions, especially aldehydes are at low level, positive influence of oxygenates on these emissions is visible at higher engine loads.

ACKNOWLEDGEMENT

This work was supported by the Slovak Research and Development Agency under the contract No. APVV-20-037105.

REFERENCES

Ali Y, Eskridge KM, Hanna MA (1995). Testing of alternative diesel fuel from tallow and soybean oil in cummins N14-410 diesel engine. Biores. Technol., 53: 243-254.

Ali Y, Hanna MA, Borge JE (1995). Optimization of diesel, methyl tallowate and ethanol blend for reducing emissions from diesel engine. Biores. Technol., 52: 237-243.

Barabas I, Todoruț A, Bǎldean D (2010). Performance and emission characteristics of an CI engine fueled with diesel-biodiesel-biethanol blends. Fuel 89: 3827-3832.

Cvengroš J, Považanec F (1996). Production and treatment of rapeseed oil methyl esters as alternative fuels for diesel engines. Biores. Technol., 55: 145-152.

EPA Air Resources Board SOP MLD 104. Standard operating procedure for the determination of aldehyde and ketone compounds in automotive source samples by HPLC. Southern Laboratory Branch, Monitoring and Laboratory Division, Apr. 2006, California, USA.

Fernando S, Hanna M (2004). Development of a novel biofuels blend using ethanol-biodiesel-diesel microemulsions: EB-diesel. Energy Fuels, 18: 1695-1703.

Gerdes KR, Suppes GJ (2001). Miscibility of ethanol in diesel fuels. Ind. Eng. Chem. Res., 40: 949-956.

Hansen AC, Zhang Q, Lyne PWL (2005). Ethanol-diesel fuel blends.

Biores. Technol., 96: 277-285.

Henein NA (1973). Diesel Engine Combustion and Emissions. In: Engine emissions: Pollutant formation and measurement. Editors: Springer GS and Patterson DJ, Ann Arbon, Michigan, Plenum Press, New York-London, 6: 211-266.

Chen H, Wang J, Shuai S, Chen W (2008). Study of oxygenated biomass fuel blends on a diesel engine. Fuel, 87: 3462-3468.

Kass MD, Thomas JF, Storey JM, Domongo N, Wade J, Kenreck G (2001). Emissions from a 5.9 liter diesel engine fueled with ethanol diesel blends. SAE Technical paper 2001-01-2018.

Lapuerta M, Garcia R, Campos-Fernández J, Dorado MP (2010). Stability, lubricity, viscosity, and cold-flow properties of alcohol-diesel blends. Energy Fuels, 24: 4497-4502.

Lapuerta M, Garcia-Contreras R Agudelo JR (2010). Lubricity of ethanol-biodiesel-diesel fuel blends. Energy Fuels, 24: 1374-1379.

Lapuerta M, Armas O, García-Contreras R (2007). Stability of diesel-bioethanol blends for use in diesel engines. Fuel, 86: 1351-1357.

Lechter TM (1983). Diesel blends for diesel engines. S. Afr. J. Sci., 79: 4-7.

Mc Cormick RL, Parish R (2001). Technical barriers to the use of ethanol in diesel fuel. NREL/MP-540-32674, Nov. 2001, Colorado, USA.

Meiring P, Hansen AC, Vosloo AP, Lyne PWL (1983). High concentration ethanol-diesel blends for compression-ignition engines. SAE Technical Paper No. 831360.

Meyer TR, Brear MJ, Jin SH, Gord JR (2010). Formation and diagnostics of sprays in combustion: Combustion diagnostics and pollutants. Handbook of combustion, edited by Lackner M and Winter F, Agarwal AK, WILEY-VCH, Chapter 11, Volume 2.

Moses CA, Ryan TW, Likos WE (1980). Experiments with alcohol/diesel fuel blends in compression-ignition engines. VI International symposium of alcohol fuels technology, Guaruja, Brazil.

Potter W, Karst U (1996). Identification of chemical interferences in aldehyde and ketone determination using dual-wavelength detection. Anal. Chem., 68: 3354-3358.

Rakopoulos DC, Rakopoulos CD, Kakaras EC, Giakoumis EG (2008). Effects of ethanol-diesel fuel blends on the performance and exhaust emissions of heavy duty DI diesel engine. Energ. Conv. Manage., 49: 3155-3162.

Shi X, Yu Y, He H, Shuai S, Wang J, Li R (2005). Emission characteristics using methyl soyate-ethanol-diesel fuel blends on a diesel engine. Fuel, 84: 1543-1549.

Schaus JE, McPartlin P, Cole RL, Poola RB, Sekar R (2000). Effect of Ethanol Fuel Additive on Diesel Emissions. Report by Argonne National Laboratory for Illinois Department of Commerce and Community Affairs and US Department of Energy.

Spreen K (1999). Evaluation of Oxygenated Diesel Fuels. Final report for Pure Energy Corporation, Southwest Research Institute, San Antonio, TX.

Torres-Jimenez E, Jerman MS, Gregorc A, Dorado MP, Kegl B (2010). Comparative study of various renewable fuels blends to run a diesel power plant. In: Int. Conf. Renew. Energies and Power Quality (ICREPQ'10), 23-25th March 2010, Granada, Spain.

Torres-Jimenez E, Jerman MS, Gregorc A, Lisec I, Dorado MP (2011). Physical and chemical properties of ethanol-diesel fuel blends. Fuel, 90: 795-802.

Wrage KE, Goering CE (1980). Technical feasibility of diesohol. Trans. ASAE, 23: 1338-1343.

23

Biodiesel production from Jatropha caucus oil in a batch reactor using zinc oxide as catalyst

B. K. Highina[1]*, I. M. Bugaje[1] and B. Umar[2]

[1]Department of Chemical Engineering, University of Maiduguri, Borno State, Nigeria.
[2]Department of Agricultural and Water Resources Engineering, University of Maiduguri, Borno State, Nigeria.

This paper presents the effect of different factors that affect the conversion of Jatropha caucus (JCO) oil to biodiesel and the optimum conditions in a batch reactor using Zinc oxide as catalyst. Four replicate transesterification experimental runs were carried out for each of the mixing duration 25, 50, 75, and 100 min under different typical transesterification reaction conditions of 9.115 g JCO, temperature (37, 57, and 67°C), ZnO concentration (0.25, 0.5, 1.0 and 1.5%) (wt% JCO), methanol to oil ratio (6:1, 8:1, 12:1 and 24:1) at a constant mixing rate. The optimum conditions were found to be 1% ZnO, 18:1 methanol to oil ratio, 67°C temperature and highest yield of biodiesel obtained was 98%.

Key words: Yield, transesterification, Jatropha oil, zinc oxide.

INTRODUCTION

Majority of the world energy needs are supplied through petrochemical sources, coal, and natural gases, with the exception of hydroelectricity and nuclear energy. Of all these sources that are finite, the current usage rates will be consumed shortly. Non-renewable energy sources such as petroleum are related to several drawbacks including; increase green house emission, high cost of processing the crude petrol and energy demand during the process, non-renewable etc. This has provided incentives to seek for alternative sources for petroleum-based fuels.

Alamu et al. (2007) has reported that Nigeria currently imports about 80% of its petroleum requirements and has been hit hard by rapidly increasing cost and uncertainty. Unfortunately, in Niger Delta region, the centre of oil extraction of the country, severe environmental impacts have been ignored in the country's haste to develop the oil industry. This has generated militancy from the local people (Ijaw) making successful oil prospecting a nearly impossible task for the multinational companies in Nigeria. As a result the cost of extracting the reserve will

go on increasing in Nigeria. Thus, there is an urgent need to find alternative renewable forms of energy before mineral oil supplies run dry. Hence, in the medium (2008 to 2015) and long term (2016 to 2025), Nigeria envisions an energy transition from crude oil to renewable energy (ECN, 2005). An alternative fuel must be technically feasible, economically competitive, environmentally acceptable and readily available. One possible alternative to fossil fuel is the use of oils of plant origin like vegetable oils and tree borne oil seeds. The alternative diesel fuel can be termed as biodiesel. The fuel is biodegradable, non-toxic and has low emission profiles as compared to petroleum-based diesel. Usage of biodiesel will allow balance to be sought between agriculture, economic development and the environment (Meyer et al., 2004).

Various edible and non edible vegetable oils, like rice bran oil, coconut oil, Jatropha caucus oil, castor oil, cottonseed oil, mahua, karanja which are either surplus and are non-edible type, can be used for the preparation of biodiesel (Malhotra and Das, 1999; Shah et al., 2004; Freedman et al., 1986). Jatropha caucus oil is considered in this research work due to its generous advantages.

Chemically the oils/fats consist of triglyceride molecules of three long chain fatty acids that are ester bounded to a single glycerol molecule. These fatty acids differ by

---
*Corresponding author. E-mail: bkhighina@gmail.com.

length of carbon chains, the number, orientation and position of the double bounds in these chains. Thus, the biodiesel refers to as lower alkyl esters of long chain fatty acids, which are synthesized either by transesterification with lower alcohols or by esterification of fatty acid (Meyer et al., 2004). Most research works seek to study the factors affecting biodiesel production by transesterification of *Jatropha caucus* oil using homogenous catalysts (Chhetri et al., 2009; 2007; 2005). However, use of homogeneous catalyst in biodiesel production process has problems of soap formation and difficulty in separations (Abdullah et al., 2007; Alcantara et al., 2000; Ma and Hanna, 1999; Ma et al., 1998). Hence, this paper is aimed to study the factor that affect biodiesel production using heterogenous catalyst which may result in no soap formation and will make easy separations, catalyst and methanol as an alcohol.

MATERIALS AND METHODS

Equipment

The reactions were carried out in a 250 ml three-necked flat bottom flask with a reflux condenser (to reduce the loss of methanol by evaporation), thermometer and a stopper to add the catalyst solution. The reaction mixture was heated and stirred by a hot plate with a magnetic stirrer (Gupta and Sastry, 2006).

Procedure

The transesterification reaction of Jatropha oil was carried out by using zinc oxide as a catalyst. 2.0 g (2.5 ml) of methanol was measured and poured into a test tube after which 0.09114 g of ZnO pellet was carefully added to the test tube. A cork was replaced tightly. The test tube was swirled round thoroughly for about two minutes repeatedly about six times for complete dissolution of ZnO pellet in the methanol. 9.115 g (10 ml) of JCO was measured out, pre-heated to 65°C in a three-necked flat bottom flask with reflux condenser using a water bath and thermometer to observe the temperature. The pre-heated JCO is poured in a 250 ml beaker placed on a magnetic stirrer. The prepared zinc methoxide from the test tube was carefully poured into the JCO. Then the beaker was secured tightly using a stopper and the magnetic stirrer switched on and moderate agitation in the beaker was maintained for 25 min. The mixture was poured from the beaker into a second test tube for settling and the top secured using cork. The reaction mixture was allowed to stand overnight while phase separation occurred by gravity settling into golden/pale liquid biodiesel on the top with the light brown glycerol at the bottom of the test tube. The JCO biodiesel was carefully decanted into a plastic test tube (pierced at the bottom) leaving the glycerol at the based. The biodiesel was washed with water as detailed in Alamu (2007).The procedure was repeated by varying parameters controlling the transesterification reaction. In each case biodiesel yield as well as glycerol yield was measured and recorded.

Factors that affect transesterification reaction

Additional experiment were conducted to study the effect of parameters such as catalyst concentration (ZnO), Methanol-oil ratio and reaction temperature at constant mixing rate on transesterification reaction with time. The concentration of catalyst used were 0.25, 0.50, 1.0, and 1.5% by weight of the JCO. The methanol-oil ratios used were 6:1, 12:1, 18:1 and 24:1, respectively. The temperature variation was 35, 55 and 65°C. ASTM standard fuel tests were subsequently carried out on the JCO biodiesel and low sulphur diesel fuel (No. 2 diesel) purchased at a fuel station in Maiduguri, Nigeria. Specific gravity and viscosity measurements were made using the thermal-Hydrometer apparatus and viscometer (Canon-Fenke calibrated, 15cSt max. range), following ASTM Standards D 1298 and D 445, respectively. The biodiesel was analyzed for cloud point and pour point using Baskeyl Sate point cloud and pour point apparatus following ASTM standard D 25100-8 and D 97, respectively (ASTM, 1995).

RESULTS AND DISCUSSION

Transesterification reactions were carried out with Jatropha oil with different reaction conditions, shown in Figures 1, 2, and 3. The effect of the different process variables; catalyst concentration, alcohol to oil ratio, temperature of the reaction and reaction time were analyzed experimentally.

The main role of catalyst in reaction kinetics is to reduce the activation energy. Four different concentrations, (0.25, 0.50, 1.0 and 1.5% w/w of oil) of sodium hydroxide were used to study the effect of catalyst on transesterification. It was observed in all experiments that the equilibrium conversions were achieved between 80 to 100 min. Moreover, the average JCO biodiesel yield of 51.93, 87.22, 93.89 and 93.89% were obtained for each of the experiment. This is shown in Figure 1. It is to be noted that at any given condition, changing catalyst concentration has no significant effect on the equilibrium conversion.

One of the most important variables affecting the conversion of triglycerides (TG) is the molar ratio of alcohol to TGs. The stoichiometric molar ratio for transesterification is three moles of alcohol (for example, methanol) to one mole of TGs, to produce three moles of alkyl esters (for example, methyl esters) and one mole of glycerol. The effect of molar ratio on transesterification reaction is associated with the type of catalyst used. It is reported in open literature, acid-catalyzed transesterification requires a molar ratio of 30:1, while alkali-catalyzed reaction requires only 6:1 molar ratio to achieve the same equilibrium conversions.

In the present study, the effects of four ratios (6:1, 12:1, 18:1 and 24:1) were investigated. Low molar ratios were considered since only alkali-catalyzed transesterification reactions with Jatropha oil were carried out. Figure 2 shows the effect of alcohol- oil ratio on the transesterification reaction at different conditions. It was observed that the equilibrium conversion increased with an increase in molar ratio resulting to average JCO biodiesel yield of 93.14, 93.46, 96.46 and 96.46%, respectively for each of the experiment. This is in

Effect of catalyst concentration on reaction

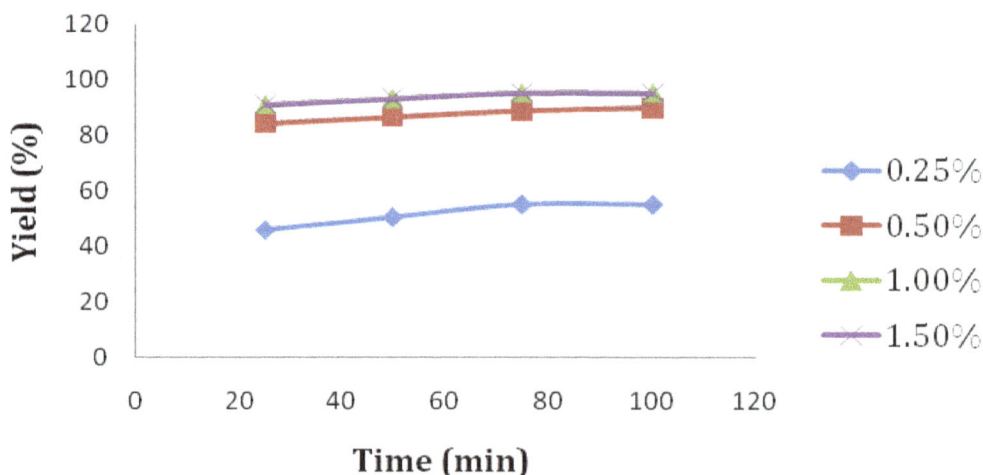

Figure 1. Effect of zinc oxide (ZnO) concentration on transesterification reaction.

Effect of MeOH:Oil ratio on reaction

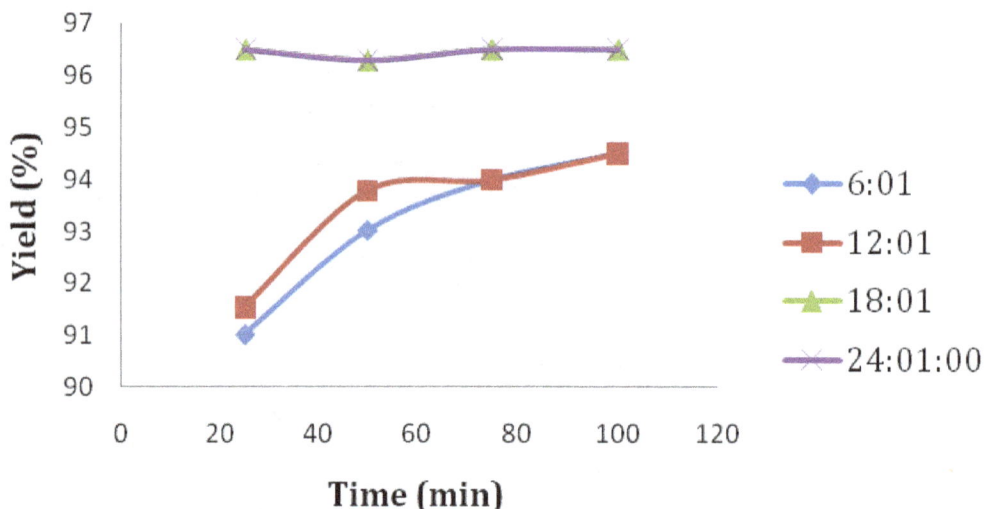

Figure 2. Effect of methanol-oil ratio on transesterification reaction.

agreement with theory (Le Chateliers principle). According to these observations, 18:1 is the optimum alcohol ratio.

The effects of three temperatures (35, 55 and 65°C) on transesterification reaction have been investigated. An increase in equilibrium conversion with an increase in temperature was observed as shown in Figure 3. Due to the sampling policy, the equilibrium conversions were observed between 85 to 100 min in experiment two and

three. The equilibrium conversions at 35, 55 and 65°C were observed to be approximately 81.75, 83.85 and 93.75%, respectively. Thus, it was observed that the equilibrium conversions increased substantially for every degree rise in temperature. Hence, rate of reaction is strongly dependent on temperature of reaction.

Based on this research work, the best reaction time to be used for transesterification reaction of JCO to biodiesel is obtained by plotting average biodiesel yield

Effect of temperature on reaction

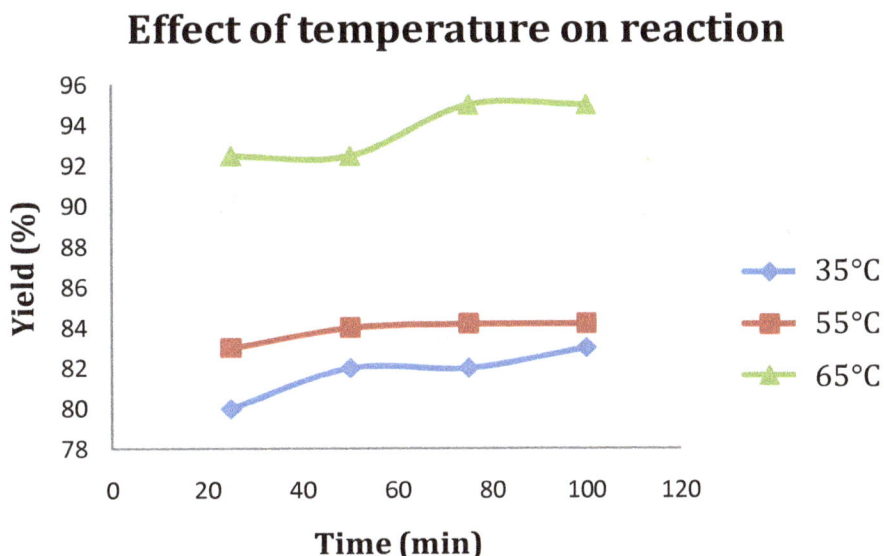

Figure 3. Effect of temperature on transesterification reaction.

against reaction time as shown in Figure 3. According to the plot, the best reaction time for transesterification reaction of JCO to biodiesel is 100 min. Dorado et al. (2002) also reported similar finding using soybeans and sunflower oil, an approximate yield of 80% was observed after 1 min with methanol to oil ration of 6:1, 1% zinc methoxide catalyst at 60°C. After one hour, the conversion was almost the same (93 to 98%).

Conclusions

The effects of different reaction factors on the production of biodiesel using zinc oxide as a hetereogeneous catalyst were studied. Four replicate transesterification experimental runs were carried out for each of the mixing duration 25, 50, 75, and 100 min under different typical transesterification reaction conditions of 9.115 g JCO, temperature (37, 57, and 67°C), ZnO concentration (0.25, 0.5, 1.0 and 1.5%) (wt% JCO), methanol to oil ratio (6:1, 8:1, 12:1 and 24:1) at a constant mixing rate. The optimum conditions were found to be 1% ZnO, 18:1 methanol to oil ratio, 67°C temperature. From this study the highest conversion is 98%.

REFERENCES

Abdullah AZ Razali N, Mootabadi H, Salamatinia B (2007). Critical Technical areas for Future Improvement in Biodiesel Technologies. IOP Publishing, Environ. Res. Lett., p.2.

Alamu OJ, Waheed MA, Jekayinfa SO (2007). Alkali-catalysed Laboratory Production and Testing of Biodiesel Fuel from Nigerian Palm Kernel Oil". Agricultural Engineering International: the CIGR Ejournal. Manuscript Number EE 07 009. Vol. IX. July.

Alcantara R, Amores J, Canaira L, Fidalgo E, Franco MJ, Navarro A (2000). Catalytic Production of Biodiesel from Soybean Oil, Using Frying Oil and Tallow, Biomass Bioener., 80: 515.

ASTM International (2005). Standard Specification for Biodiesel Fuel-Blend Stock (B100) for Distillate fuel (West Conshohocken, PA: ASTM).

Chhetri AB, Martin ST, Suzanne MB, Chris WK, Rafiqul IM, (2009). Non-Edible Plant Oils as New Sources for Biodiesel Production, Int. J. Mol. Sci., 9: 169-180.

Chhetri V, Islam AB, Mann P (2007). Zero-waste multiple uses of jatropha and dandelion. J. Nat. Sci. Sust. Tech., 1: 75-99.

Chitra P, Venkatachalam P, Sampathrajan A (2005). Optimisation of experimental conditions for biodiesel production from alkali-catalysed transesterification of Jatropha Caucus oil. Energy Sustain. Dev., 9(3): 13-18.

Dorado MP, Ballesteros E, Almeida JA, Schellet C, Lohrlein HP, Krause R (2002). An alkali-catalyzed 855 transesterification process for high free fatty acid oils. Trans ASAE, 45(3): 525-529.

Freedman B, Butterfield RO, Pryde EH (1986). Transesterification kinetics of soybean oil. J. Am. Oil Chem. Soc., 63(10): 1375-1380.

Gupta AK, Sastry SR (2006). Developing The Rate-Equation For Biodiesel Production Reaction, Department of Chemical Engineering, Indian Institute of Technology, New Delhi-110016, E-mail: svarsastry@yahoo.com. Advances in Energy Research (AER).

Ma F, Hanna MA (1999). Biodiesel Production: A Review, Biores. Technol., 70: 1-15.

Ma F, Clements LD, Hanna MA (1998). The effect of catalyst, free fatty acids, and water on transesterification of beef tallow. Trans ASAE, 41(5): 1261-1264.

Malhotra RK, Das LM (1999). Biofuel as blending components for Motor Gasoline and Diesel fuels. J. Sci. Ind. Res., 61: 91.

Meher LC, Sagar DV, Naik SN (2004). Technical Aspect of Biodiesel Production by Transesterification - A Review. Renew. Sustain. Energy Rev., p. 3.

Shah S, Sweta S, Gupta MN (2004). Biodiesel Preparation by Lipase Catalysed Transesterification of Jatropha oil, India Energy and Fuels, 18: 154-159.

24

Agricultural waste plastics conversion into high energy liquid hydrocarbon fuel by thermal degradation process

Moinuddin Sarker*, Mohammad Mamunor Rashid and Md Sadikur Rahman

Natural State Research Inc., 37 Brown House Road, Stamford, CT-06902, USA.

For worldwide cultivation purposes farmers use numerous plastics. To cultivate watermelon, grapes, pumpkin, cucumber, potatoes, etc. they use polyethylene poly bags or polyethylene sheets. Polyethylene poly bags parts and polyethylene sheets are used to protect Crops from adverse situation such as natural disaster, muddy water and some external parasites insects. After that, the used plastics become a waste and they are non-biodegradable. Most common way to deal with those agricultural waste plastics is landfill/incineration. However this method is not a sustainable solution in the long run, because it will create ecological problems. So the problems of waste agriculture plastics cannot be solved by landfilling/incineration. Because safe depots are expensive and incineration stimulates the growing emission of harmful green house gases for example, NOx, SOx, COx etc.; Natural State Research, Inc. (NSR) can resolve the problems of agricultural waste plastics by converting it into high energy liquid hydrocarbon fuel. Produced fuel can be use alternative of Gasoline, Diesel, Heating oil, and Aviation fuel. To mitigate the present market demand we can substitute the method as a potential source of energy.

Key words: Cultivation, non-biodegradable, parasites, atmosphere, incineration, contamination, conversion, fuel, thermal degradation, agriculture plastic.

INTRODUCTION

The American Environmental Protection Agency (EPA) has struggled to resolve the problems posed by massive amount of waste material produced by feedlots, slaughterhouses, agricultural, food processing factories, etc. Organic waste is generated by the agricultural and food production industries at the rate of billions of tons each year and that's just in the USA. Problems with the disposal of waste sludge from these facilities are receiving a tremendous amount of attention (www.pmcbiotec.com/downloads/agriocultural.pdf). Each year about 550,000 of non-natural wastes are generated on farms in England, of which it is estimated that about

85,000 are waste plastics (http://www.agriculturewasteplastics.org.uk/agri/about.html). Over 90% of holding produce plastic waste, with packaging waste estimated to be about 21000 tons each year and non-packaging plastics waste estimated at about 65,000 tons per year. (http://www.agriculturewasteplastics.org.uk/agri/about.html). Governments around the world have funded hundreds of research projects to find an efficient waste treatment technology to handle this astonishing amount of material. As waste plastic treatment NSR is able to convert those wastes into liquid hydrocarbon fuel and this will be profitable. An agricultural waste plastics is a polymer of High Density Polyethylene (HDPE-2), Low Density Polyethylene (LDPE-4) and imitative of other, consisting of a chain of organic units joined by $-CH_2-$ links. Numerous $-CH_2-$ monomer units are linked together to form long chain hydrocarbon by Polymerization Reaction. Polyethylene including other plastics has enormous use in farmer's field, such Polyethylene's are: watermelon and bread bags, frozen food bags, grocery bags and etc. The Synthesis of Polyethylene can be

---

*Corresponding author. E-mail: msarker@naturalstateresearch.com.

Abbreviations: NSR, Natural state research; **EPA,** environmental protection agency; **HDPE-2,** high density polyethylene; **LDPE-4,** low density polyethylene; **DSC,** differential scanning calorimeter; **EA,** elemental analyzer; **FTIR,** fourier transforms infra-red spectroscopy.

Table 1. Experiment yield percentage calculation.

Production yield (Fuel)(%)	Residue yield (%)	Gas yield (%)
60.46	32.56	6.98

Calculations of table-1 shows that the total experiment yield is 100% (Liquid fuel is 60.46, solid residue is 32.56 and light gas (C_1-C_4) is 6.98.

derived from Ethylene. Polymerization of Ethane into Polyethylene reaction is given below.

Polymerization reaction

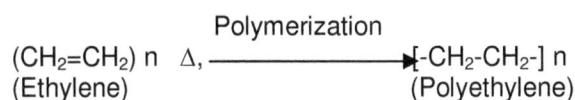

$$(CH_2=CH_2) n \quad \Delta, \xrightarrow{\text{Polymerization}} [-CH_2-CH_2-] n$$
$$(\text{Ethylene}) \qquad\qquad\qquad (\text{Polyethylene})$$

Both high and low density polyethylene has many types of uses in our daily life including various agricultural aspects. Polyethylene is made by polymerization of ethylene. To save fertility of land and restrain environmental balance in the atmosphere waste agricultural plastics need to be cleaned up from the land, otherwise remaining waste plastics in the land is non-rotten able. It makes unsaturated mixture in the soil and turns it into worst and complex blend that inhibit the crops growth which leads to terrible impact in total cultivation process. When the seeds are cultivated in the land with such unsaturated blend, some seeds are seeding on remaining waste polyethylene surface, as a result of this after blooming, the plants soft root fails to go through the soil to complete photosynthesis mechanism due to polyethylene layer and ultimately plants dies due to the lack of glucose. Many of researchers and experts have done lots of research and worked on agricultural waste plastics: thermal degradation process (Aguado et al., 2007; Marcilla et al., 2007), pyrolysis (Cozzani et al., 1997; Marcilla et al., 2008), and the catalytic conversion (Manos et al., 2000).

EXPERIMENTAL

Sample preparation

Agricultural waste plastics samples (watermelon bag, grapes bag, wheat Bag, etc.) NSR used is collected from California, USA. Prior to the experiment each unclean waste agricultural plastics samples were washed, dried and torn manually by scissors.

Sample loading and set- up

NSR carried out the experiment in laboratory scale and inside of labconco protector laboratory fume hood in glass reactor (Chemglass Round Bottom Boiling Flask one neck, KIMAX Brand Liebig Condenser, Heat Mantle and Variac Glascol and Chemglass Round Bottom Boiling Flask 2 neck). To perform the experiment

NSR took the volume of the boiling flask which is 1000 ml. Due to experimental purpose we poured 86 g of waste agricultural sample in the boiling flask without any catalyst. Then the boiling flask filled with sample is placed on the heat mantle and then we connected the variac meter with the heat mantle. During experiment also, attached distillation adapter, clump joint, condenser and collection flask with high temperature apiezon grease. Subsequently, to measure electricity consumption electricity logger was connected.

Condensation process

When the experiment started the variac percent was 70 (315ºC) for quick melting, after the sample melted the variac percent is decreased to 50 (225ºC) due to smoke formation. The average (optimum) used variac percent in this experiment is 70 (315ºC). Gradually, temperature contour was maintained by variac meter, with proper monitoring. Produced vapor travel through the condenser, due to cooling water, the vapor turns into liquid fuel and confirmed that the condensation process has been completed properly. The process performed without used any vacuum, chemical or catalysts.

Fuel calculation and residue collection

The complete process took about 2 h and 36 min and the electricity used is about 0.346 kWh. Based on our laboratory scale, 1gallon of fuel production purposed electricity used 13.3 kWh and 1 Metric Ton of agricultural waste plastics total electricity would be 4030.30 kWh. The obtained fuel is 52 g that is 69 ml and derived density is 0.76 g/ml that is compared to gasoline fuel and it has burning properties. At the end of the state we cooled down the experiment and collected the ash type carbon residue 28 g from the boiling flask for further analysis. This residue will be less when NSR will go for commercialization plant.

If NSR proceed further fractional distillation process of the produced fuel, the probable product would be Aviation, Naphtha, Diesel and Fuel oil. Ultimately, without having any problem, NSR technology can convert all of agriculture sample into liquid hydrocarbon fuel. The materials of liquid Hydrocarbon fuel are concluded by hydrocarbon functional group analysis by FT-IR data. Fuel Production, Residue and Gas Yield Percentages are given below in Table 1. NSR laboratory scale one gallon production cost is $1.46; While NSR will go commercialization plant cost will be less than dollar (Table 2).

RESULTS AND DISCUSSION

Fuel characterization by fourier transforms infra-red spectroscopy (FTIR analysis)

Agricultural watermelon plastics fuel was analyzed by FTIR Spectrum - 100 (Fourier Transform Infra- Red

Table 2. Comparison data with agriculture waste plastic to fuel and other fuel production cost.

Feedstock	Fuel name	Commercial cost /gallon ($)	NSR Lab scale/ gallon ($)
Mineral	Diesel	2.45	-
Sugar cane	Bio-fuel	1.9	-
Agriculture plastic	Liquid fuel	-	1.46

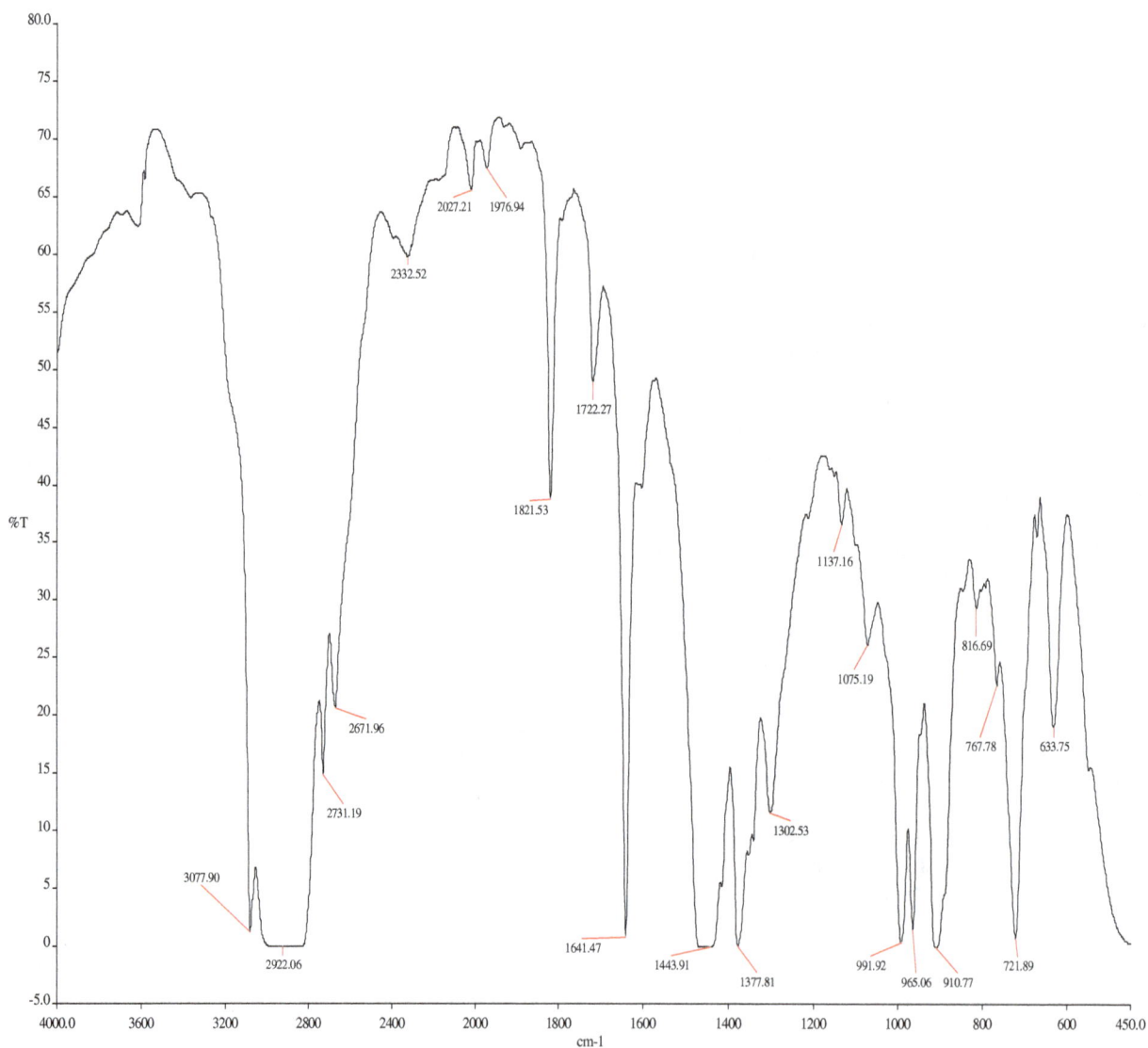

Figure 1. FT-IR spectrum of agricultural watermelon plastic fuel.

Spectroscopy). FTIR analysis indicated the following types of functional groups are present into liquid fuel; such groups are: H bonded NH, CH_2, C-CH_3, Non-conjugated, CH_3,-CH=CH_2, -CH=CH- (trans) and -CH=CH- (cis) etc. (Figure 1). By FTIR analysis we calculated energy of each spectrum band functional group. In the functional group H bonded NH the energy is 6.11×10^{-20} J; functional group C-CH_3, the energy is 5.80×10^{-20} J and in Non-conjugated the energy is 3.6×10^{-20} joule. Also in functional group CH_3, the energy is 2.87×10^{-20} J and the functional group -CH=CH_2, the energy is 1.97×10^{-20} Joule. In the functional group -CH=CH- (trans) the energy is 1.93×10^{-20} J and in functional group -CH=CH- (cis) the energy is 1.91×10^{-20} J. In FTIR library search found out that the following compounds are present in Agricultural watermelon waste plastics fuel. 0.394 F06640 4-Amino - Acetaphenone, 0.393 F37460 2, 5-Dihydroxy Acetaphenone, 0.302 F00508 Ethyl

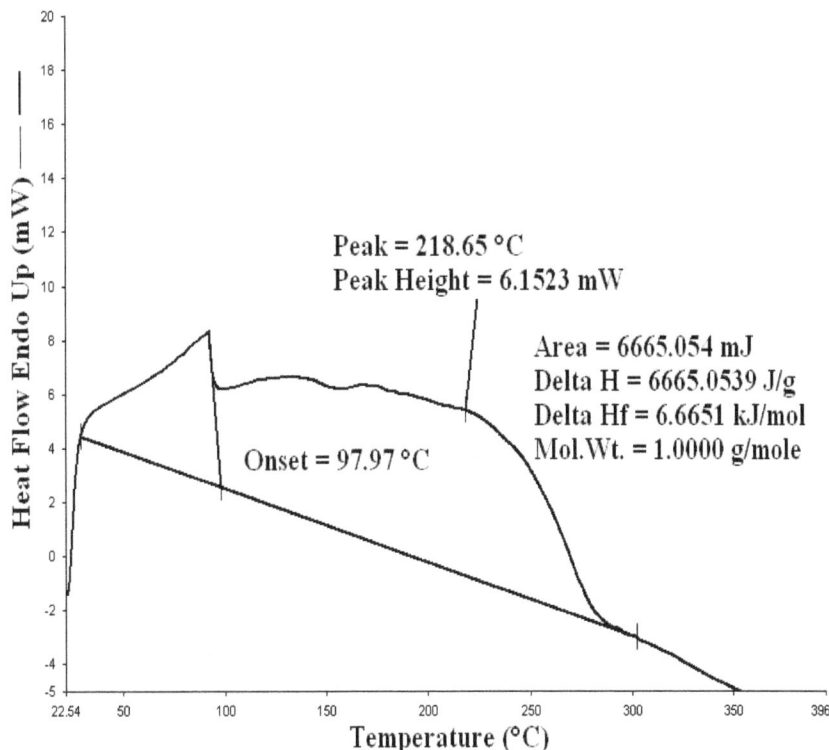

Figure 2. DSC graph of agriculture watermelon plastic fuel.

Table 3. Elemental analyzer data for residue analysis.

Carbon (%)	Hydrogen (%)	Nitrogen (%)	Sulfur (%)
48.26	6.85	0.04	1.11

Acetohydroxamate, 0.272 F54150 2-Hydroxy Acetaphenone, 0.260 F65470 3-Methyl Acetophenone, 0.249 F22850 4- Chloro Acetophenone 0.245 F91080 Trichloro Acetonitrile, 0.236 F65155 2- Methoxy Phenyl Acetonitrile, 0.220 F80072 BIS (3, 5, 5-Trimethyl Hexyl) Phthalate, 0.214 F35038 1, 1-Dichloro Acetone.

Fuel characterization by differential scanning calorimeter (DSC analysis)

Agricultural watermelon waste plastics fuel ran by Differential Scanning Calorimeter (DSC). From DSC analysis we found onset temperature is 97.97 °C (Figure 2). Onset temperature represents boiling point of the fuel and at this point derived compound is Heptane (C_7H_{16}) with Molecular Weigh of 100.21 g/mole, and the density is 0.683 g/mole with flash point -4 °C. (Reference: Hydrocarbon Standard Data Table).

DSC Graph postulates the following items (Figure 2): Peak = 218.65 °C, Peak Height = 6.1523 Mw, Area =

6665.0539 mJ, Delta H = 6.6651 kJ/mol, Mol.Wt. = 1.000 g/mol

Residue characterization by elemental analyzer (EA-2400)

California Waste Watermelon plastics sample solid residue ran by Elemental Analyzer (EA-2400). From EA-2400 result found the following composition of Carbon and Hydrogen ratio as well as Sulfur and Nitrogen availability as a trace quantity that can be ignorable.

EA-2400 data table (residue)

From EA-2400 (CHNS Mode) residue analysis data shown (Table 3), we noticed that Carbon and Hydrogen percentages are only 48.26 and 6.85%, respectively and other impurities under consideration. During fuel production period 6.98% light gas produced and NSR did not analyzed yet is under consideration.

Conclusion

The conversion of agricultural waste plastics to liquid hydrocarbon fuel was carried out in thermal degradation process and in absence of a catalyst. NSR executed the experiment by washing out agricultural waste plastics. Experiment procedures are followed by NSR's own process. By utilizing NSR's technology can solve all agricultural waste plastics problems by saving environment, landfill cost, etc. The produced fuel from agricultural waste plastics can be used as feedstock of refinery because it has high content of energy value or can be produce electricity.

ACKNOWLEDGEMENTS

The author acknowledges the support of Dr. Karin Kaufman, the founder and the sole owner of Natural State Research (NSR), Inc. The authors also acknowledge the valuable contribution of NSR's laboratory team members during the preparation of this manuscript.

REFERENCES

Convert Agricultural Waste into Sustainable Profitable Green Energy, Available online at www.pmcbiotec.com/downloads/agriocultural.pdf.

Available online at http://www.agriculturewasteplastics.org.uk/agri/about.html.

Available online at http://www.ethicalcorp.com/content.asp.

Aguado J, Serrano DP, Vicente G, Sanchez N (2007). Enhanced Production of r-Olefins by Thermal Degradation of High-Density Polyethylene (HDPE) in Decalin Solvent: Effect of the Reaction Time and Temperature. Ind. Eng. Chem. Res., 46: 3497-3504.

Marcilla A, Garcia AAN, Hernandez MR (2007). Thermal Degradation of LDPE-Vacuum Gas Oil Mixtures for Plastic Wastes Valorization. Energy Fuels, 21: 870-880.

Cozzani V, Nicolella C, Rovatti M, Tognotti L (1997). Influence of Gas-Phase Reactions on the Product Yields Obtained in the Pyrolysis of Polyethylene. Ind. Eng. Chem. Res., 36: 342-348.

Marcilla A, Beltran MI, Navarro R (2008). Evolution with the Temperature of the Compounds Obtained in the Catalytic Pyrolysis of Polyethylene over HUSY. Ind. Eng. Chem. Res., 47: 6896-6903.

Manos G, Garforth A, Dwyer J (2000). Catalytic Degradation of High-Density Polyethylene on an Ultrastable-Y Zeolite. Nature of Initial Polymer Reactions, Pattern of Formation of Gas and Liquid Products, and Temperature Effects. Ind. Eng. Chem. Res., 39: 1203-1208.

Manos G, Garforth A, Dwyer J (2000). Catalytic Degradation of High-Density Polyethylene over Different Zeolitic Structures. Ind. Eng. Chem. Res., 39: 1198-1202.

25

Determination of the catalytic converter performance of bi-fuel vehicle

Sameh M. Metwalley, Shawki A. Abouel-seoud and Abdelfattah M. Farahat

Faculty of Engineering, Helwan University, Cairo, Egypt.

The reduction of pollutant emission from spark ignition engines is desirable in order to reduce the highly impact on the green environment, produced from transport release like trains, trucks, traveler vehicles and others. However, modern vehicles are equipped with catalytic converters. A bi-fuel vehicle that has been retrofitted for both fuel systems: namely, compressed natural gas (CNG) and base fuel gasoline. A locally produced three-way catalytic converter (TWC) fitted on the exhaust system. The objective of this investigation is to evaluate the effectiveness of TWC in reducing vehicle exhaust emissions. In addition, the individual conversion efficiency of the vehicle-out emissions have been calculated and presented. Operations under idle state and on-road emission test procedures were carried out on a newly registered gasoline/CNG bi-fuel vehicle in Egypt market (Hyundai-star) where is assessed against the European standard urban driving cycle (ECE-15). Two different fuel injection systems are used; namely multi-point (MPI) and venture (mixer) closed-loop. The emission results such as CO, CO$_2$ and THC were measured and compared between the earlier mentioned two fuels. The results show that the arrangement of TWC and operation in idle state is very effective to reduce exhaust emissions than that in transient state. Moreover, the results of this investigation will be used to develop CNG emissions based TWC.

Key words: Vehicle engine emissions, bi-fuel vehicle, idle measurement, on-road measurements, fuel injection systems, air index.

INTRODUCTION

The exhaust emissions and performance were evaluated for a computer integrated bi-fuel spark ignition engine that has been retrofitted for two fuel system: namely, compressed natural gas (CNG) and base fuel gasoline. Operations under steady state with lean burn condition. The used engine was a Proton Magma 4-cylinders spark ignition engine. The emission results such as CO, HC and NO$_x$ were measured and compared between the earlier mentioned two fuels. A three way catalytic converter (TWC) was used to assess the emissions. The results show that the arrangement of retrofitting catalytic converter and operation with lean burn condition is very effective to reduce exhaust emissions. From the investigation, it is found that CNG produced 15% less

brake power, 15 to 18% less specific fuel consumption (SFC) and 10% higher thermal efficiency than gasoline fuel. The emission results showed at the entrance of TWC that CNG produced 30% higher than NO$_x$ emissions and lower 12 and 90% HC and CO, respectively. The details about the emissions management system together with the catalytic and engine performance results have been presented and discussed (Yaacob et al., 2002).

Motor vehicle exhaust emissions could be sharply reduced by the introduction of exhaust after-treatment systems, modern engine control concepts and cleaner fuels. However, it has not been possible to significantly improve air quality in cities with regard to particulates and ozone in the past 10 years. It appears that the reduction in vehicle exhaust emissions achieved is offset by the growth in traffic as well as by changes in the composition of exhaust emissions and the corresponding reactivity in the environment. The reduction of selected pollutants

*Corresponding author. E-mail: metwalley1965@hotmail.com

therefore remains an important issue alongside greenhouse gas reduction. The present investigation illustrates the emissions of actual gasoline, diesel and natural gas passenger cars in the official European driving cycle and in the real-world driving cycle Artemis. The natural gas vehicles show the lowest impact on air quality (Yaacob and Rahman, 2003; Burch et al., 1996).

Uniform flow distribution inside a catalytic converter is highly desirable in order to enhance converter efficiency and extend catalyst durability. Mal-distributed flow constantly results in a penalty of deterioration performance and reduced lifetime. In addition, the formation of recirculation zone inside the catalytic sensor installed in the converter diffuser, especially with closed coupled design. In such a case, information from the oxygen sensor may not represent the overall combustion characteristics of all the cylinders, which may lead to a false feedback to the engine control system. Generally, velocity distribution is not uniform inside the catalytic converter of production engines. This because the flow distribution within the converter is a momentum transfer process and sensitive to different boundary conditions and momentum sources inside the system (Kalam et al., 2001).

Previous studies reported that uniform flow distribution at the catalyst improves the conversion efficiency and durability of the catalytic converter (Burch et al., 1996; Kalam et al., 2001; Weaver, 1989). Uniform flow distribution at the monolith inlet lowers local peak velocities and temperature gradients in the catalytic converter, and delays aging of the catalytic converter, It is possible for a reduced-volume of close-coupled catalytic converter (CCC) to have the same durability if flow distribution becomes uniform. This leads to a reduction in cost and mass of the CCC. The key factor in designing a CCC is to optimize the exhaust manifold and the CCC inlet. If they are poorly designed, flow at the monolith inlet becomes highly non-uniform, resulting in poor catalyst conversion efficiency and durability. The continuing global enthusiasm for mobility and the resultant ever increasing burden placed upon the environment lead to co-political discussions and further on to more stringent emission standards for all combustion engines (Weaver, 1989; Tong et al., 2000).

A compressed natural gas (CNG) or gasoline is being developed for the transition to alternative fuel usage in a spark ignition bi-engine. Individual conversion efficiency of the emissions have been measured for the vehicle's engine in steady-state and transient-state conditions, where their results indicate that regulated total hydrocarbon (THC), nitrogen oxides (NO_x), carbon monoxides (CO) and carbon dioxides (CO_2) were characterized. Moreover, the equivalence ratio range over which the catalyst can relatively reduce NO_x, CO and CO_2 plays an important role in reduction (Pipitone and Beccari, 2007; Al-Shemmeri, 1993).

More stringent emissions standards around the world

are challenging the auto industry to advance emissions control technology. Engine management strategies have been introduced to achieve faster exhaust heat up and tighter closed-loop operation, which results in faster converter light-off. Catalysts have achieved lower light-off temperature and higher conversion efficiency after the vehicle is at operating temperature. New exhaust manifold design and dual wall exhaust piping has been utilized for better thermal management to achieve faster light-off. Catalyst supports have been developed to achieve lower mass and higher surface area for faster light-off and higher conversion efficiencies (Kalam et al., 2004).

A driving cycle is a time series of vehicle speeds recorded at successive (equally spaced) time points USEPA (1993). It represents a typical driving pattern for the population of a city. For emission testing, a test driving cycle in the most general case, attempts to synthesize real driving conditions with respect to a number of measures, including speed, acceleration, specific power, trip patterns, road grade, and temperature. Driving cycles have been developed to provide a single speed-time profile that is representative of urban driving. Standard driving cycles have a wide range of uses (Tong et al., 1999). Vehicle manufacturers need these cycles to provide a long term basis for design, tooling and marketing. Traffic engineers require driving cycles in the design of traffic control systems and simulation of traffic flows and delays. Environmentalists are concerned with the performance of the vehicle in terms of the pollutants generated, while negotiating specific driving patterns. Furthermore, a speed-time trace can provide a convenient laboratory-based means to estimate fuel consumption and emissions of vehicles within the respective urban areas.

As known gaseous fuels, such as liquefied petroleum gas (LPG) and natural gas (NG), thank to their good mixing capabilities, allow complete and cleaner combustion than normal gasoline, resulting in lower pollutant emissions and particulate matter. Moreover, the use of natural gas, mainly constituted by methane, whose molecule has the highest hydrogen/carbon ratio, leads also to lower CO_2 equivalent emissions. Some of the automobile producers already put on the market "bi-fuel" engines, which may be fed either with standard gasoline or with natural gas. These engines, endowed of two separate injection systems, are originally designed for gasoline operations, hence they do not fully exploit the good qualities of methane, such as its high knocking resistance (Mah, 1993), which would allow higher compression ratios. Moreover, when running with gasoline at medium high loads, the engine is often operated with rich mixture and low spark advance in order to prevent from dangerous knocking phenomena: this produces both high hydrocarbon and carbon monoxide emissions (also due to the low catalyst efficiency caused by the rich mixture) and high fu

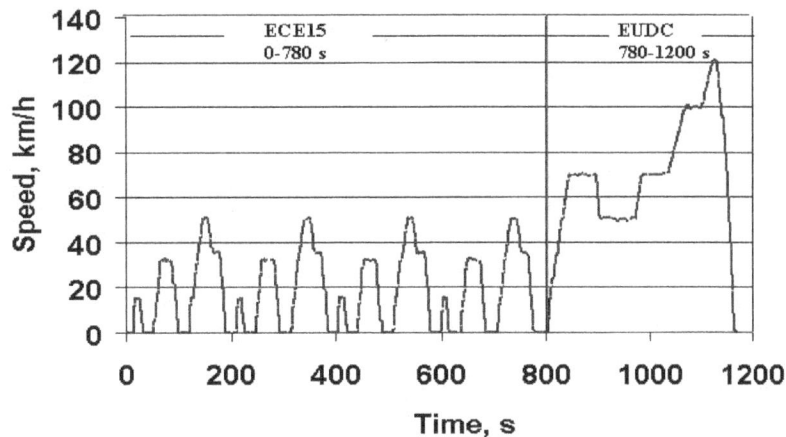

Figure 1. The European driving cycle ECE-15.

consumption.

The objective of this investigation is to evaluate the effectiveness of TWC in reducing vehicle exhaust emissions. In addition, the individual conversion efficiency of the vehicle-out emissions have been calculated and presented. Operations under idle state and on-road emission test procedures were carried out on a newly registered gasoline/CNG bi-fuel vehicle in Egypt market (Hyundai-star) where is assessed against the European standard urban driving cycle (ECE-15). Two different fuel injection systems (that is, Multi-point MPI-sequential and Closed-loop venturi-continuous) are used.

EUROPEAN DRIVING CYCLE (ECE-15)

In order to investigate the amounts of exhausted gas emissions and fuel consumption rates of vehicles traveling in Egypt, a generic driving characteristic or pattern for any vehicle traveling in the traffic of the city under consideration must be established. So far there is no such driving cycle officially developed for representing Egypt traffic. The driving cycle used for the assessment of the exhaust emissions of newly registered automobiles in Egypt is based upon the standard driving cycle of the European Community (called ECE-15 cycle) where the driving conditions are not the same. Furthermore, it is modal driving cycle which derived from various representative constant acceleration and speed driving modes contrast to the cycle that constructed from the real micro trips obtained from actual on-road driving data such as the US75 cycle and Melbourne peak cycle.

The European test driving cycle is based on Euro III and it is presented in Figure 1. The driving cycle consists of two parts, ECE15 and EUDC, that correspond to urban and highway (extra-urban) driving conditions in that order. ECE15 test cycle simulates a 4.052 km urban trip at an average speed of 18.7 km/h and at a maximum

speed of 60 km/h. Its duration is 780 s. The same part of the ECE15 driving cycle is repeated four times to obtain an adequate driving distance and temperature (Figure 2). The EUDC cycle instead illustrates the aggressive, high speed driving at a maximum speed of 120 km/h. Its duration is 400 s and 6.955 km at an average speed of 62.6 km/h. In this work, only part of urban cycle with the duration of 280 s is used (Figure 3).

CATALST TECHNOLOGY

The catalytic converter

Most modern vehicles are equipped with three-way catalytic converters. Three-way refers to the three regulated emission components. The converter uses two different types of catalysts, a reduction catalyst and an oxidation catalyst. Both types consist of a ceramic structure coated with a metal catalyst, usually platinum, rhodium and/or palladium. The idea is to create a structure that exposes the maximum surface area of catalyst to the exhaust stream, while also minimizing the amount of catalyst required. There are two main types of structures used in catalytic converters, honeycomb and ceramic beads.

A locally produced three-way catalytic converter (TWC) is shown in Figure 4, and has been used to reduce and assess the unwanted pollutant gases like CO, HC and NOx from the exhaust gas stream. The catalytic material consists of platinum (Pt), palladium (Pd), rhodium (Rh) and cerium oxide (CeO_2). The Pt and Pd are used to oxidize CO and HC to CO_2 and H_2O. The Rh is used to reduce NOx to N_2 and O_2. However, the efficiency of the three-way catalytic converter depends on the availability of O_2 and the temperature in the exhaust gas stream. The role of CeO_2 in TWC is to afford as an O_2 storage capacity (OSC) which liberates or adsorbs O2 if the air to fuel ratio is perturbed. The details about work function of

ECE-Cycle

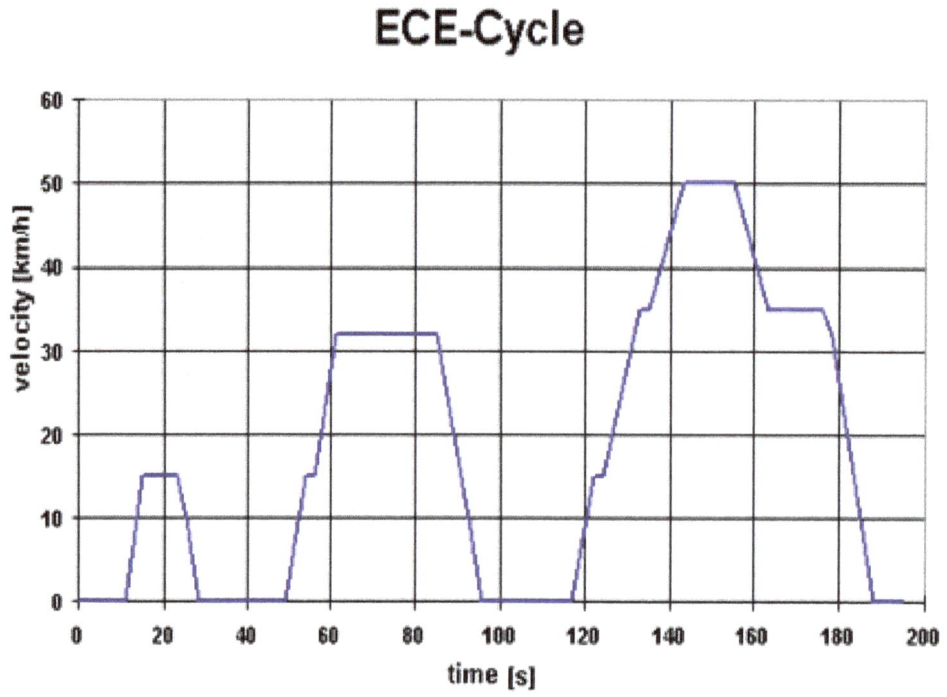

Figure 2. Part of the ECE-Cycle.

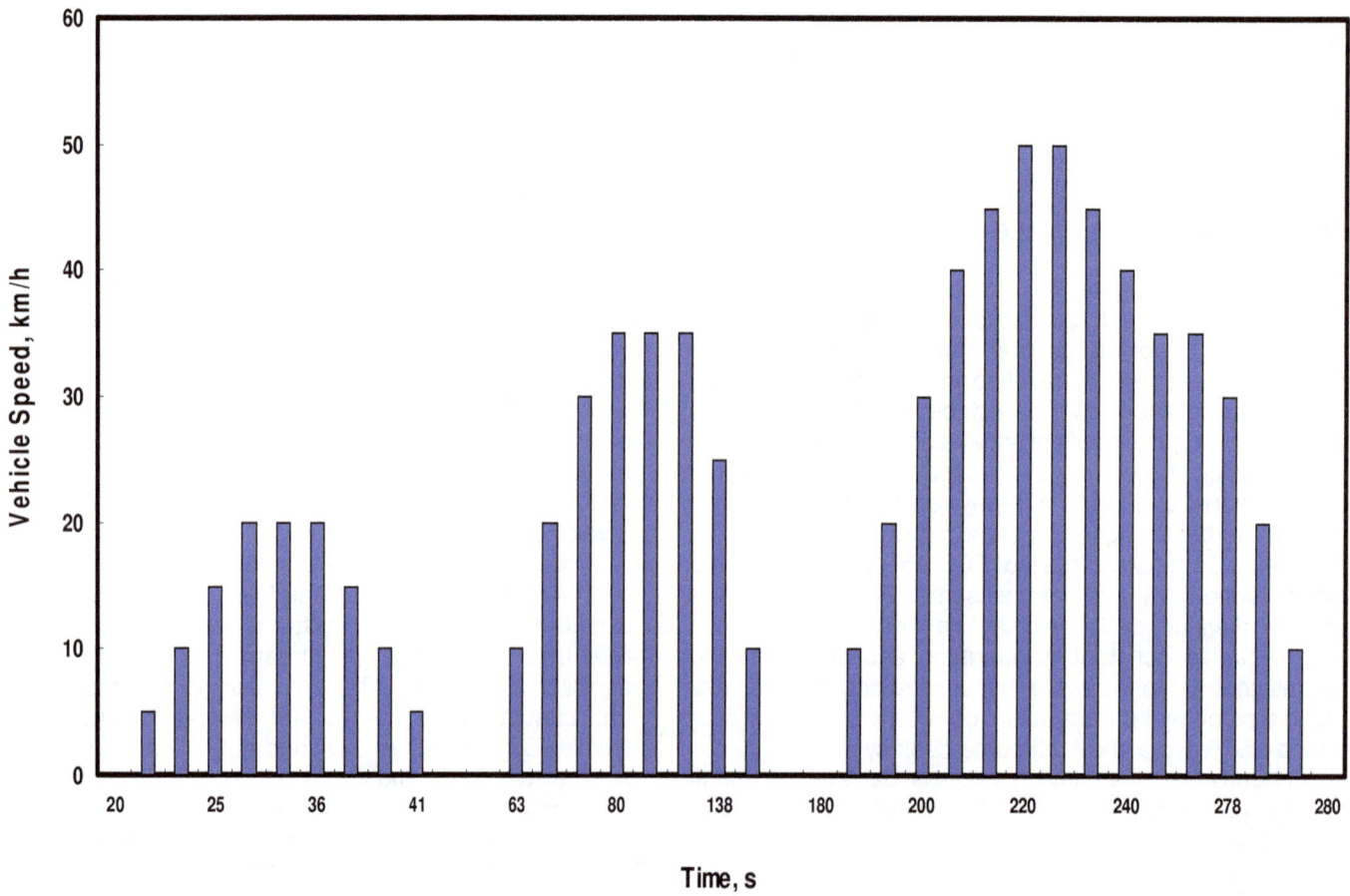

Figure 3. Part of the ECE-Cycle used.

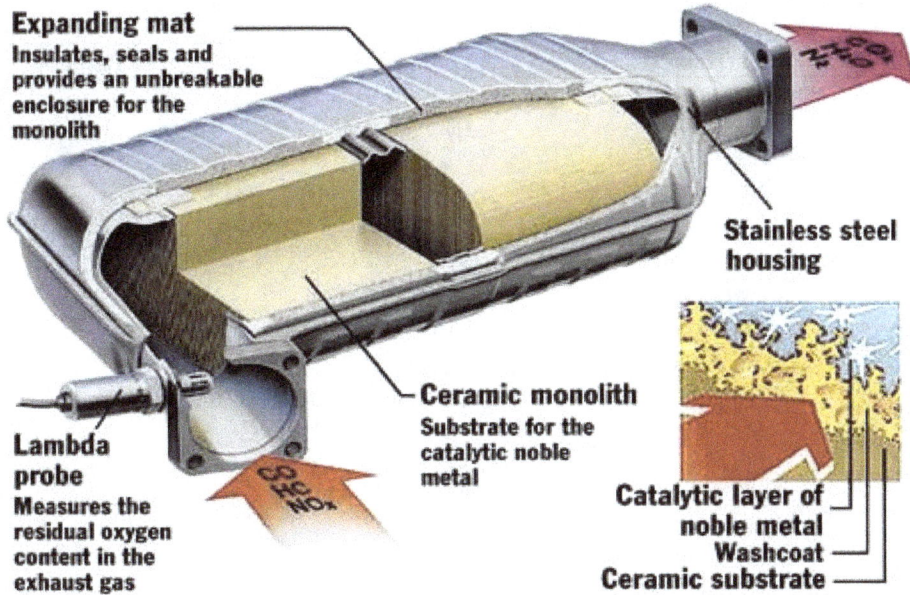

Figure 4. Structure of three-way catalytic converter.

Table 1. Types of the fuel injection system.

No	Injection system	Fuel	Remarks
1	Multi-point (MPI-sequential)	Gasoline Compressed natural gas (CNG)	
2	Closed-loop(venturi-continuous)	Compressed natural gas (CNG)	

CeO_2 in the TWC can be found elsewhere (Yaacob et al. (2002); Yaacob and Rahman (2003).

The catalytic bed was placed in the exhaust pipe (near manifold) where the maximum temperature (catalytic bed temperature) reaches about 560 °C. This high temperature is required to oxidize THC (as well as un-burn CH_4) in the exhaust gas from natural gas combustion. The details about the catalytic bed placement in the exhaust system can be found elsewhere (Burch et al., 1996).

The reduction catalyst

The reduction catalyst is the first stage of the catalytic converter. It uses platinum and rhodium to help reduce the NO_x emissions. When an NO or CO_2 molecule contacts the catalyst, the catalyst rips the nitrogen atoms out of the molecule and holds on to it, freeing the oxygen in the form of O_2. The nitrogen atoms bond with other nitrogen atoms that are also stuck to the catalyst, forming N_2.

Catalyst conversion efficiency

Catalyst system conversion efficiencies were calculated for each emission constituent using the following equation:

(1- (Without Catalytic Converter)/(With Catalytic Converter)) × 100

(1)

EXPERIMENTAL METHODOLGY

The experimental work is carried out on specified vehicle to avoid vehicle-to-vehicle variation which results from different designs, operating conditions and maintenance processes. According to the aim of this study, the idle and on-road test modes are employed and applied on the vehicle that has been retrofitted for both fuels namely compressed natural gas (CNG) and base fuel gasoline together with a locally produced three-way catalytic converter (TWC) fitted on the exhaust manifold to reduce its emissions. The fuel injection system used in this work is tabulated in Table 1. The vehicle is equipped with infrared gas analyzer and rpm pickup transducer. The exhaust gas concentration, engine rotational speed and vehicle speed are recorded during the test. A newly registered

Table 2. Some of the technical data for vehicle used in the present experimental study.

S/No.	Parameters	Values	Remarks
I	Vehicle		
	Type	Hyundai-star	
II	Engine		
	Type	SOHC, 4 cyl, 4 stroke	
	Fuel type	Gasoline/CNG	
	Swept volume	1598 (cm^3)	
	Fuel supply	MPI / Closed Loop (venturi)	
	ignition system	Electronic	
	Exhaust system	Catalytic converter	
	Maximum power	106 HP@ 4300 rpm	
	Maximum torque	143 Nm @ 3000 rpm	
III	Performance		
	0-60 mph	13.2 s	
	Quarter mile	18.5 s	
	Top speed	156 km/h	

gasoline/CNG bi-fuel vehicle in Egypt market (Hyundai-star) was used. The vehicle is Verna-star, 1600 cc. The vehicle engine is spark ignition, four strokes, four cylinders in-line and water cooled. It is transversally mounted with front wheel drive technique. The vehicle is equipped with manually operated gear box mounted transversally and sharing the same oil sump of the engine. The gear box offers 4 forward speeds and single reversal speed. Some of the technical data for the vehicle used are tabulated in Table 2.

Portable version of infrared gas analyzer is used during the experimental work. The gas analyzer is equipped with gas sampling probe to collect the exhaust gas from the muffler. The gas is then filtered and dried before entering the analyzer. Magnetic inductive pickup transducer is used also to measure the vehicle speed in km/h. Figure 5 represents the layout of the gas analyzer used in the present study, while the installation of instruments and probes at different measuring points in the vehicle are shown in Figure 6.

Intensive measurements program were done at different operating condition. The selected vehicle is equipped with previously mentioned measuring instruments. The gas analyzer and its accessories are mounted in the rear seat of the passenger cabinet. Rechargeable power supply and printer are the most important attachment to the analyzer. Gas sampling probe with 3 m long inserted inside the muffler. Its other terminal is connected to the gas analyzer through the window of the rear door. Magnetic inductive transducer is clipped also to the spark plug cable to measure the engine rotational speed. Before starting the measurements, the catalyzer either removed (without) or leaved (with), and the following precautions are taken into account:

(i) The vehicle engine is warm enough before starting the measurement and runs steadily at standard idling configuration.
(ii) All electric accessories like electric fan and radio-cassette are off.
(iii) All the windows of the cabinet are closed except one of the rear windows, which is partially opened (to permit the gas sampling connection). This is to keep the drag effect within the standard value.

During the idle or on-road test, two persons are required to carry out the experimental work. The first person is the vehicle driver, which perform the test program with certain sequence. The driver is responsible to drive the vehicle steadily for enough periods required to obtain steady measurements. The second person is the instruments operator, which is responsible to review the test procedure with the driver and observe the output readings. When the signals become steady, the output readings are recorded and next step of vehicle speed is performed.

RESULTS AND DISCUSSION

The effectiveness of the catalytic converter (catalyzer) results

In Figures 7 to 9, the effectiveness of catalytic converter on the vehicle exhaust gas emissions components in the idle state is shown, where the values of carbon monoxide (CO), carbon dioxide (CO_2) and total hydrocarbon (THC) measured for the bi-fuel vehicle in steady-state when operated by multi-point injection (MPI) system are shown respectively. The fuel is being gasoline and the results are presented with and without catalytic converter (catalyzer). A substational reduction is observed for the levels of the exhaust gas emissions components of CO, CO_2 and THC when the catalyzer is used.

In Figures 10 to 12, the effectiveness of catalytic converter on the vehicle exhaust gas emissions components in the idle state is shown, where the values of carbon monoxide (CO), carbon dioxide (CO_2) and total hydrocarbon (THC) measured for the bi-fuel vehicle in steady-state when operated by multi-point injection (MPI) system are shown respectively. The fuel is being compressed natural gas (CNG) and the results are

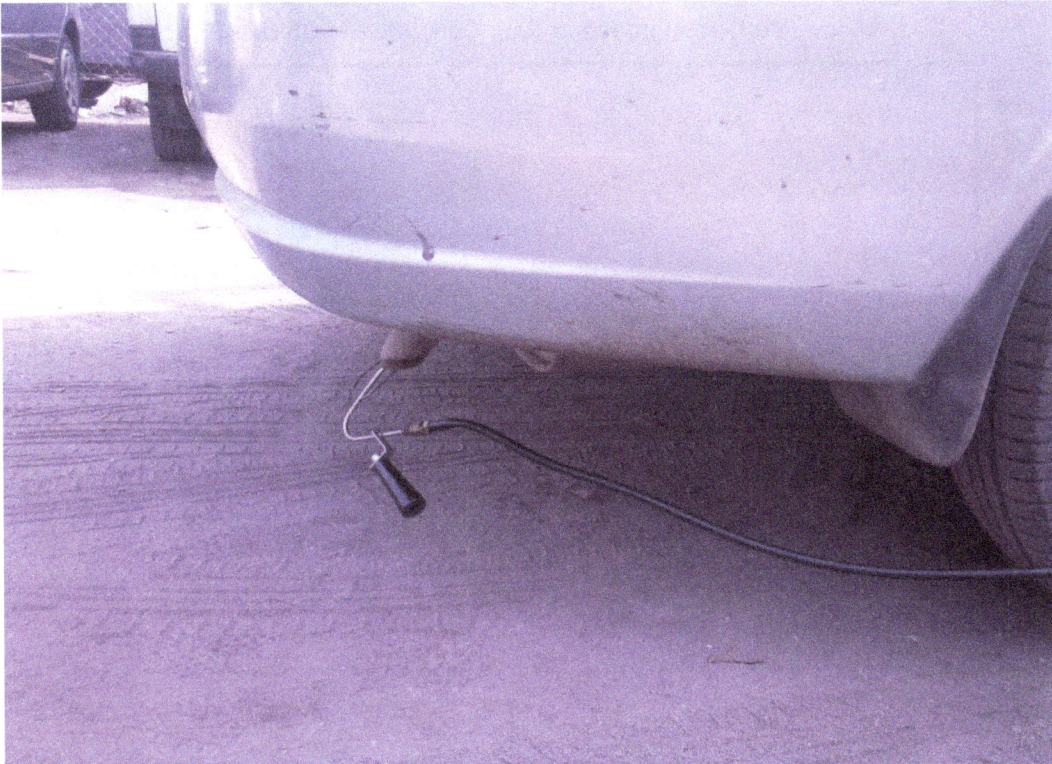

Figure 6. The probe inside the exhaust tailpipe.

Multi-point injection (MPI Gasoline - sequential)

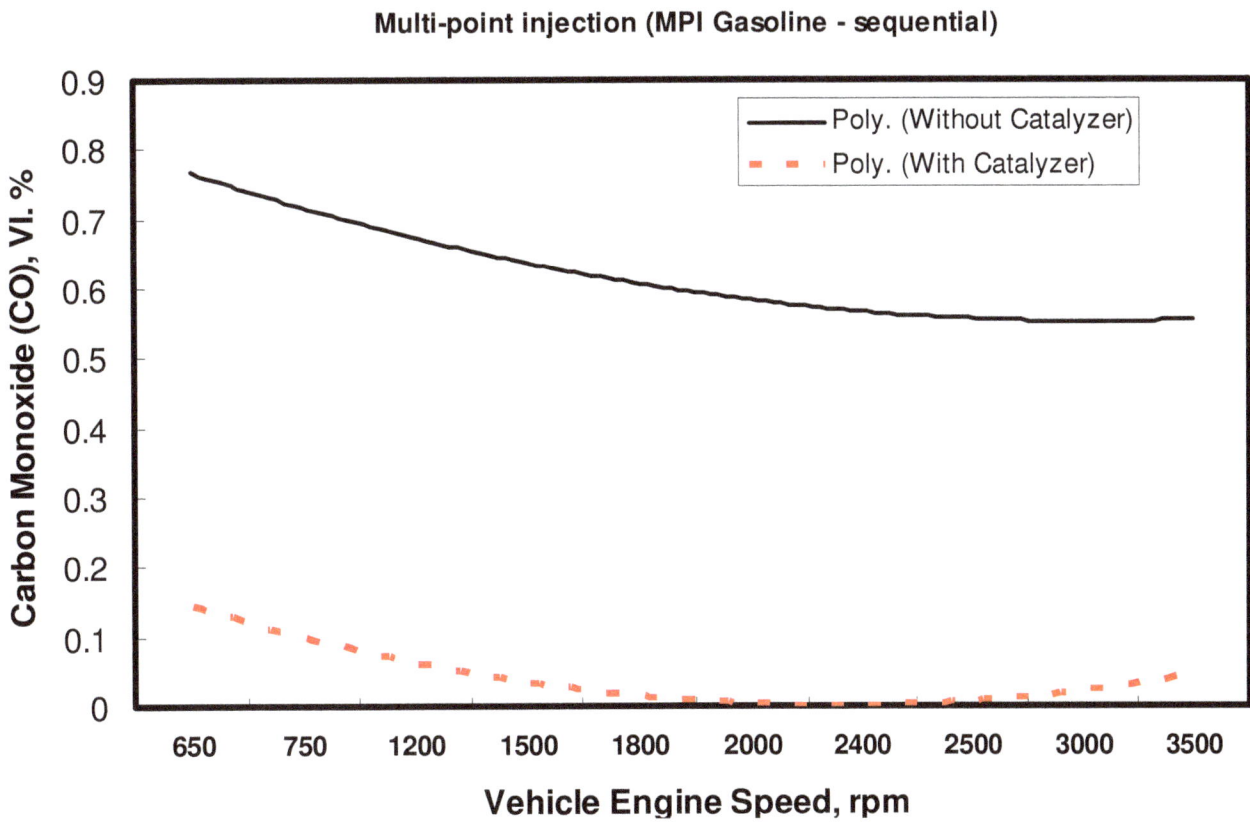

Figure 7. The influence of using catalyzer on carbon monoxide (CO).

Multi-point injection (MPI Gasoline - sequential)

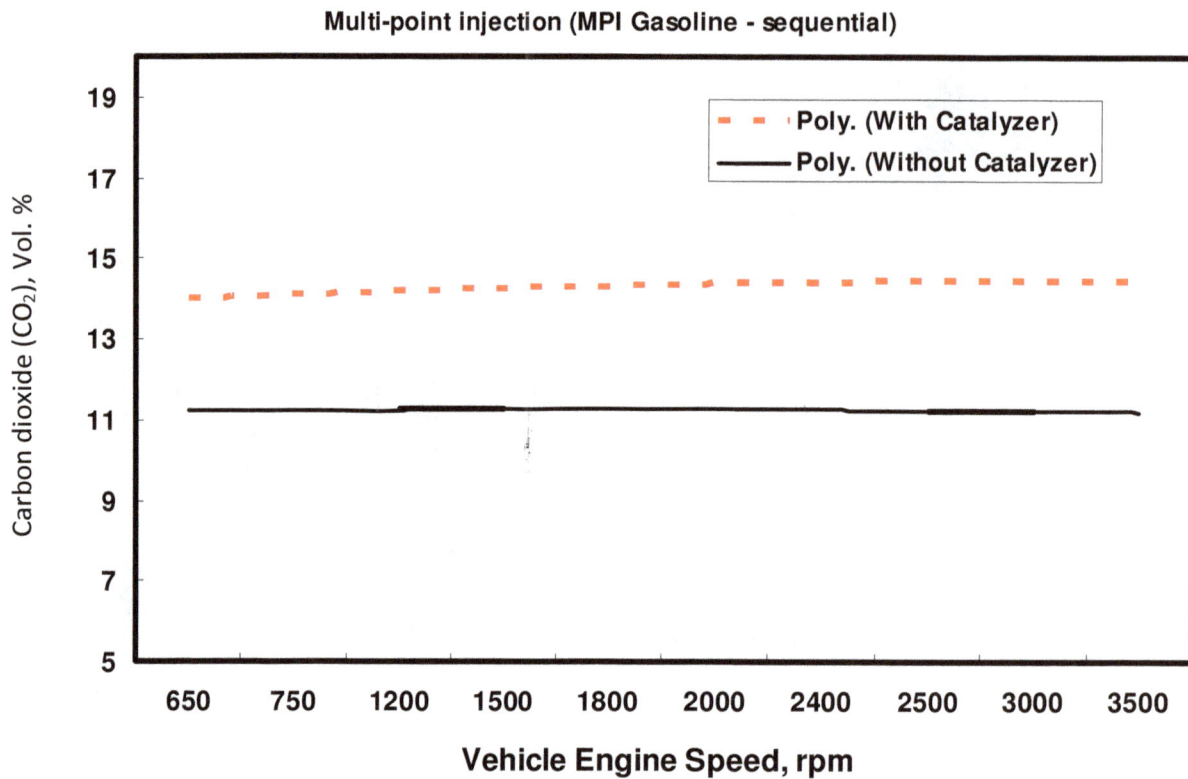

Figure 8. The influence of using catalyzer on carbon dioxide (CO_2).

Multi-point injection (MPI Gasoline - sequential)

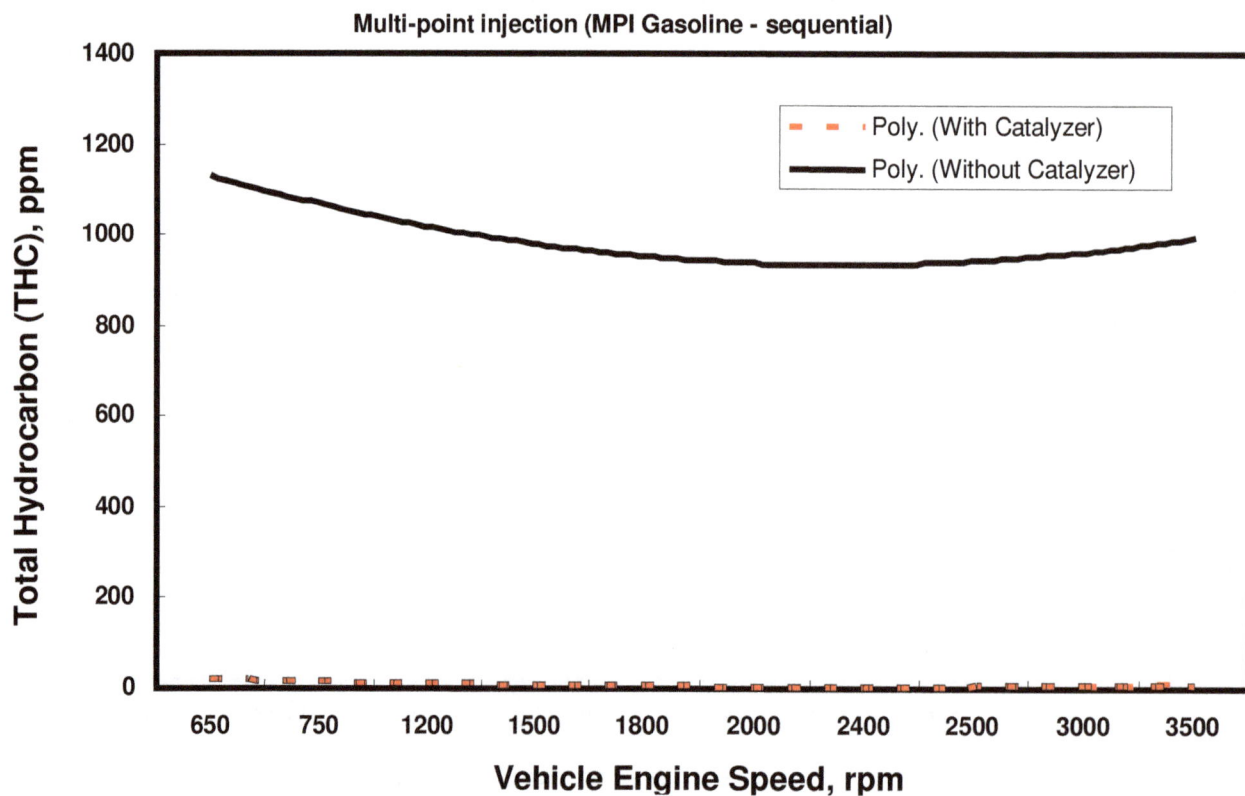

Figure 9. The influence of using catalyzer on total hydrocarbon (THC).

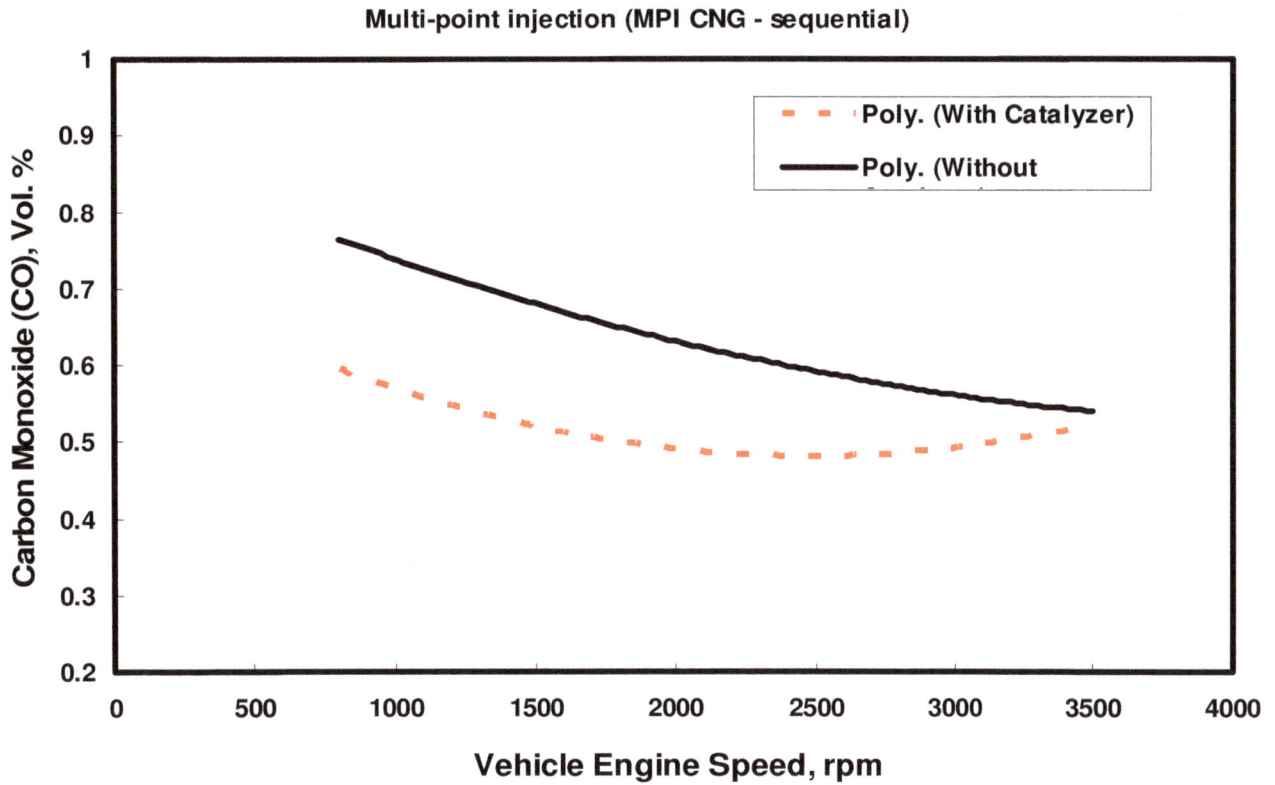

Figure 10. The influence of using catalyzer on carbon monoxide (CO).

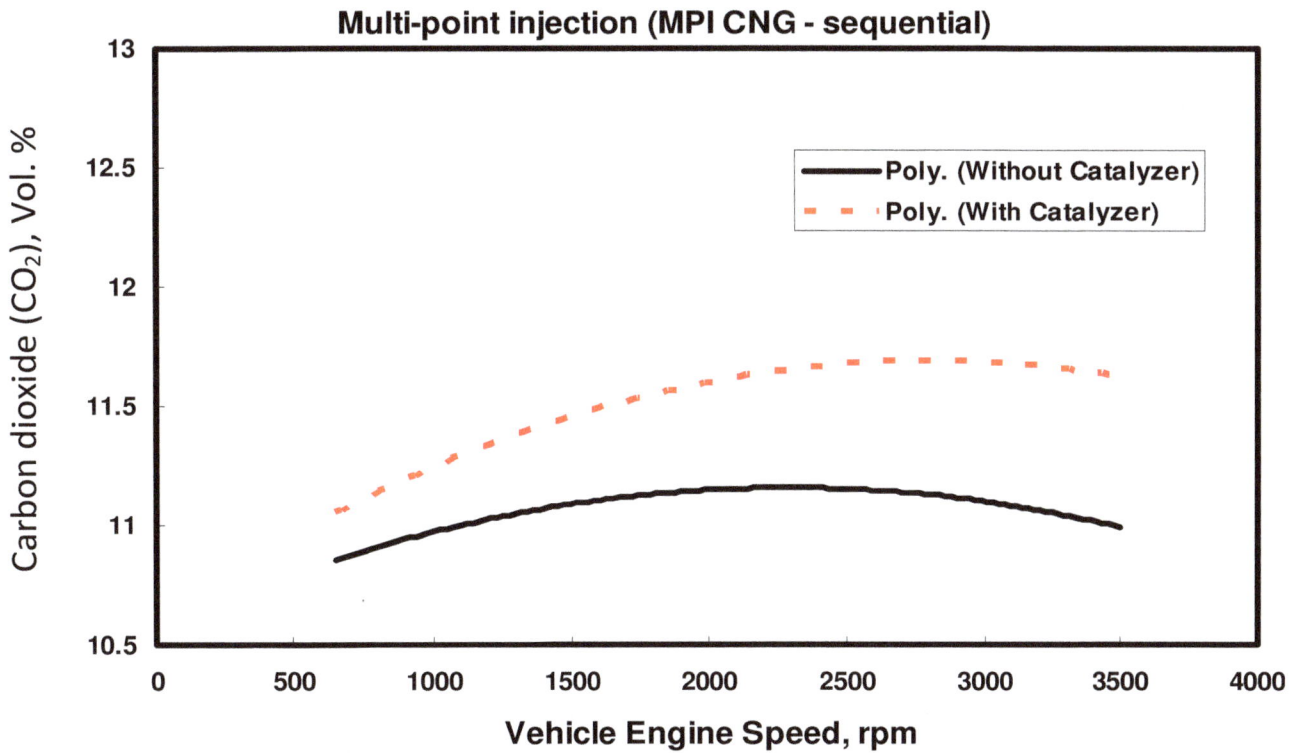

Figure 11. The influence of using catalyzer on carbon dioxide (CO_2).

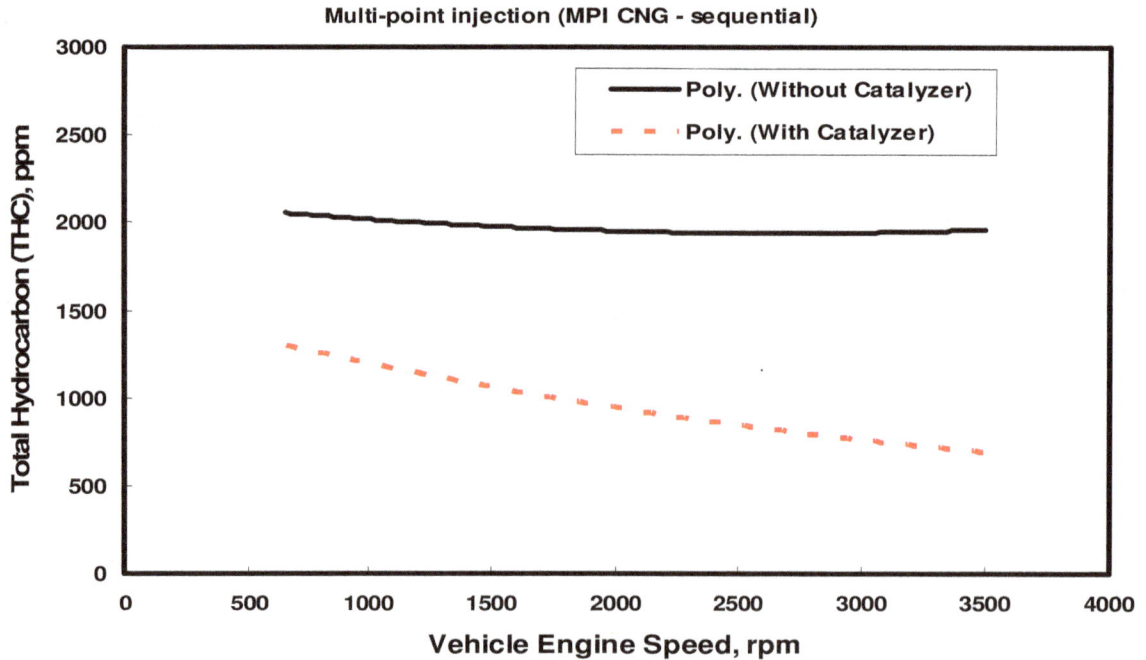

Figure 12. The influence of using catalyzer on total hydrocarbon (THC).

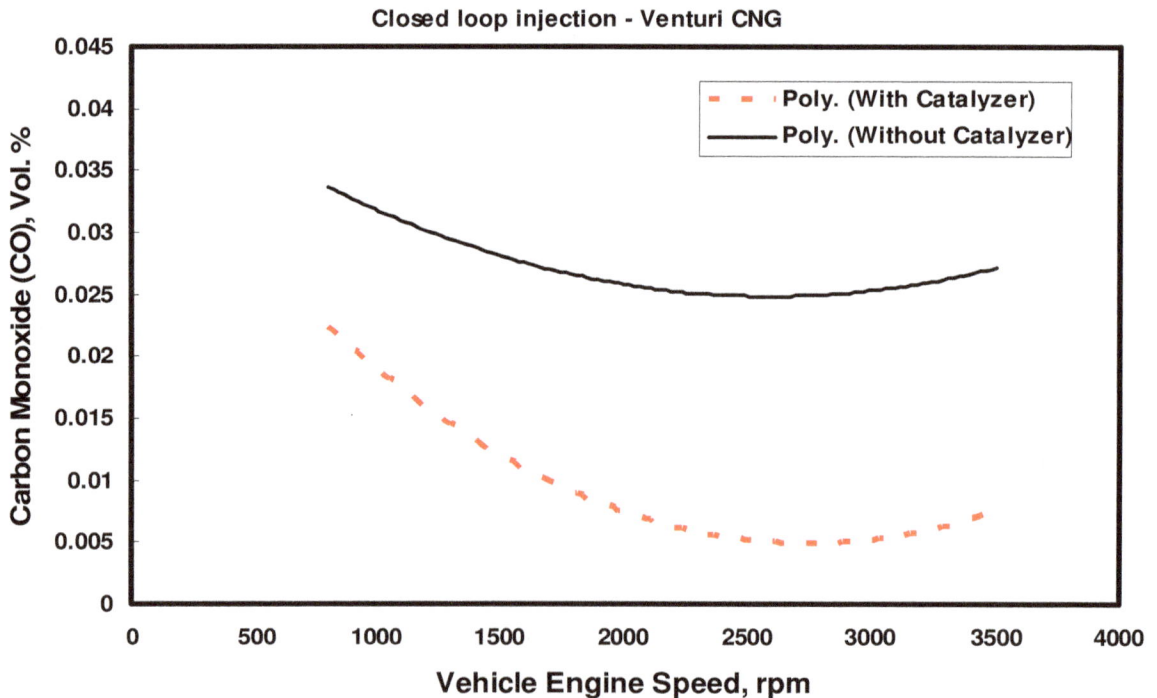

Figure 13. The influence of using catalyzer on carbon monoxide (CO).

presented with and without catalytic converter (catalyzer). A substational reduction is observed for the levels of the exhaust gas emissions components of CO, CO_2 and THC when the catalyzer is used.

In Figures 13 to 15, the effectiveness of catalytic converter on the vehicle exhaust gas emissions components in the idle state is shown, where the values of carbon monoxide (CO), carbon dioxide (CO_2) and total

Closed loop injection - Venturi CNG

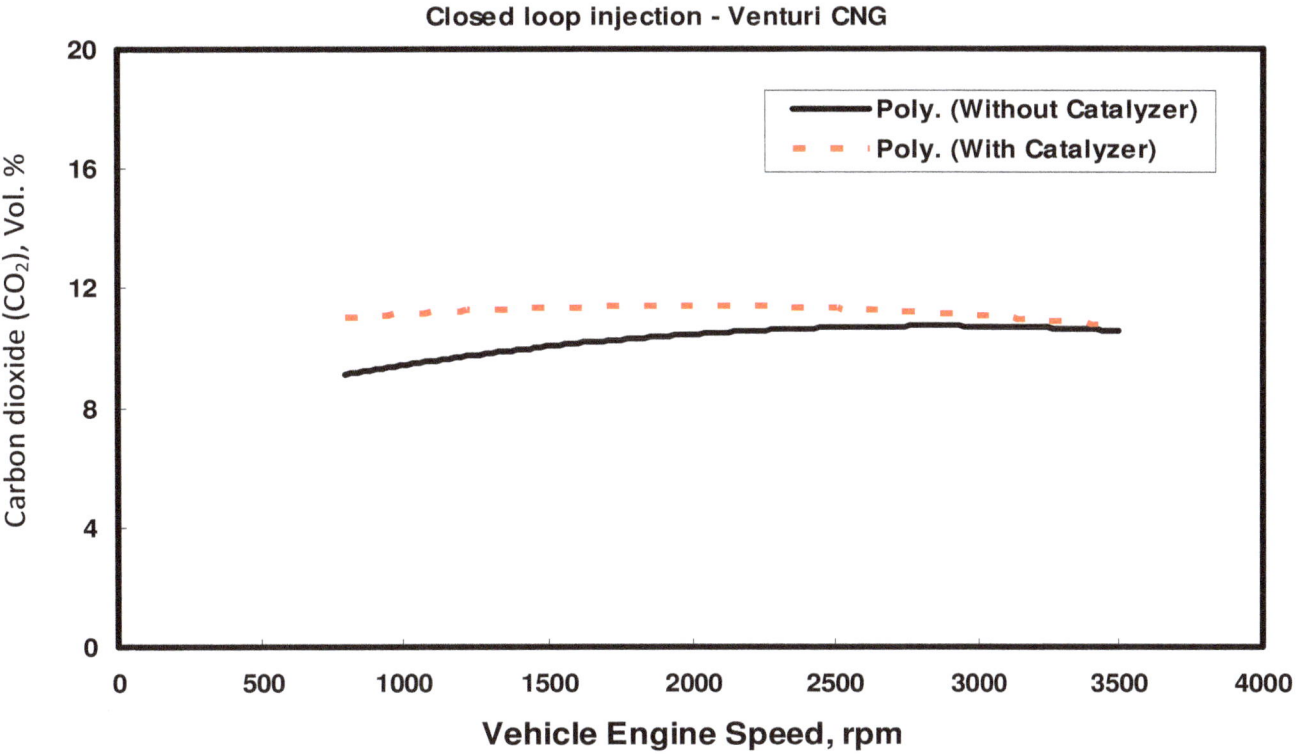

Figure 14. The influence of using catalyzer on carbon dioxide (CO_2).

Closed loop injection - Venturi CNG

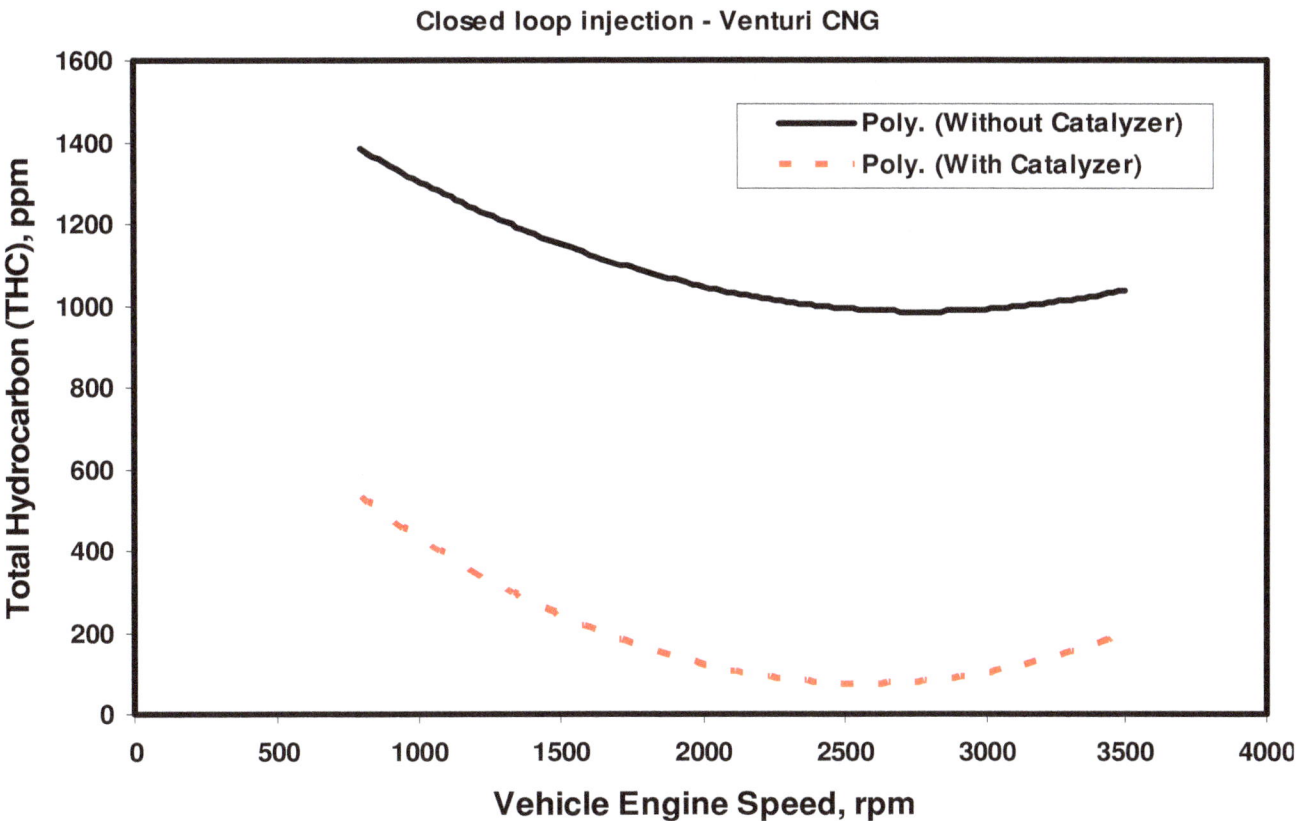

Figure 15. The influence of using catalyzer on total hydrocarbon (THC).

Multi-point injection (MPI Gasoline - sequential

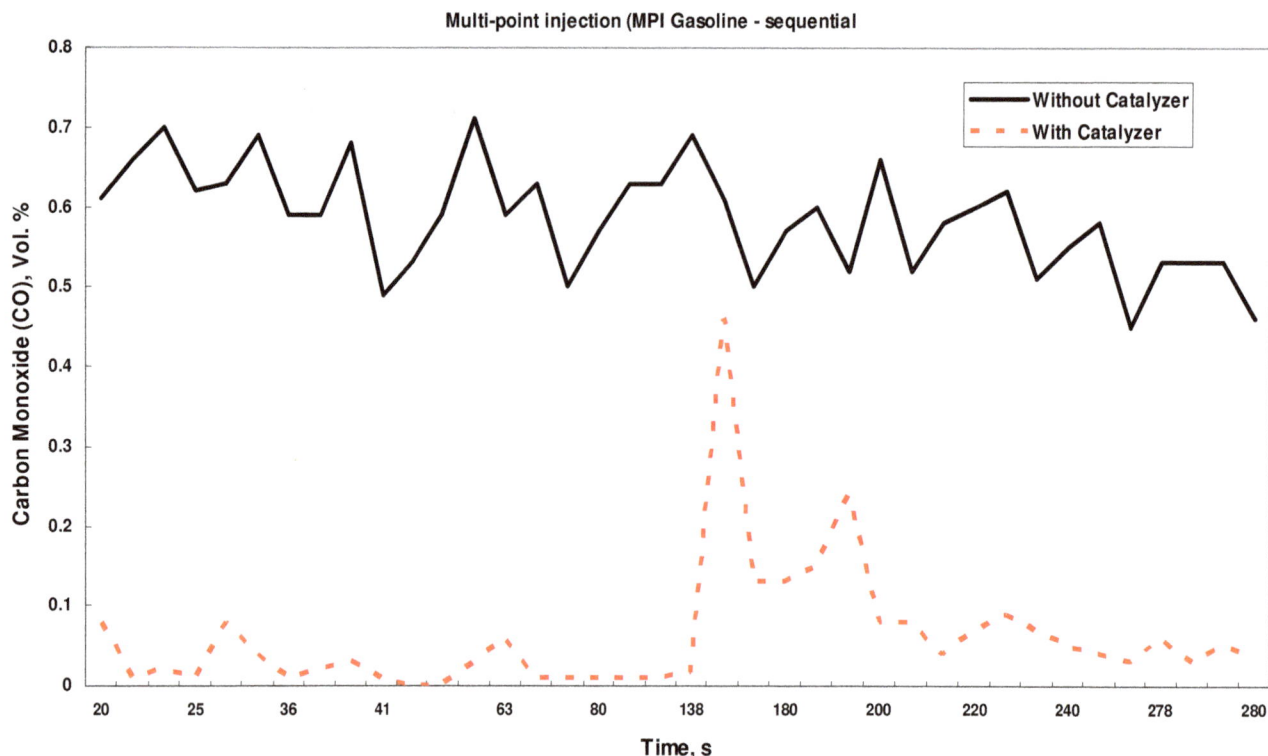

Figure 16. The influence of using catalyzer on carbon monoxide (CO).

hydrocarbon (THC) measured for the bi-fuel vehicle in steady-state when operated by closed loop (venturi) injection system are shown respectively. The fuel is being compressed natural gas (CNG) and the results are presented with and without catalytic converter (catalyzer). A substational reduction is observed for the levels of the exhaust gas emissions components of CO, CO_2 and THC when the catalyzer is used.

In Figures 16 to 18, the effectiveness of catalytic converter on the vehicle exhaust gas emissions components in the transient state based on the European Driving Cycle (ECE-15), the values of carbon monoxide (CO), carbon dioxide (CO_2) and total hydrocarbon (THC) measured for the bi-fuel vehicle when operated by multi-point injection (MPI) system are shown respectively in Figure 3. The fuel is being gasoline and the results are presented with and without catalytic converter (catalyzer). A substational reduction is observed for the levels of the exhaust gas emissions components of CO, CO_2 and THC when the catalyzer is used.

In Figures 19 to 21, the effectiveness of catalytic converter on the vehicle exhaust gas emissions components in the transient state based on the European Driving Cycle (ECE-15), the values of carbon monoxide (CO), carbon dioxide (CO_2) and total hydrocarbon (THC) measured for the bi-fuel vehicle when operated by multi-Figure 3. The fuel is being compressed natural gas (CNG) and the results are presented with and without

catalytic converter (catalyzer). A substational reduction is observed for the levels of the exhaust gas emissions components of CO, CO_2 and THC when the catalyzer is used.

In Figures 22 to 24, the effectiveness of catalytic converter on the vehicle exhaust gas emissions components in the transient state based on the European Driving Cycle (ECE-15), , the values of carbon monoxide (CO), carbon dioxide (CO_2) and total hydrocarbon (THC) measured for the bi-fuel vehicle when operated by closed loop (venturi) injection system are shown respectively in Figure 3. The fuel is compressed natural gas (CNG) and the results are presented with and without catalytic converter (catalyzer). A substational reduction is observed for the levels of the exhaust gas emissions components of CO, CO_2 and THC when the catalyzer is used.

Catalyst conversion efficiency results

In idle state testing, Figures 25 to 27 show the catalyst conversion efficiencies for the CO, CO_2 and THC components resulted from the bi-fuel vehicle, where the point injection (MPI) system are shown respectively invehicle is being equipped by multi-point injection (MPI) system (gasoline), multi-point injection (MPI) system (gasoline) (CNG) and closed loop (venturi) injection

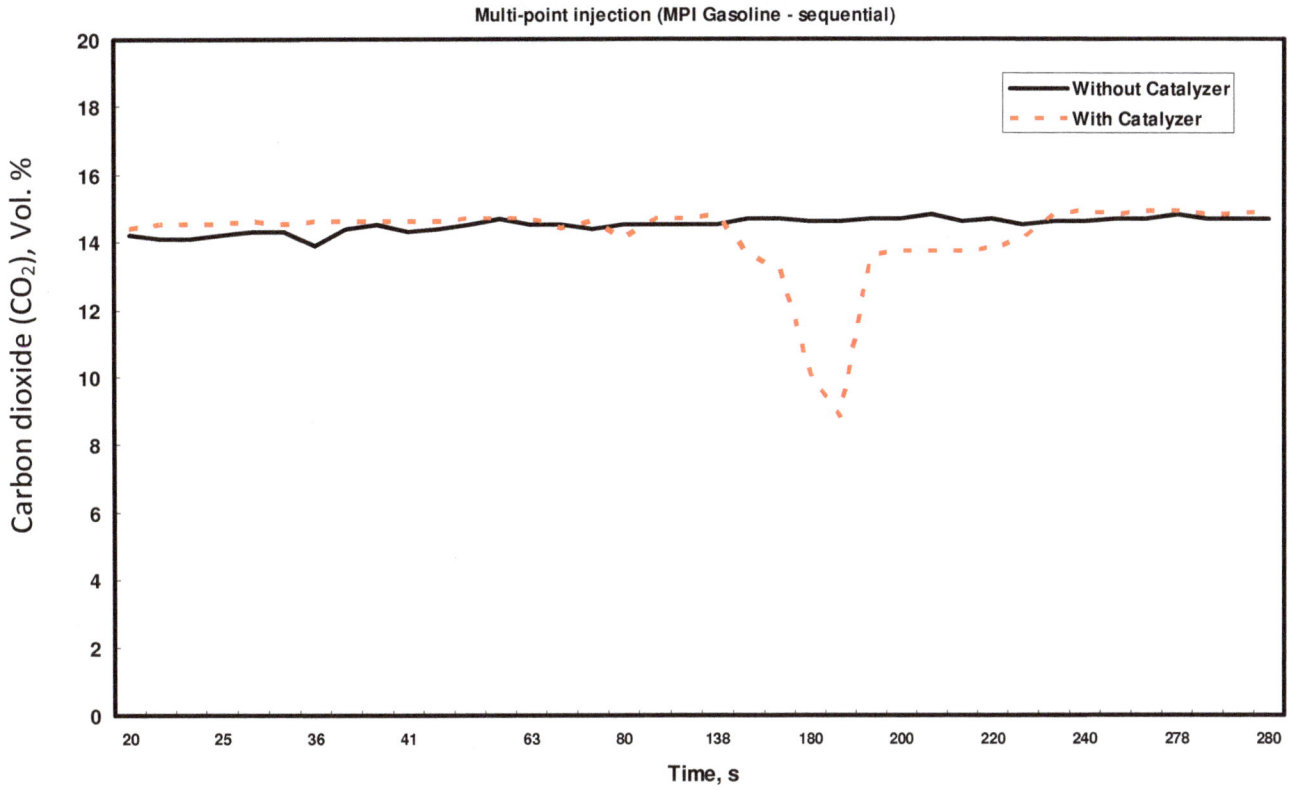

Figure 17. The influence of using catalyzer on carbon dioxide (CO_2).

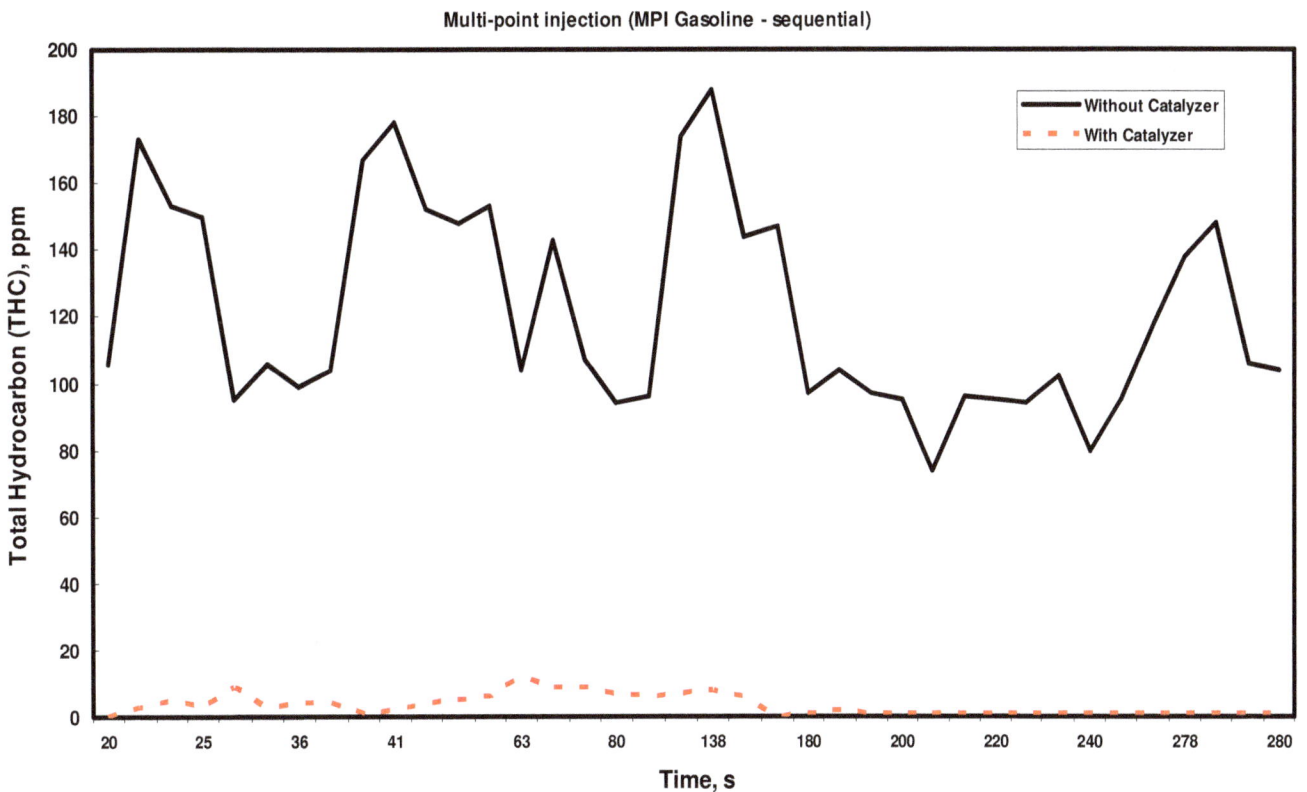

Figure 18. The influence of using catalyzer on total hydrocarbon (THC).

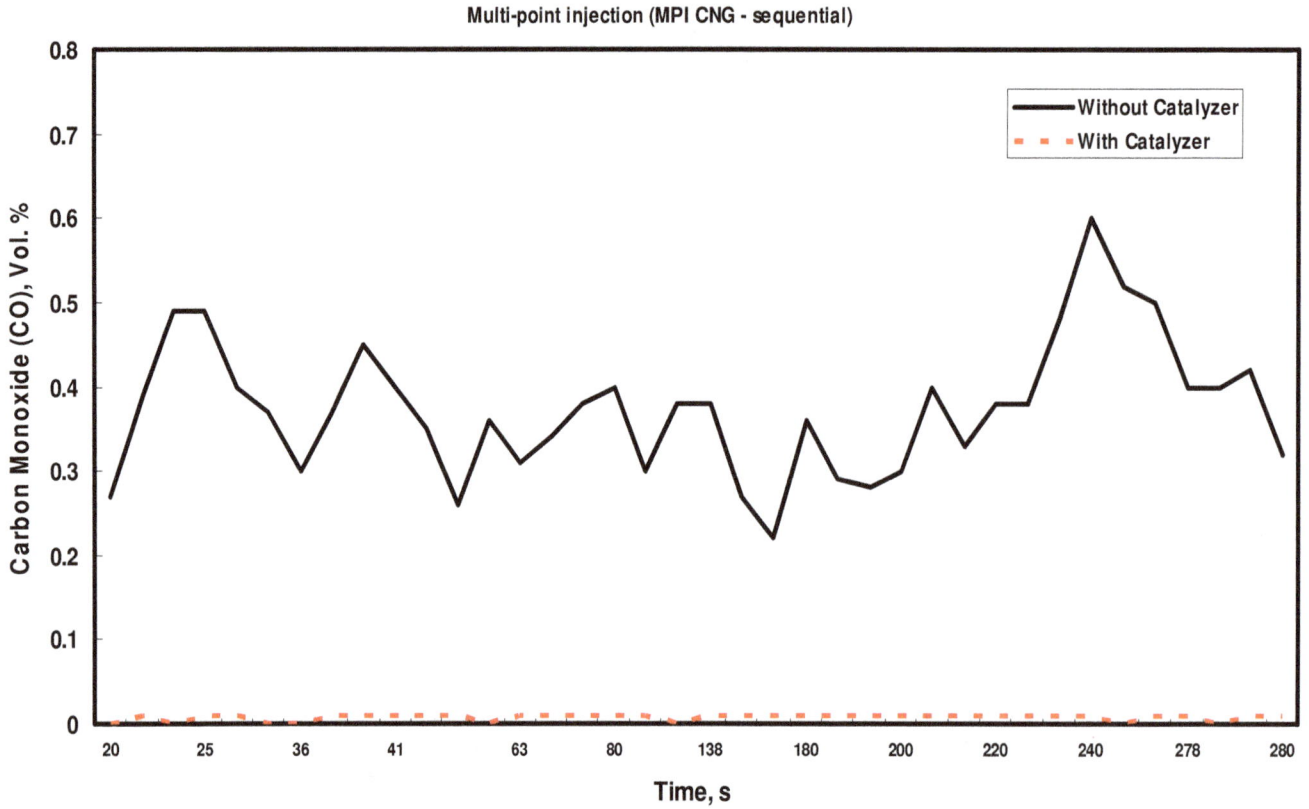

Figure 19. The influence of using catalyzer on carbon monoxide (CO).

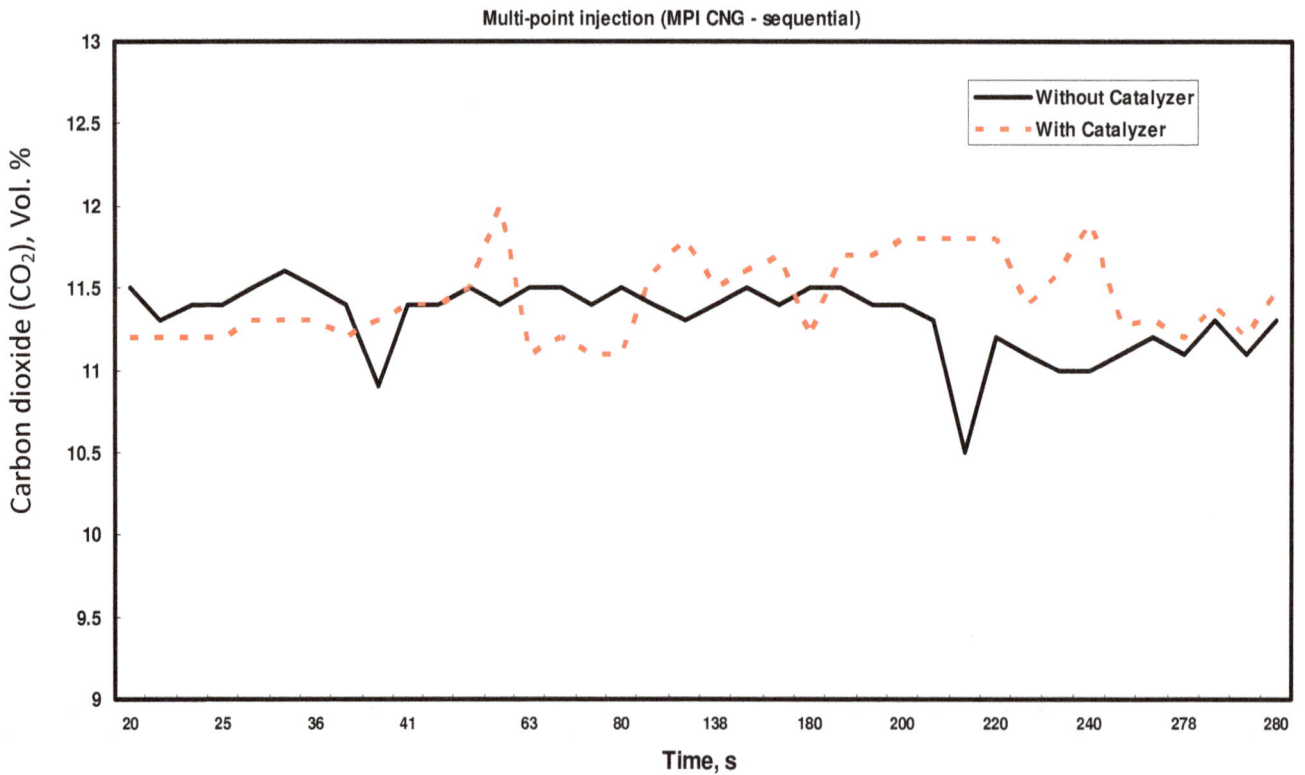

Figure 20. The influence of using catalyzer on carbon dioxide (CO_2).

Multi-point injection (MPI CNG - sequential)

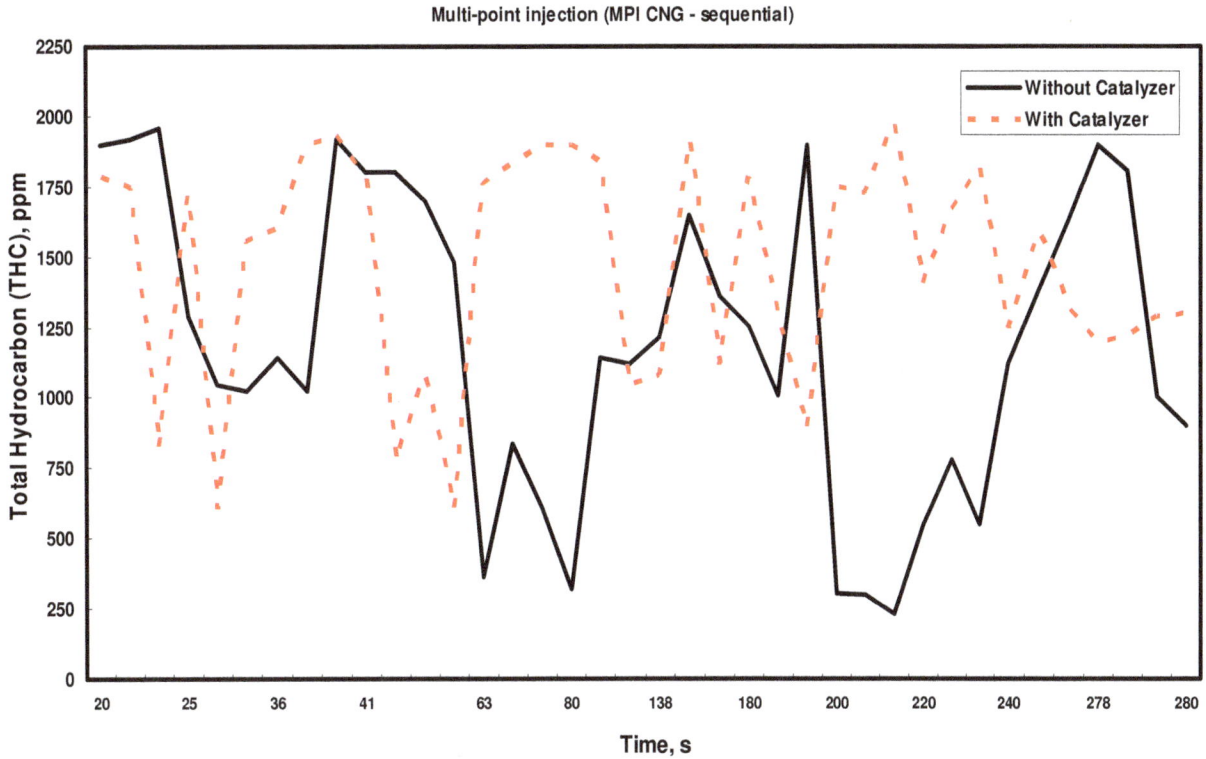

Figure 21. The influence of using catalyzer on total hydrocarbon (THC).

Closed loop injection - Venturi CNG

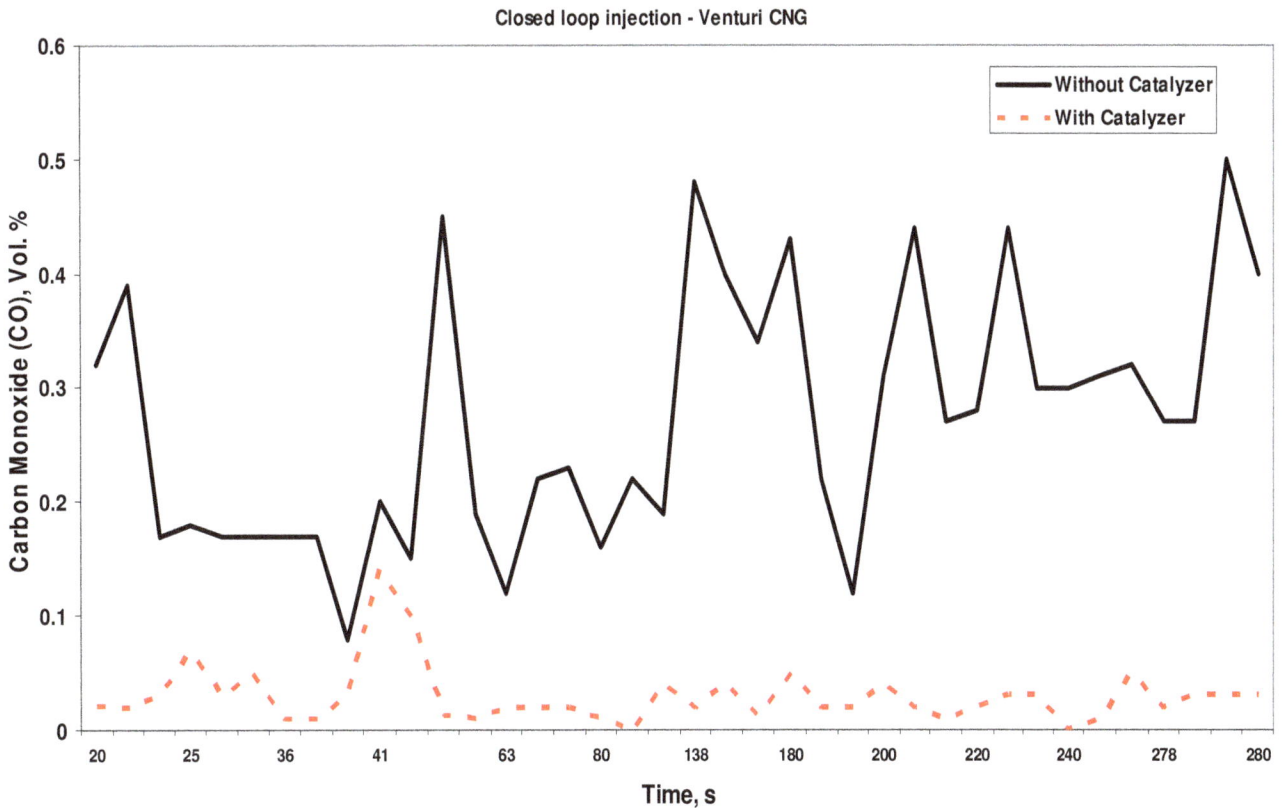

Figure 22. The influence of using catalyzer on carbon monoxide (CO).

Closed loop injection - Venturi CNG

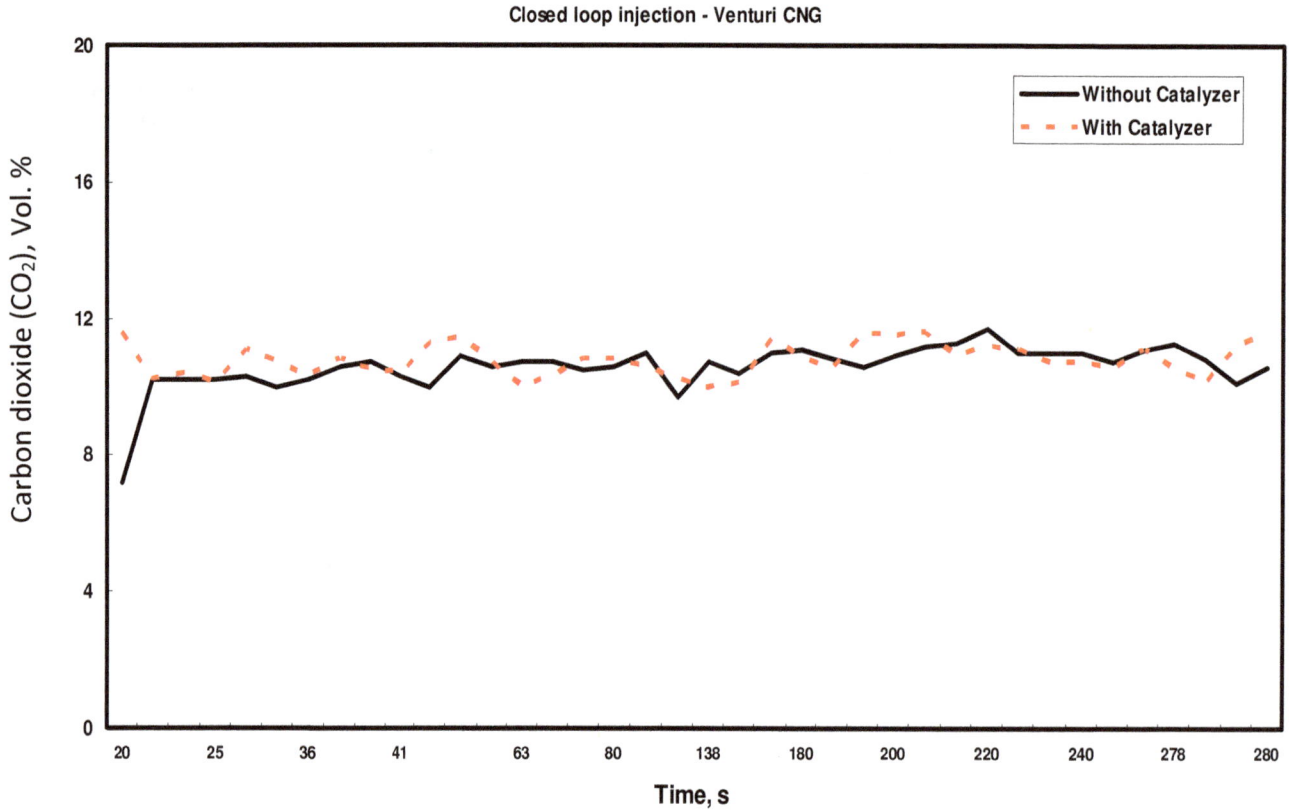

Figure 23. The influence of using catalyzer on carbon dioxide (CO_2).

Closed loop injection - Venturi CNG

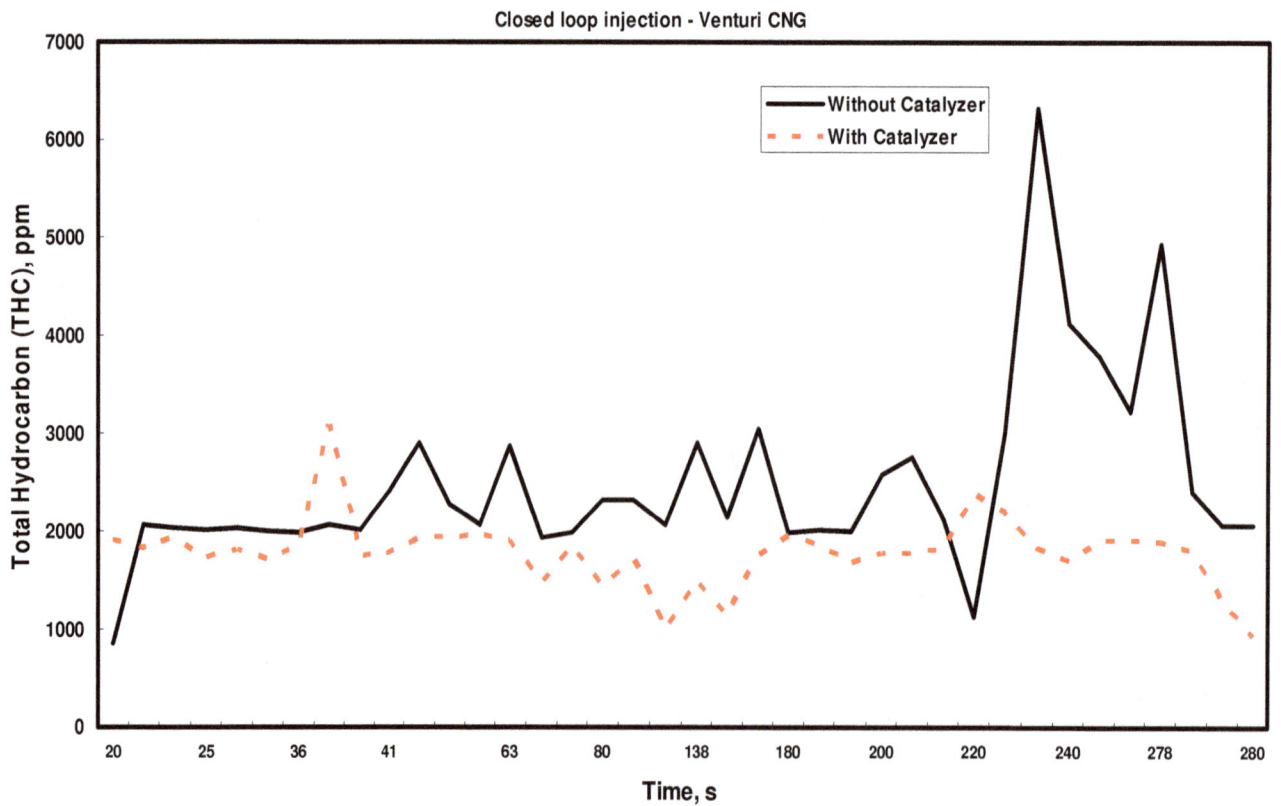

Figure 24. The influence of using catalyzer on total hydrocarbon (THC).

Carbon Monoxide (CO) - Catalyst Conversion Efficiency

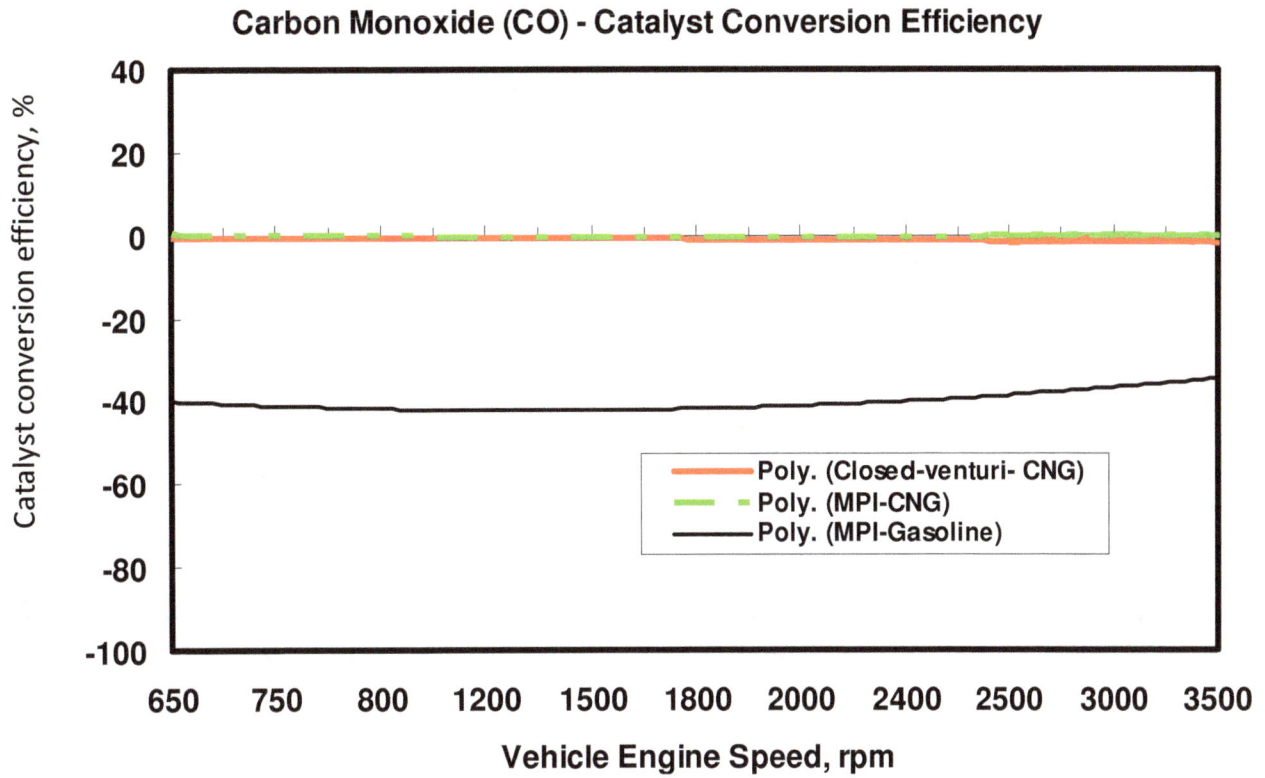

Figure 25. Catalyst conversion efficiency of carbon monoxide (CO).

Carbon dioxide (CO_2) - catalyst conversion efficiency

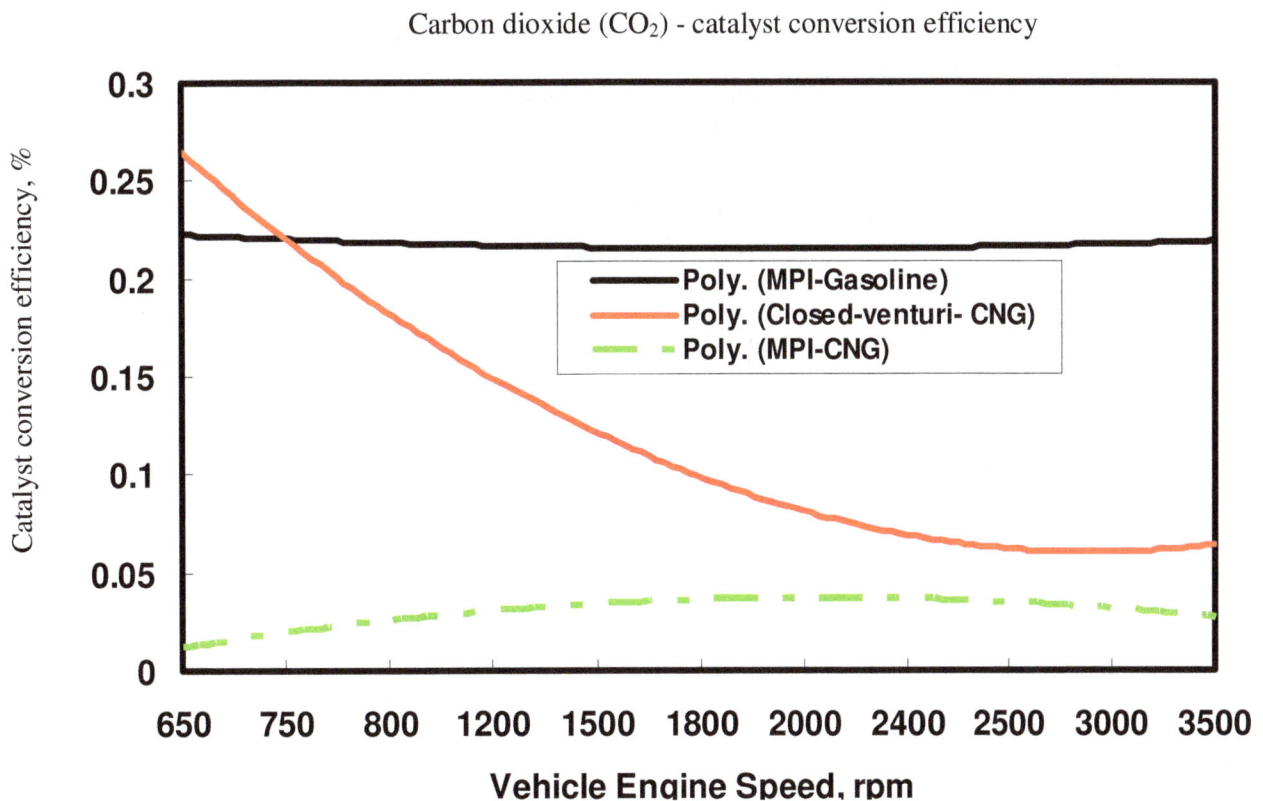

Figure 26. Catalyst conversion efficiency of carbon dioxide (CO_2).

Total Hydrocarbon (THC) - Catalyst Conversion Efficiency

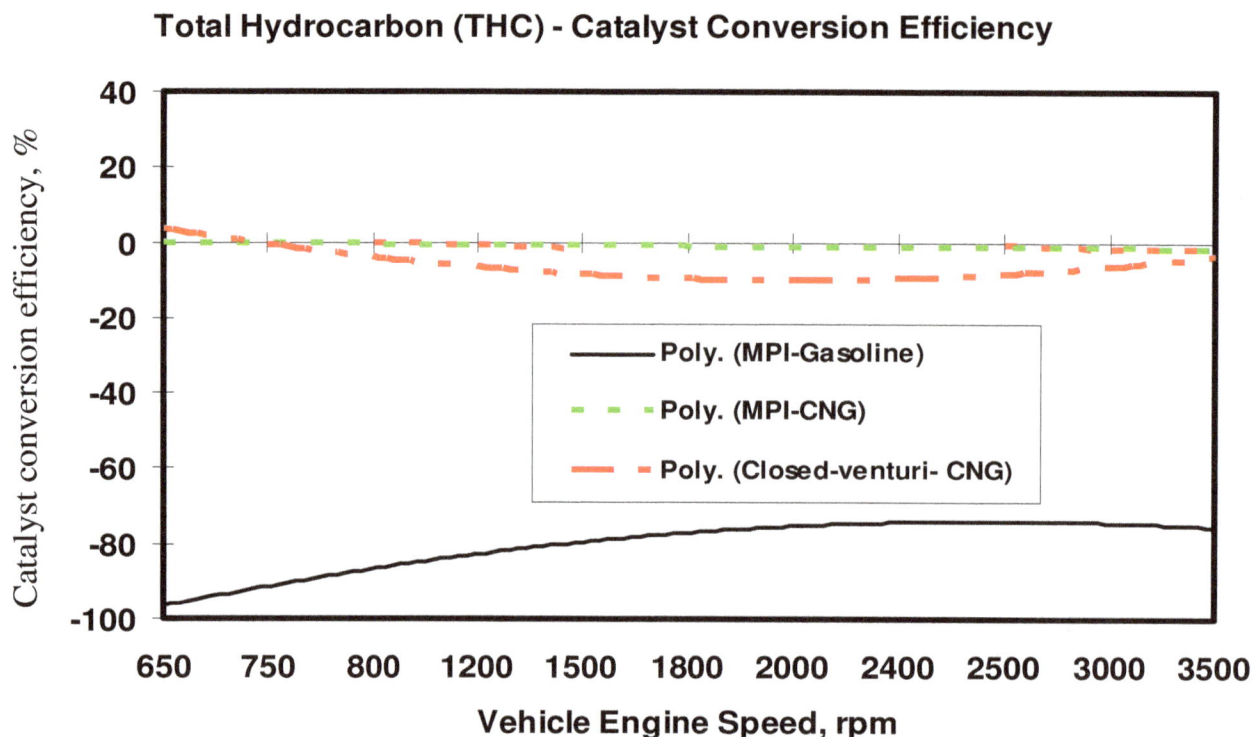

Figure 27. Catalyst conversion efficiency of total hydrocarbon (THC).

(CNG) system, respectively. The catalyst conversion efficiencies are calculated based on Equation 1. It is clearly seen that the catalyst conversion efficiencies (CCE) gained due to the use of the catalytic converter and multi-point injection (MPI) system in the gasoline fuel phase that the CCE is more effective in carbon monoxide (CO) and total hydrocarbon (THC), where they reaches over -40% and over -90% respectively. In the case of carbon dioxide (CO_2), the CCE indicates the value of about 0.22%. The use of multi-point injection (MPI) system in the CNG fuel phase is less effective, where the reduction in CO and THC are about -0.1% and -0.15% respectively, while increase in CO_2 is about 0.05%. In closed loop (venturi) injection system in the CNG fuel phase, the reduction in CO and THC are -0.0 and -10% respectively, while increase in CO_2 is ranged from 0.06 to 0.28%. However, the catalytic converter reduces the amount of both CO and THC, while increases the amount of CO_2

In on-road (transient state) testing and based on the European driving cycle (ECE-15) shown in Figure 3, Figures 28 to 30 show the catalyst conversion efficiencies for the CO, CO_2 and THC components resulted from the bi-fuel vehicle, where the vehicle is being equipped by multi-point injection (MPI) system (gasoline), multi-point injection (MPI) system (gasoline) (CNG) and closed loop (venturi) injection (CNG) system respectively. The catalyst conversion efficiencies are calculated based on Equation 1. It is observed that the catalyst conversion

efficiencies (CCE) for carbon monoxide (CO) gained due to the use of the catalytic converter and multi-point injection (MPI) system in the gasoline fuel phase, multi-point injection (MPI) system in the CNG fuel phase and closed loop (venturi) injection system in the CNG fuel phase, where the values of CO for all the injection systems considered vary against the testing time within the side of reduction with the variation values for MPI-gasoline is bigger followed with MPI-CNG with the lest for closed-loop system (Figure 28). In the case of carbon dioxide (CO_2) values for all the injection systems considered only MPI- gasoline and MPI-CNG vary against the testing time in both sides (decrease and increase) with the variation values for MPI-CNG is bigger than that for MPI-gasoline with almost tiny reduction values for closed-loop system (Figure 29). In the case of total hydrocarbon (THC) values for all the injection systems considered only MPI- gasoline varies against the testing time within the side of reduction, while the other two injection systems exhibit almost tiny reduction values (Figure 30). However, the catalytic converter is more effective in MPI-gasoline than those for the other two injection systems.

It is well known that the formation of CO in internal combustion engine is a result of incomplete combustion of the fuel. This occurs when there is insufficient oxygen near the hydrocarbon (fuel) molecule during combustion. Moreover, incomplete combustion could be caused by the quenching of the hydrocarbon oxidation near a

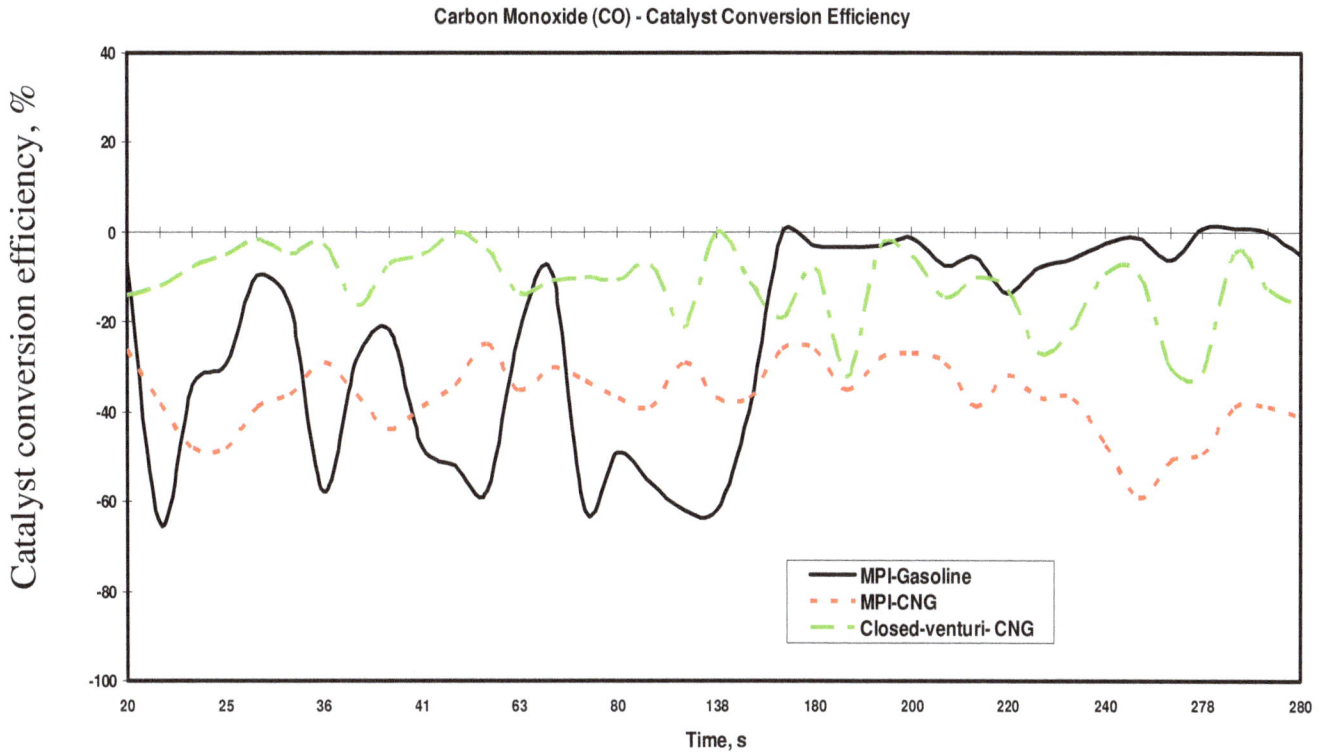

Figure 28. Catalyst conversion efficiency of carbon monoxide (CO).

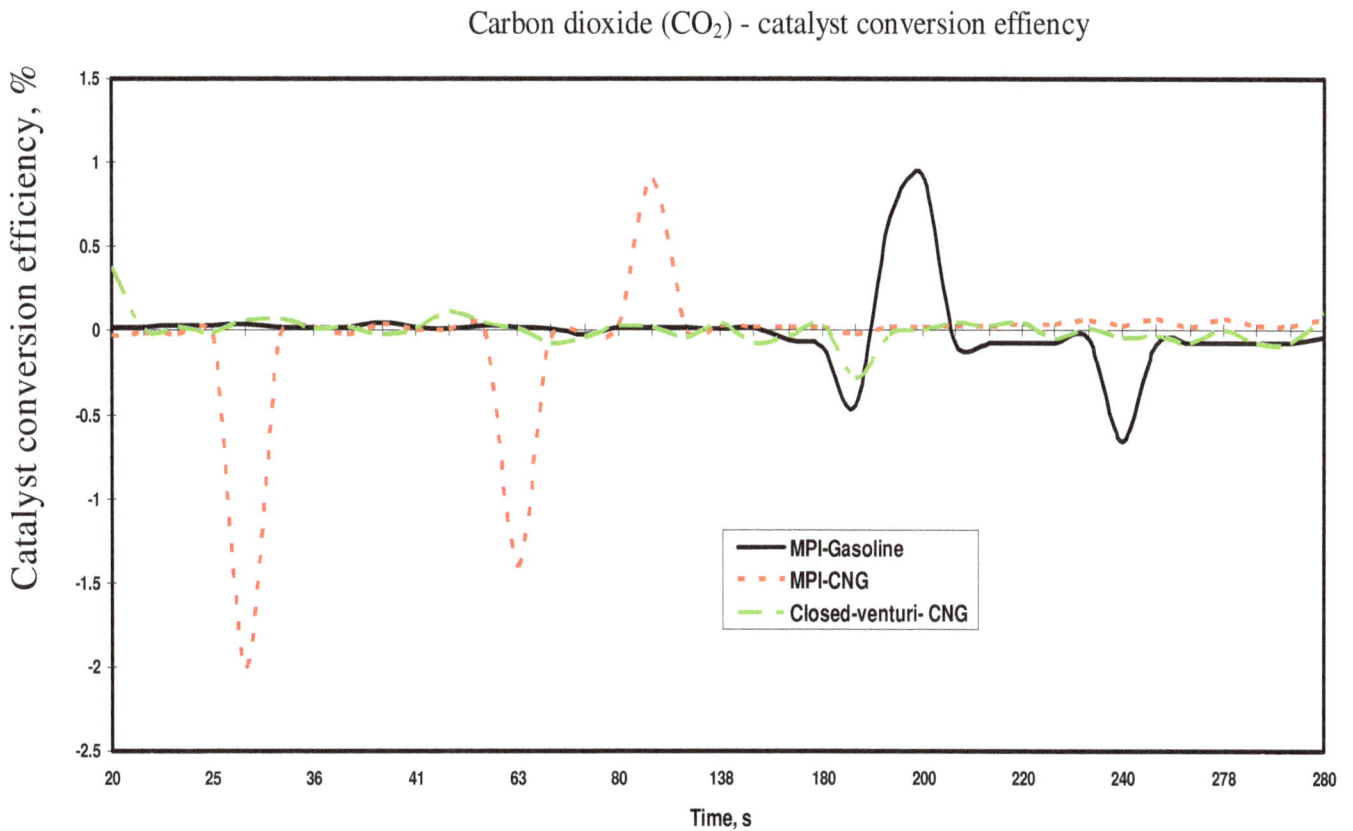

Figure 29. Catalyst conversion efficiency of carbon dioxide (CO_2).

Total Hydrocarbon (THC) - Catalyst Conversion Efficiency

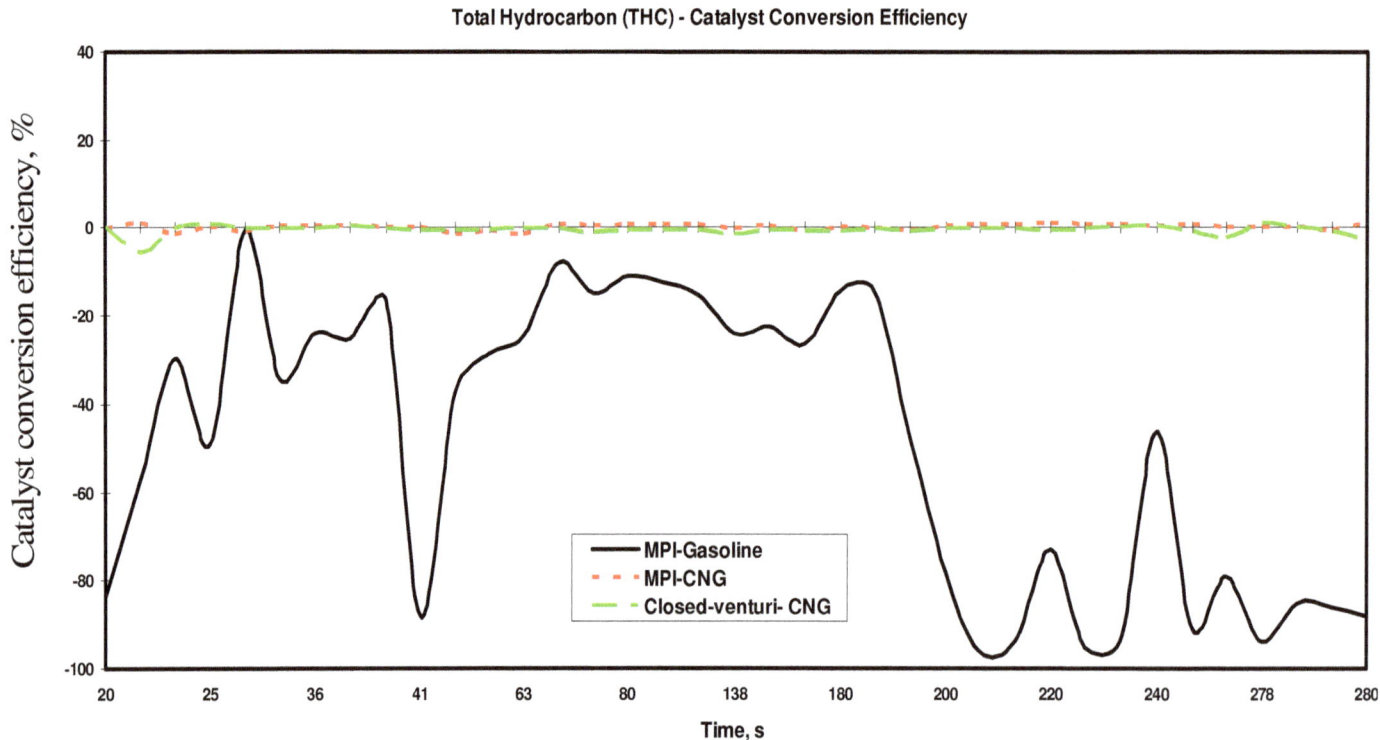

Figure 30. Catalyst conversion efficiency of total hydrocarbon (THC).

cold surface in the combustion chamber. However, in either steady or transient state and with MPI system, it is found that CNG produces a lower level of CO emissions due to lean conditions as compared to gasoline fuel. This result reveals that CeO_2 is effective as an OSC that supports platinum (Pt) to oxidize CO into CO_2 at low to high temperature. On the other hand, it is found that CNG reduces lower THC emission (non methane THC) mainly due to lean burn condition than gasoline. The catalytic material is effective to oxidize HC with increasing speed or catalytic temperature, where the conversion efficiency of CNG is lower than gasoline fuel.

Conclusions

1- It is found that the reason behind the increase of total hydrocarbon (THC) in compressed natural gas (CNG) operated vehicle over that produced in gasoline operated vehicle is due to the difficulty in oxidizing the unburned hydrocarbons in the exhaust gases, where the oxidization of hydrocarbons is one of the functions of the three–way catalyzer. The exhaust hydrocarbons of a gas-operated vehicle have a significantly different composition to those of a gasoline-operated vehicle.

2- The effectiveness of catalytic converter (catalyzer) in the condition of idle state is much better than that for transient state based on the European Driving Cycle (ECE-15) shown in Figure 3, where in the case of idle

state, the catalyzer is fully warm and hence operating at it's maximum efficiency, while for the case of transient state condition the effectiveness is not very well, where the catalyzer was faced by air flow and temperature changes rapidly.

3- In the case of catalyst conversion efficiency, the catalytic converter reduces the amount of both CO and THC, while increases the amount of CO_2 (idle state). This is attributed to the fact that the vehicle emission levels are quite low and the noble metal catalyst required for the THC emissions would easy remove most of the CO under these lean conditions. Furthermore, the catalytic converter is more effective in MPI-gasoline than those for the other two injection systems (transient state).

4- It is hope that the results in this study can help to adopt and developed Cairo driving cycle, which is more realistic to represent Cairo traffic conditions, and can be used for tests of vehicles running in Cairo in future in order to report the real world performance of vehicles in service. Thus providing information for Egyptian's energy and environmental ministry on how to set up proper national standards for the motor vehicles fuel consumption and exhaust emissions. It is also hoped that the proposed methods could be used for other big cities as well as for other types of vehicle.

REFERENCES

Yaacob II, Hadi A, Rahman NLA (2002). Evaluation of the catalytic converterperformance for car engines," Proc., 6th Asia-Pacific int. symposium on combustionand energy utilization. Kuala Lumpur, Malaysia, pp. 262-267.

Yaacob II and Rahman NLA (2003). Zeolite catalyst synthesis: Characterization and their catalytic activity reduction of gasoline exhaust gases. Materials science forum 43.

Burch DS, Keyser A, Colucci MPC, Potter FT, Benson KD, Biel PJ (1996). Applications and benefits of catalytic converter thermal management. J. Fuels Lubricants, SAE 961134, 4(105): 839-844.

Kalam MA, Masjuk HH, Maleque MA (2001). Gasoline engine operated on compressed natural gas. Proc., Adv. Malaysian Ener. Res., pp. 307-316.

Weaver S (1989). Natural Gas Vehicles: A Review of the State of the Art. SAE, 892133.

Tong HY, Hung WT, Cheung CS (2000). On-Road Motor Vehicle Emissions and Fuel Consumption in Urban Driving Conditions. J. Air Waste Manage. Assoc., 50: 543-554.

Pipitone E, Beccari A (2007). A Study on the Use of Combustion Phase Indicators for MBT Spark Timing on a Bi-Fuel Engine. SAE 2007-24-0051.

Al-Shemmeri TT (1993). Experimental and computational study of a spark ignition engine fuelled with petrol or natural gas. IMechE 1993, C465/015/93, pp. 79-83.

Kalam MA, Masjuki HH, Maleque MA, Amalina M, Abdesselam AH, Mahlia TM (2004). Air-fuel ratio calculation for a natural gas fuelled spark ignition Engine. SAE No. 01-0640.

US (1993). Environmental Protection Agency (US.EPA). Federal Test Procedure Review Project. Preliminary Technical Report, EPA 420-R-93-007. USA.

Tong HY, Hung WT, Cheung CS (1999). Development of A Driving Cycle for Hong Kong. Atmospheric Environ., 33: 2323-2335.

Mah TA (1993). Modal assessment of exhaust emissions potential of a lean burn petrol engine vehicle. In Proc., IMechE, C465/030/93, London, pp. 89-196.

26

Evaluation of homogeneous catalysis and supercritical methodology for biodiesel production from *Jatropha curcas, Azadirachta indica and Hevea brasiliensis* oil seeds

C. S. Ezeanayanso[1]*, E. M. Okonkwo[1], E. B. Agbaji[2], V. I. Ajibola[2] and O. J. Okunola[1]

[1]National Research Institute for Chemical Technology, P. M. B. 1052, Basawa, Zaria, Kaduna State, Nigeria.
[2]Department of Chemistry, Ahmadu Bello University, Samara, Zaria, Kaduna State, Nigeria.

The present study investigated the production of biodiesel from *Jatropha curcas, Azadirachta indica, Hevea brasiliensis* oil seed via transesterification methods: base catalysed and supercritical methanol, a catalyst-free process, as a cleaner alternative to conventional chemically catalyzed process. In biodiesel production using base catalyzed transesterification, temperature, methanol: oil molar ratio, catalyst type, catalyst concentration and agitation were found to have significant role on ester conversion. For supercritical transesterification method, reaction was performed in a tubular reactor, at 250°C, with 200 rpm, varying the pressure in the range from 1 to 100 bars. The results obtained showed that non-catalytic supercritical methanol technology required 3, 4, 5 min reaction time to produce 98, 97.17 and 87.1% biodiesel from *J. curcas, A. indica and H. brasiliensis* seed oil, respectively. Compared to conventional catalytic methods, which required at least 1 h reaction time to obtain similar yield, supercritical methanol technology has been shown to be superior in terms of time and energy consumption. Apart from the shorter time, it was found that separation and purification of the products were simpler since no catalyst is involved in the process. Hence, formation of side products such as soap in catalytic reactions does not occur in the supercritical methanol method.

Key words: Biodiesel, *Jatropha curcas, Azadirachta indica, Hevea brasiliensis,* supercritical.

INTRODUCTION

Developing alternative energy source to replace traditional fossil fuels has recently become more and more attractive due to the high energy demand, the limited resource of fossil fuel and environmental concerns (Ramadhas et al., 2005). Biodiesel fuel derived from vegetable oils or animal fats is one of the promising possible sources to be substituted for conventional diesel fuel and produces favorable effects on the environment (Shieh et al., 2003). However, in spite of favorable impact, the economic aspect of biodiesel production is still a barrier for its development, mainly due to the lower price of fossil fuels (Antolin et al., 2002). The cost of biodiesel production is dependent on two factors; the

costs of feedstocks and production process cost. Although, the cost of biodiesel production is highly dependent on the costs of feedstock which affect the cost of the finished product up to 60 to 75% (Cetinkaya and Karaosmanoglu, 2004). The 25 to 40% cost of production also dependent on the production process using catalyzed transesterification including homogeneous and hetero-geneous catalyst. The transesterification reaction is affected by alcohol type, molar ratio of glycerides to alcohol, type and amount of catalyst, reaction temperature, reaction time and free fatty acids and water content of vegetable oils or animal fats (Vieitz et al., 2008). Nevertheless, the usage of catalyst complicates the transesterification reaction mainly in the need to separate the catalyst from the product mixture.

Recently, new production processes have been developed among which supercritical process is one, an alternative catalyst-free method for transesterification

*Corresponding author E-mail: chikaeze2@yahoo.com, chikaeze2@gmail.com.

which uses supercritical methanol at high temperatures and pressures. Although, high temperatures and pressures are required for this process, however, energy costs of production are similar to catalytic production routes. In studies, it has been reported that with methanol at supercritical stage, a single phase of methanol-oil mixture can be obtained instead of two phase methanol-oil mixture observed at room conditions. This is because at supercritical stage, the dielectric constant of liquid methanol and enables the transesterification reaction to complete in a very short reaction time. Since catalyst is not being used, this process is much simpler and superior as compared to conventional catalytic transesterification process in terms of cost saving and purification of the product mixture (Demirbas, 2006).

Currently, partially or fully refined and edible-grade vegetable oils, such as soyabean, rapeseed and sunflower, are the predominant feedstock for biodiesel production (Hass, 2005), which obviously results in the high price of biodiesel. Therefore, exploring ways to reduce the cost of raw materials is of much interest in recent biodiesel research.

With the foregoing information, in the present study, *Jatropha curcas*, *Azadirachta indica,* and *Hevea brasiliensis* seed oil, was chosen as the raw material to produce biodiesel using supercritical condition of methanol. The process can tolerate water in the feedstock, free fatty acids are converted into fatty acid methyl esters with similar or higher rates than the corresponding triglycerides, and therefore, the process can use high FFA, cheaper feedstocks such as rubber seed oil. Also, the catalyst removal step is eliminated making this process as environmental friendly.

MATERIALS AND METHODS

Dried *J. carcus* and *A. indica* seeds were obtained from the National Research Institute for Chemical Technology, Zaria (Nigeria), while the dried *H. brasiliensis* seeds were obtained from the Rubber Research Institute Benin Nigeria. The seeds were cleaned by removing debris using hand picking method, dried to constant mass in an oven at 50°C for 72 h, dehauled to remove the seed coat (with little or no oil), and the seeds dried at 50°C for another 48 h. The oil was extracted by oil expeller. The seeds were fed to a series of expellers to receive a mild pressing on continuous basis, as each screw press gradually increased the pressure on the incoming material through the interior of a closed barrel.

The extracted oil were drained out through a small gap between positioned hardened steel bar in the barrel cage. The oils were dark Yellow colour, with neem oil being more viscose and denser. The oils were filtered using muslin cloth to remove dirt and other inert materials. The oils were heated in flasks up to near boiling point to remove water contaminant, allowed to cool to room temperature (27°C) and taken for biodiesel production.

Esterification of seed oil

The free fatty acid (FFA) content of the biodiesel was determined using ASTM standard methods by initially determination of acid value by neutralization reaction using 0.1 M KOH, hence the FFA is evaluated by dividing the acid value by 2. The Jatropha and neem oil had 0.62 and 0.78% free fatty acid (FFA) content respectively and were found suitable for direct transesterification. The rubber seed oil had high FFA content of 16.58% and was refined, as the yield of esterification process decreases considerably where FFA value > 2% (Ramadhas et al., 2005). The FFA of rubber seed oil was reduced below 2% using paratoluene sulphonilic acid as catalyst prior to transesterification. A round bottom flask was used as laboratory scale reactor vessel, and a hot plate with magnetic stirrer was used for the heating. The reaction/heating time, agitation speed and temperature were measured for the test runs of combination of reactants and process conditions.

Homogeneous catalyzed transesterification (alkaline esterification)

The homogeneous catalyzed transesterification of jatropha, neem and rubber oil was carried out using methanol with potassium hydroxide (KOH) catalyst to obtain monoesters of fatty acids. The controlling parameters for the alkaline esterification were methanol to oil molar ratio (3:1 to 12:1), catalyst concentration (0.25 to 1.50% wt/wt of oil), reaction temperature (50 to 65°C) and rate of agitation (150 to 450 rpm). The wide range of reaction conditions were selected to obtain higher conversion to optimize the parameters. Each experiment was carried out in duplicate.

Supercritical transesterification

The reaction of transesterification of triglycerides with methanol is carried out in supercritical conditions by adopting reaction temperature of 250°C, higher than the critical temperature of methanol (240°C) and agitation of 200 rpm. The effect of molar ratio of supercritical methanol to rapeseed oil (Kusdiana and Saka, 2001) and cottonseed oil (Demirbas, 2006) has been investigated and it was determined that the ratio of 42:1 and 41:1, respectively, was the optimum ratio for the conversion to methyl esters. Therefore, a molar ratio of 40:1 for alcohol to oil was chosen in this study. The initial amounts of methanol and oil were chosen such that when the desired reaction temperature (250°C) was reached, the pressure in the system was 1 to 100 bar. The oil and methanol were in a single phase, and reaction occured spontaneously and rapidly. When the desired temperature was reached, the process remained for a set time. The temperature of the reaction vessel was measured with a thermocouple. Then, the vessel was transferred to an ice water bath to quench the reaction.

Although, some initial experimental work was carried on biodiesel production at elevated pressure in this study. During the reaction, the effect of pressure on reaction time was studied for all three feedstocks. A Shimadzu (Kyoto, Japan) gas chromatograph, GCMS-QP2010 PLUS, coupled with a flame ionization detector (FID) was used to analyze the fatty acid composition and ester content of the oils and biodiesel obtained from the transesterification reaction.

RESULTS AND DISCUSSION

Homogeneous catalyzed transesterification (alkaline esterification)

Effect of molar ratio of methanol to oil

Figures 1 to 3 show the effect of alcohol/oil molar ratio on

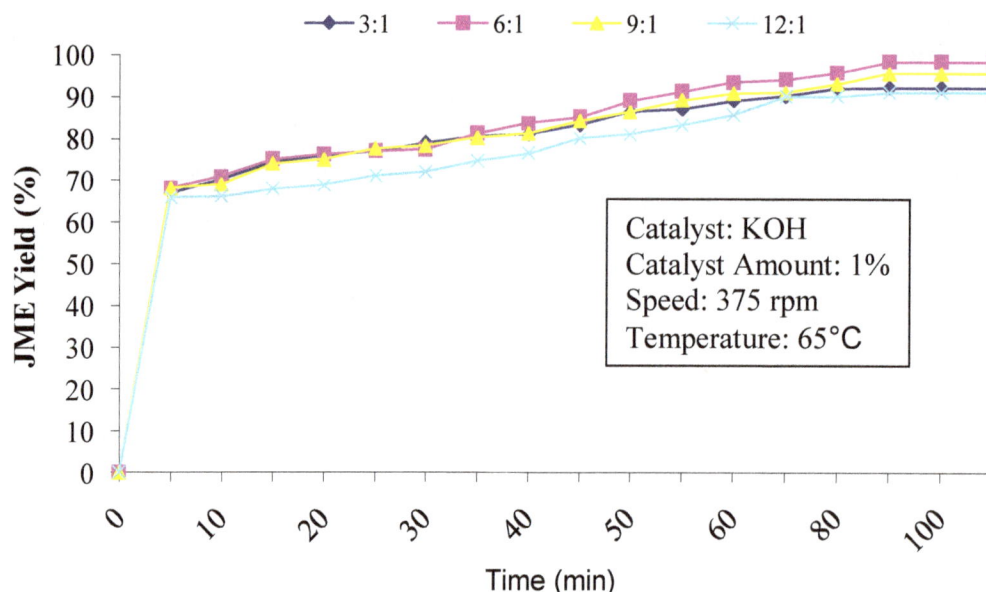

Figure 1. Effect of molar ratio on methyl ester yield for based catalysed transesterification of *Jatropha curcas* oil at varying reaction time showing 6:1 the best.

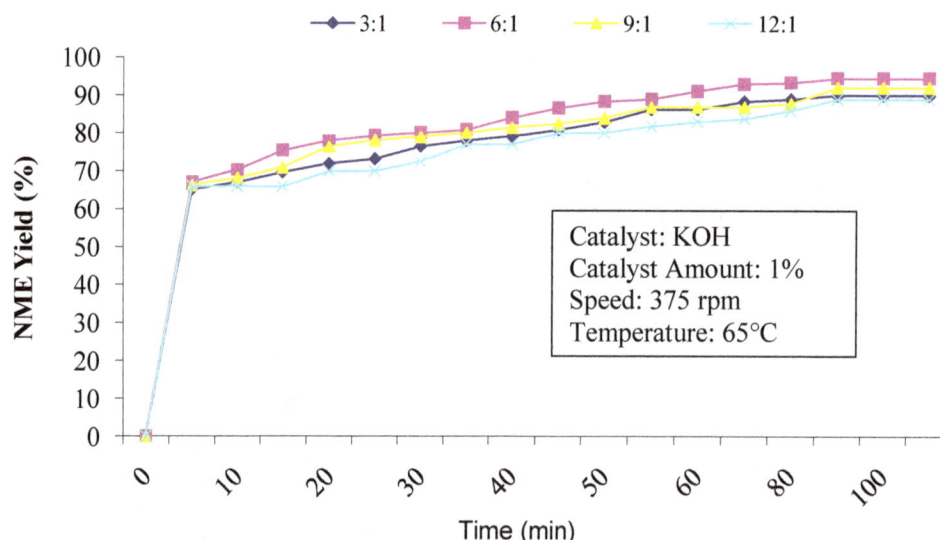

Figure 2. Effect of molar ratio on methyl ester yield for based catalysed transesterification of *Azadirachta indica* seed oil at varying reaction time showing 6:1 the best.

J. curcas methyl ester (JME), *A. indica* methyl ester (NME) and *H. brasiliensis* methyl ester (RME) yield, respectively using KOH as a catalyst. Concentration of catalyst, speed and the temperature were maintained constant. From the graph, highest yield for JME and NME was observed with respect to a molar ratio of 6:1 whereas that for the RME was 9:1. Analysis of variation the yield at 6:1 and 9:1 when compared to other molar ratio using SPSS version 17 showed no significant (p > 0.05) among the yields. This indicates that despite the

highest yield of JME and NME at 6:1 and JME at 9:1 was not significantly different from 3:1 and 12:1. However, the higher molar ratios than the stoichiometric value of 3:1 resulted in a greater ester conversion and could ensure complete reaction (Leung and Guo, 2006). This result obtained for JME and NME is in line with reports of many investigations based on Canola oil (Leung and Guo, 2006; Zhang et al., 2003; Boocock et al., 1996).

For RME, the larger ratio 9:1 indicated that more methanol were required for achieving the maximum yield

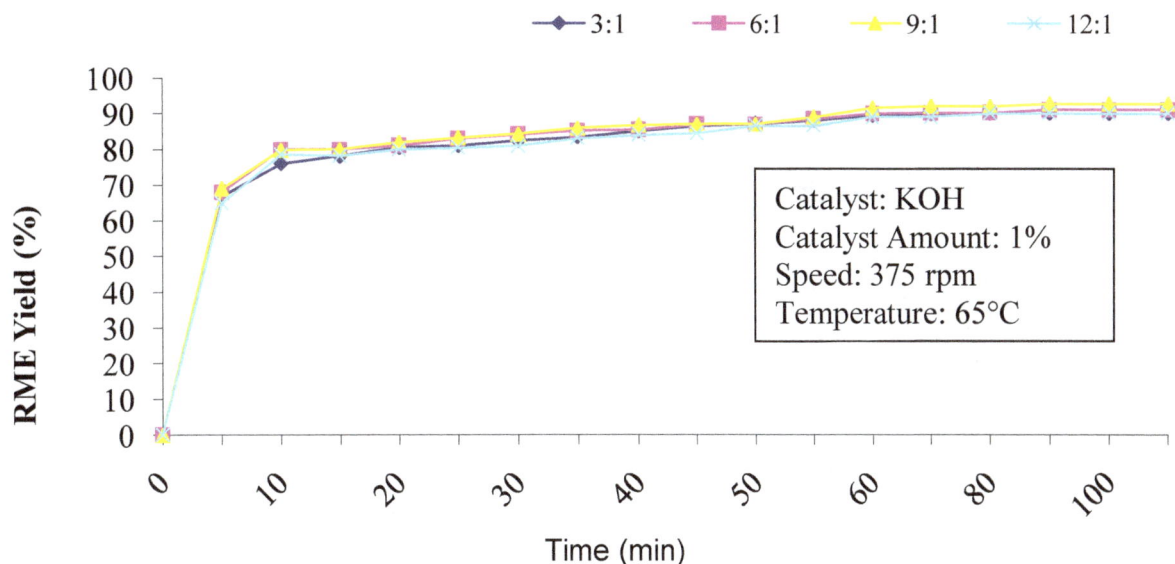

Figure 3. Effect of molar ratio on methyl ester yield for based catalysed transesterification of *Hevea brasiliensis* seed oil at varying reaction time showing 9:1 the best.

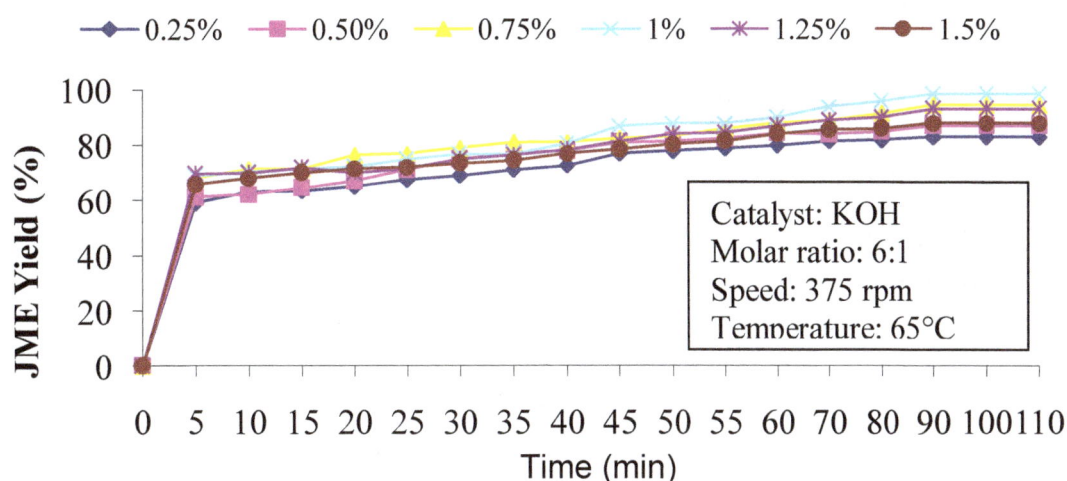

Figure 4. Effect of catalyst concentration on methyl ester yield for base catalyzed transesterification of *Jatropha curcas* seed oil at varying reaction time showing 1% is the best.

in the transesterification of rubber oil. The reason for this difference is likely due to the fact that *H. brasiliensis* oil was more viscose than the *J. curcas* and *A. indica* oils. More methanols increase the solubility of the oil in the methanol and improves the contact between oil and methanol molecules, thereby maximizes the conversion.

Effect of catalyst concentration

Figures 4 to 6 show the effect of catalyst concentration on JME, NME and RME yield, respectively using KOH as a catalyst. Transesterification reaction was carried out under optimal conditions obtained in the previous section (that is, 6:1 methanol/oil molar ratio for JME and NME and 9:1 methanol/oil molar ratio for RME) while temperature, rate of agitation and alcohol/oil molar ratio were kept constant. From the graphs, highest yield of all the biodiesel fuel has been observed with respect to catalyst concentration of 1%. Analysis of variation among the yield for different catalyst concentration showed no significant difference ($p > 0.05$). It was observed in the course of the experiment that addition of excess amount of catalyst leads to the formation of emulsion, which increased the viscosity of the mixture and led to the formation of gel. It was also observed that excess

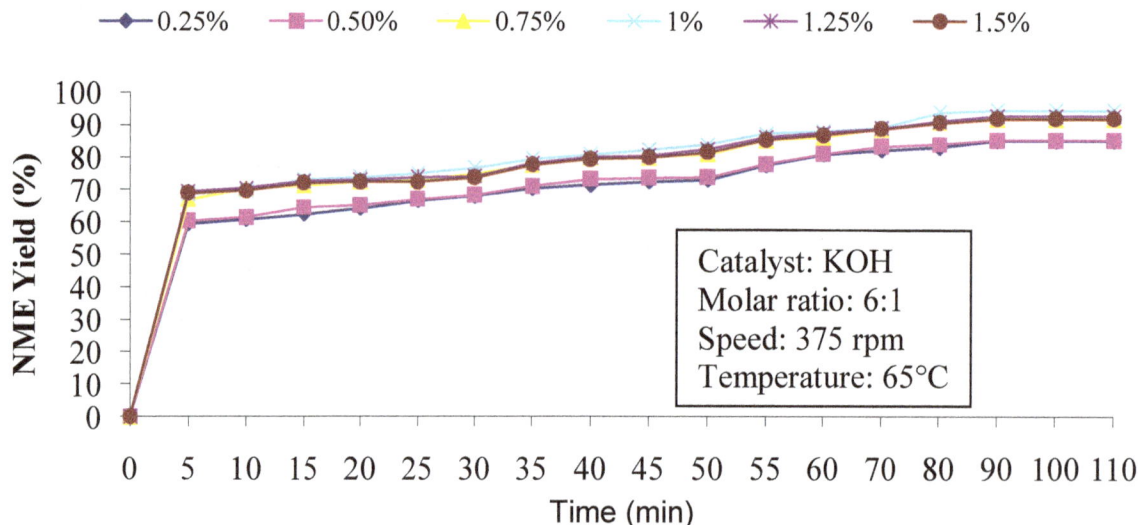

Figure 5. Effect of catalyst concentration on methyl ester yield for base catalyzed transesterification of *Azadirachta indica* seed oil at varying reaction time showing 1% is the best.

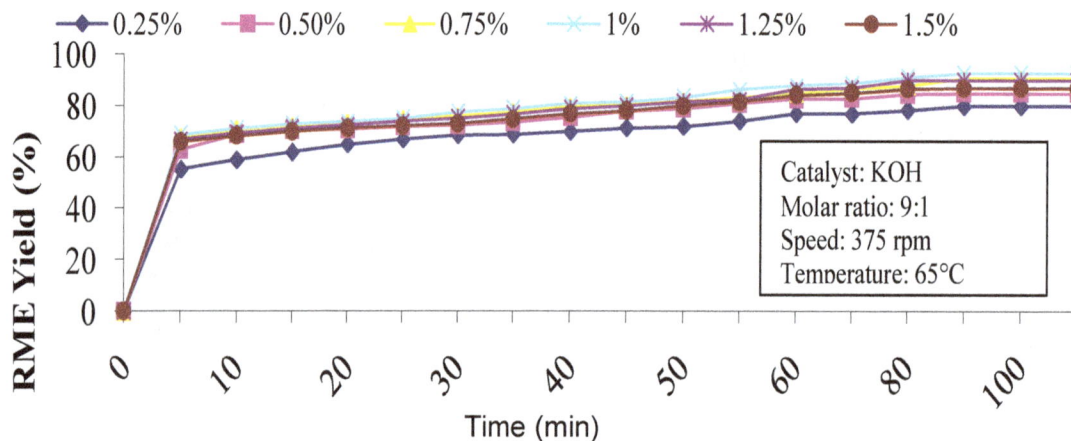

Figure 6. Effect of catalyst concentration on methyl ester yield for base catalyzed transesterification of *Hevea brasiliensis* seed oil at varying reaction time showing 1% is the best.

catalyst also reduced the total yield.

Effect of reaction temperature

Figures 7 to 9 show the effect of temperature on JME, NME and RME yield, respectively using KOH as a catalyst. Transesterification reaction was carried out under optimal conditions obtained in the previous section (that is, 6:1 methanol/oil molar ratio and 1% catalyst for JME and NME, and 9:1 methanol/oil molar ratio and 1% catalyst for RME) while speed maintained constant. As showed in the graph, highest yield of all biodiesel fuels were observed with respect to reaction temperature of 65°C, though, analysis of variation among the yield for

different temperature showed no significant difference ($p > 0.05$). The results indicated that the transesterification reaction could proceed within the temperature range studied but the reaction time to complete the reaction varied significantly with reaction temperature. The result in this study was however different from previous studies on optimization of Jatropha oil in which 55°C was reported as optimal temperature (Eevera et al., 2008).

According to Leung and Guo (2006), higher temperature could accelerate the saponification of triglycerides, and have negative effect on the product yield, it could also lead to a drastic decrease in viscosity of the oil that is favourable to increase the solubility of the oil in the methanol and improve the contact between oil and methanol molecules, thereby reaching a better

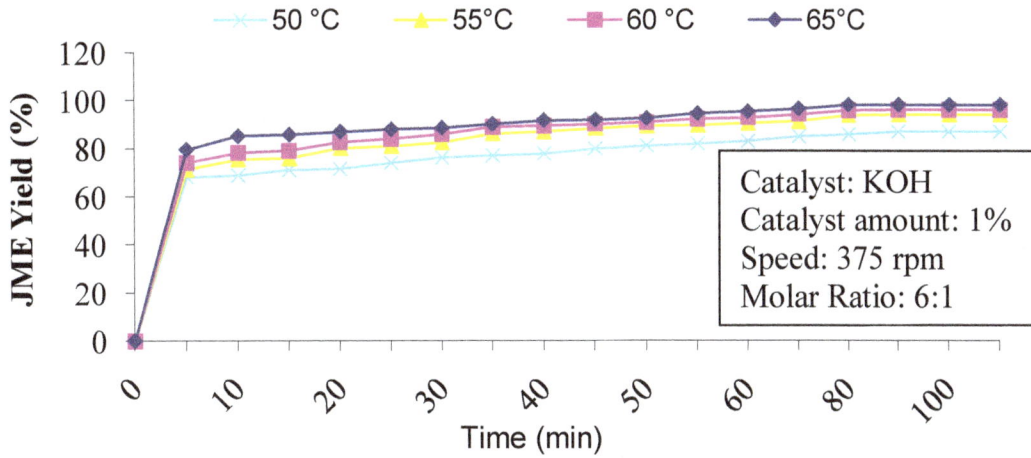

Figure 7. Effect of temperature on methyl ester yield for base catalyzed transesterification of *Jatropha curcas* oil at varying reaction time showing 65°C is the best.

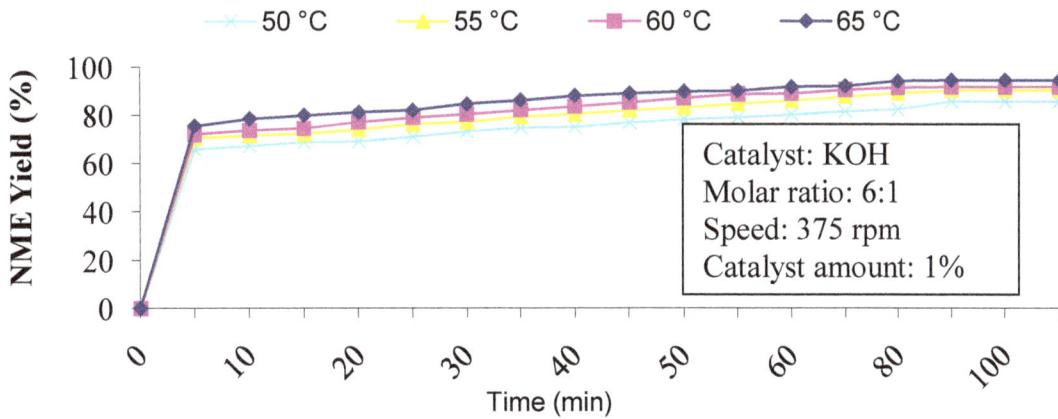

Figure 8. Effect of temperature on methyl ester yield for base catalyzed transesterification of *Azadirachta indica* seed oil at varying reaction time showing 65°C is the best.

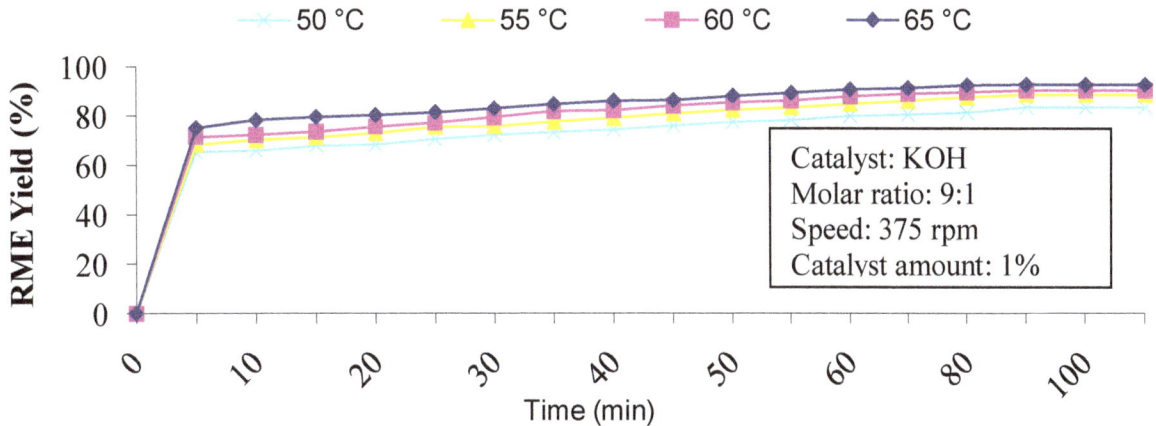

Figure 9. Effect of temperature on methyl ester yield for base catalyzed transesterification of *Hevea brasiliensis* seed oil at varying reaction time showing 65°C is the best.

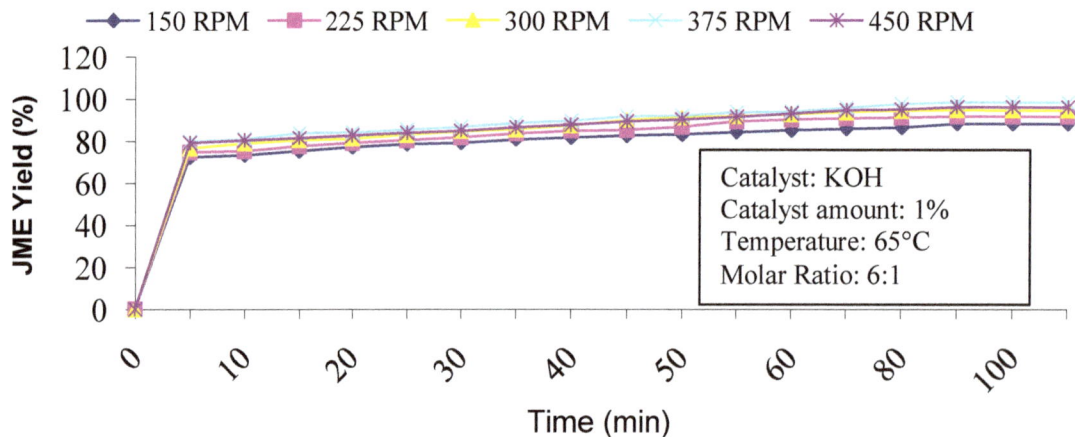

Figure 10. Effect of stirring speeds/mixing intensity on methyl ester yield for base catalyzed transesterification of *Jatropha curcas* oil at varying reaction time showing 375 rpm is the best.

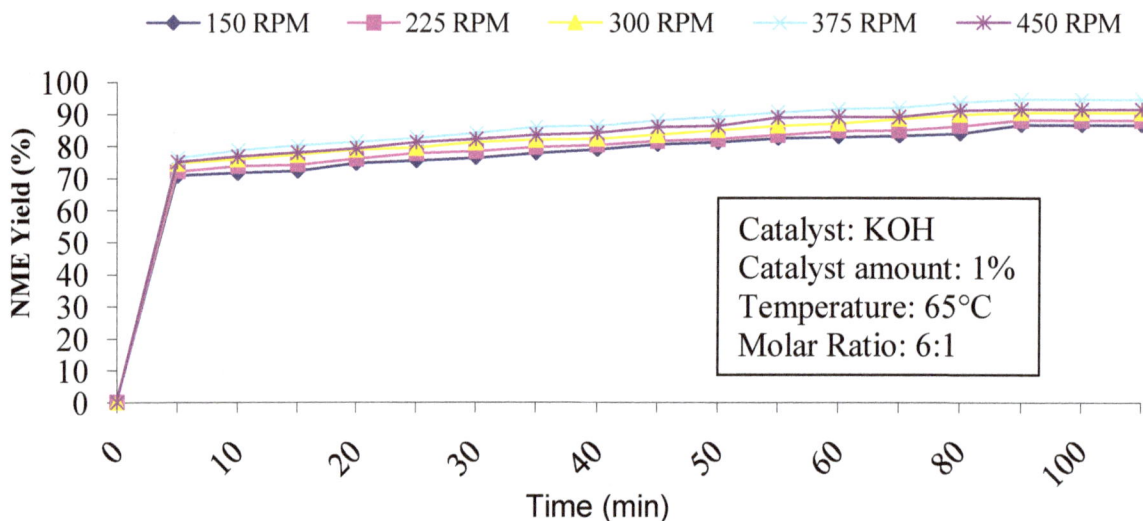

Figure 11. Effect of stirring speeds/mixing intensity on methyl ester yield for base catalyzed transesterification of *Azadirachta indica* seed oil at varying reaction time showing 375 rpm is the best.

conversion of triglycerides. This implies that optimal temperature of 65°C obtained in this study could be as a result of the competition between the main transesterification reaction and the side saponification reaction.

Effect of rate of agitation

Transesterification reaction was carried out under optimal conditions obtained in the previous section (that is, 6:1 methanol/oil, 1% catalyst and 65°C temperature for JME and NME, 9:1 methanol/oil molar ratio, 1% catalyst and 65°C temperature for RME) while rate of stirring was varied. It was observed from Figures 10 to 12 that the

ester yield for all studied oil increased with increase in rate of stirring. Results obtained from the present experiments with *J. curcas*, *A. indica* and *H. brasiliensis* seed oils reveal that 375 rpm was sufficient for the completion of the catalyzed transesterification. Analysis of variation among the yield for different catalyst concentration showed no significant difference ($p < 0.05$).

Table 1 summarized the optimal conditions of the biodiesel produced from jatropha, neem and rubber seed oil. The present optimization studies on the transesterification of the oils showed that highest conversion might be achieved at 6:1 methanol/oil nolar ratio, 1% catalyst 65°C temperature and 375 rpm for *J. curcas* methyl and *A. indica* 9:1 methanol/oil molar ratio 1% catalyst 65°C,

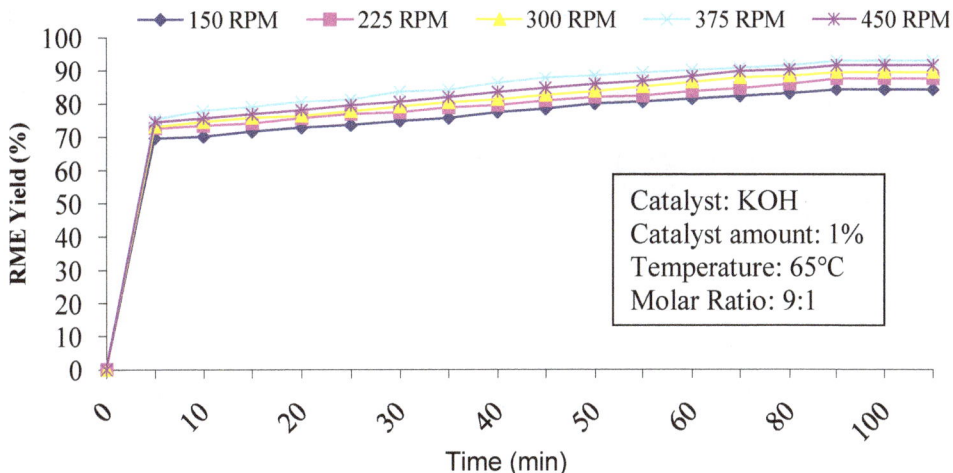

Figure 12. Effect of stirring speeds/mixing intensity on methyl ester yield for base catalyzed transesterification of *Hevea brasiliensis* seed oil at varying reaction time showing 375 rpm is the best.

Table 1. Optimized process parameters for base catalysed transesterification process.

Methyl ester	Alcohol to oil molar ratio	Catalyst (%)	Temperature	RPM	Yield (%)
Jatropha	6:1	1	65	375	98.4
Neem	6:1	1	65	375	94.6
Rubber	9:1	1	65	375	92.8

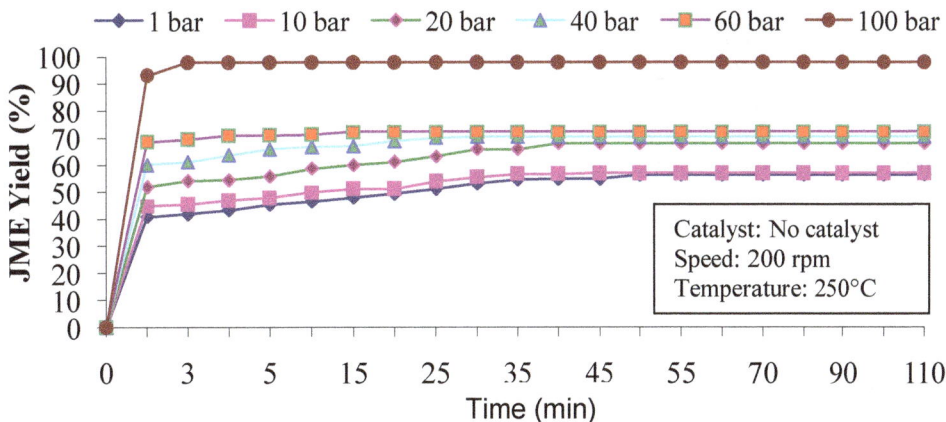

Figure 13. Variation of reaction time with increase in pressure for JME.

and 375 rpm temperature for *H. brasiliensis* seed oil. However, the maximum yield of biodiesel revealed the following trends; JME (98.4%) > NME (94.6%) > RME (92.8%).

Supercritical transesterification

An alternative, catalyst-free method for transesterification uses supercritical methanol at high temperatures and pressures. The process can tolerate water in the feedstock, free fatty acids are converted into fatty acid methyl esters with similar or higher rates than the corresponding triglycerides, and therefore, the process can use high FFA, cheaper feedstocks such as rubber seed oil. Also, the catalyst removal step is eliminated making this process environmental friendly. Some initial experimental work was carried out on biodiesel

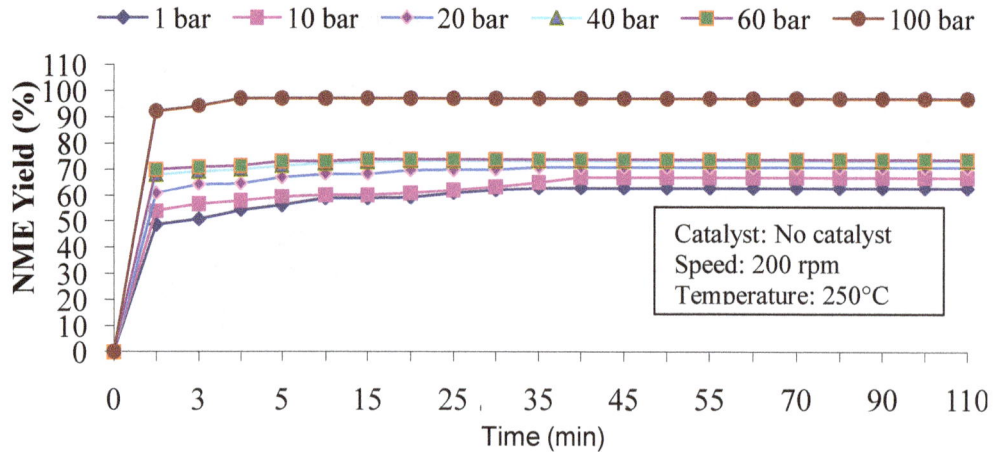

Figure 14. Variation of reaction time with increase in pressure for NME.

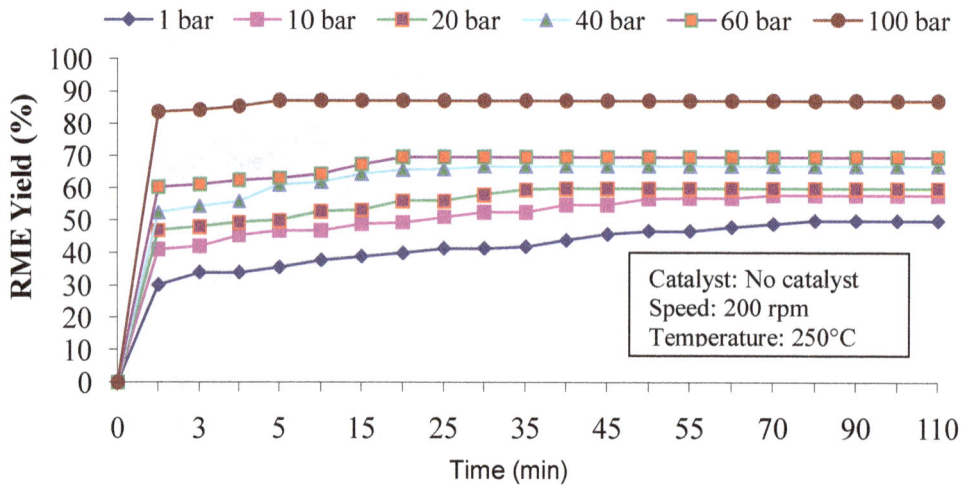

Figure 15. Variation of reaction time with increase in pressure for RME.

production at elevated pressure in this study. The effect of pressure on reaction time was studied for all three feedstocks which are summarized in Figures 13 to 15.

As shown in the figures, the effect of increase in pressure on reaction time of JME, NME and RME yield without using any catalyst. Speed and temperature were maintained constant. As can be seen in the figures, the highest yield of 98% for JME was observed at a pressure of 100 bar in 3 min, yield of 97.1% for NME was observed at a pressure of 100 bar in 4 min and highest yield of 87.1% for RME at a pressure of 100 bar in 5 min. The variation in time could be due to different FFA content of the oil.

Comparison between supercritical methanol treatment with catalytic methods

Table 2 shows the comparison between supercritical

methanol treatment with catalytic methods in terms of reaction conditions and biodiesel yield. As can be seen, supercritical methanol is superior to catalytic methods because no catalyst is needed and a relatively lower reaction time is required to obtain a very high yield of 98, 97.1 and 87.1% for JME, NME and RME. On the other hand, a longer reaction time of 65 and 120 min are necessary to obtain similar results for homogeneous and heterogeneous reactions, respectively. Apart from these, the presence of catalyst in the reaction mixture makes the purification and separation process complicated and energy consuming (Cao et al., 2005). Hence, these processes become unreactive in terms of time and economic consideration.

In another study (Kusdiana and Saka, 2004), it was observed that the presence of free fatty acids and water in palm oil did not affect the yield of biodiesel in the supercritical methanol method as no unwanted saponified product was produced as in catalytic reactions. In fact,

Table 2. Comparison between supercritical methanol with catalytic transesterification reaction.

Parameter	Catalytic	Supercritical methanol
Reaction time	> 1 h	3-5 min
Reaction pressure and temperature	1 bar, 338 K	100 bar, 523 K
Catalyst	NaOH, KOH	None
Effect of presence of FFA	Soapy products with problems of separation	No separation problems
Yield	<100%	100%
Purification needs of removal of	Methanol, catalyst	Methanol only
Process	Complicated	Simple

the free fatty acids were also esterified during the transesterification reaction and subsequently increased the total yield of biodiesel. Such studies have shown that non-catalytic supercritical methanol transesterification technology has been superior to conventional catalytic reactions in terms of cost and energy consumption. The amount of insignificant reaction time needed and the absence of a catalyst make this technology very attractive compared to other processes. Hence, it has huge potential to be explored and developed in the near future in order to fulfill the escalating demand of renewable energy sources in the world.

Conclusion

Results obtained through the study suggest that supercritical methanol method has a high potential for transesterification of triglycerides to biodiesel fuel.

REFERENCES

Antolin G, Tinaut FV, Briceno Y, Castano V, Perez C, Rarnirez AL (2002). Optimisation of biodiesel production by sunflower oil transesterification; Biores. Technol., 83: 111-114.
Boocock DGB (1996). Process of producing lower alkyl fatty acid esters. CA 0112581.
Cao W, Han H, Zhang J (2005). Preparation of biodiesel from soybean oil using supercritical methanol and co-solvent. Fuel, 84: 347-351.
Centinkaya M, Karaosmanoglu F (2004). Optimization of base-catalyzed tranesterification reaction of used cooking oil. Energy Fuels, 18: 1888-1895.
Demirbas A (2006). Biodiesel production via catalytic and no-catalytic supercritical methanol transesterification methods. Progress Energy and Combustion, 31: 466- 487.
Eevera T, Rajendran K, Saradha S (2008). Biodiesel production process optimization and characterization to assess the suitability of the product for varied environment conditions. Renewable Energy. Doi:l010 Lirnenen. 2008.04016.
Haas MJ (2005). Improving the economics of biodiesel production through the use of low value lipids as feedstocks: vegetable oil soapstocks. Fuel Process Technol., 86: 1087-1096.
Kusdiana D, Saka S (2004). "Two Step Preparation for Catalyst-Free Biodiesel Fuel Production". Appl. Biochem. Biotechnol., 115: 781-791
Leung DYC, Guo Y (2006). Transesterification of neat and used frying oil: optimization for biodiesel production. Fuel Processing Technol., 87: 883-890.
Ramadhas AS, Muraleedharan C, Jayarey S (2005). Performance emission Waluatron of a diesel engine fueled with methyl esters of rubber seed oil. Renew Energy, 30: 1789-1800.
Shieh CJ, Liao HF, Lee CC (2003). Optimization of lipase-catalysed biodiesel by response surface methodology. Biores. Technol., 88: 103-106.
Zhang Y, Dube MA, Mclean DD, Kates M (2003). Biodiesel production from waste cooking oil: 2 Economic assessment and sensitivity analysis. Biores. Technol., 90: 229-240.

Permissions

All chapters in this book were first published in JPTAF, by Academic Journals; hereby published with permission under the Creative Commons Attribution License or equivalent. Every chapter published in this book has been scrutinized by our experts. Their significance has been extensively debated. The topics covered herein carry significant findings which will fuel the growth of the discipline. They may even be implemented as practical applications or may be referred to as a beginning point for another development.

The contributors of this book come from diverse backgrounds, making this book a truly international effort. This book will bring forth new frontiers with its revolutionizing research information and detailed analysis of the nascent developments around the world.

We would like to thank all the contributing authors for lending their expertise to make the book truly unique. They have played a crucial role in the development of this book. Without their invaluable contributions this book wouldn't have been possible. They have made vital efforts to compile up to date information on the varied aspects of this subject to make this book a valuable addition to the collection of many professionals and students.

This book was conceptualized with the vision of imparting up-to-date information and advanced data in this field. To ensure the same, a matchless editorial board was set up. Every individual on the board went through rigorous rounds of assessment to prove their worth. After which they invested a large part of their time researching and compiling the most relevant data for our readers.

The editorial board has been involved in producing this book since its inception. They have spent rigorous hours researching and exploring the diverse topics which have resulted in the successful publishing of this book. They have passed on their knowledge of decades through this book. To expedite this challenging task, the publisher supported the team at every step. A small team of assistant editors was also appointed to further simplify the editing procedure and attain best results for the readers.

Apart from the editorial board, the designing team has also invested a significant amount of their time in understanding the subject and creating the most relevant covers. They scrutinized every image to scout for the most suitable representation of the subject and create an appropriate cover for the book.

The publishing team has been an ardent support to the editorial, designing and production team. Their endless efforts to recruit the best for this project, has resulted in the accomplishment of this book. They are a veteran in the field of academics and their pool of knowledge is as vast as their experience in printing. Their expertise and guidance has proved useful at every step. Their uncompromising quality standards have made this book an exceptional effort. Their encouragement from time to time has been an inspiration for everyone.

The publisher and the editorial board hope that this book will prove to be a valuable piece of knowledge for researchers, students, practitioners and scholars across the globe.

List of Contributors

Praveen K. S. Yadav
Department of Oil and Paint Technology, Harcourt Butler Technological Institute, Kanpur, India

Onkar Singh
Department of Mechanical Engineering, Harcourt Butler Technological Institute, Kanpur, India

R. P. Singh
Department of Oil and Paint Technology, Harcourt Butler Technological Institute, Kanpur, India

K. Nanthagopal
Automotive Research Centre, SMBS, VIT University, Vellore-14, Tamilnadu, India

R. Thundil Karuppa Raj
Automotive Research Centre, SMBS, VIT University, Vellore-14, Tamilnadu, India

T. Vijayakumar
Automotive Research Centre, SMBS, VIT University, Vellore-14, Tamilnadu, India

P. M. Ejikeme
Industrial and Biomass Research Laboratory, Department of Pure and Industrial Chemistry, University of Nigeria, Nsukka, Enugu State, Nigeria

I. D. Anyaogu
Department of Science Laboratory Technology, Federal Polytechnic Nasarawa, Nasarawa State, Nigeria

C. A. C. Egbuonu
Nutritional and Toxicological Biochemistry Unit, Department of Biochemistry, University of Nigeria, Nsukka, Enugu State, Nigeria

V. C. Eze
Industrial and Biomass Research Laboratory, Department of Pure and Industrial Chemistry, University of Nigeria, Nsukka, Enugu State, Nigeria

I. A. Mohammed-Dabo
Chemical Engineering Department, Ahmadu Bello University Zaria, Nigeria

M. S. Ahmad
Center for Renewable Energy Research, Umaru Musa Yarádua University, Katsina, Nigeria

A. Hamza
Chemical Engineering Department, Ahmadu Bello University Zaria, Nigeria

K. Muazu
National Research Institute for Chemical Technology Zaria, Nigeria

A. Aliyu
Chemical Engineering Department, Ahmadu Bello University Zaria, Nigeria

S. Sundarapandian
Department of Mechanical Engineering, Sethu Institute of Technology, Kariapatti, Tamil Nadu State, India

M. H. Shagal
Department of Chemistry, Modibbo Adama University of Technology, P. M. B. 2076, Yola, Adamawa State, Nigeria

J. T. Barminas
Department of Chemistry, Modibbo Adama University of Technology, P. M. B. 2076, Yola, Adamawa State, Nigeria

B. A. Aliyu
Department of Chemistry, Modibbo Adama University of Technology, P. M. B. 2076, Yola, Adamawa State, Nigeria

S. A. Osemeahon
Department of Chemistry, Modibbo Adama University of Technology, P. M. B. 2076, Yola, Adamawa State, Nigeria

A. Y. J. Akossou
Département d'Aménagement et de Gestion des Ressources Naturelles, Faculté d'Agronomie, Université de Parakou, BP 123, Parakou Bénin

E. Gbozo
Département d'Aménagement et de Gestion des Ressources Naturelles, Faculté d'Agronomie, Université de Parakou, BP 123, Parakou Bénin

A. E. Darboux
Département d'Aménagement et de Gestion des Ressources Naturelles, Faculté d'Agronomie, Université de Parakou, BP 123, Parakou Bénin

K. Kokou
Laboratoire de Botanique et d'Ecologie, Faculté des Sciences, Université de Lomé, BP 1515 Lomé Togo

E. Ramjee
Department of Mechanical Engineering, JNTUH College of Engineering, Kukatpally, Hyderabad, Andhra Pradesh, India

K. Vijaya Kumar Reddy
Department of Mechanical Engineering, JNTUH College of Engineering, Kukatpally, Hyderabad, Andhra Pradesh, India

J. Suresh Kumar
Department of Mechanical Engineering, JNTUH College of Engineering, Kukatpally, Hyderabad, Andhra Pradesh, India

Sachin Kumar
Sardar Swaran Singh National Institute of Renewable Energy, Kapurthala-144 601, India

Surendra P. Singh
Department of Paper Technology, Indian Institute of Technology, Roorkee, Saharanpur Campus- 247 001, India

Indra M. Mishra
Department of Chemical Engineering, Indian Institute of Technology, Roorkee- 247 667, India

Dilip K. Adhikari
Biotechnology Area, Indian Institute of Petroleum, Dehradun- 248 005, India

K. M. Oghenejoboh
Department of Chemical Engineering, Delta State University, Oleh Campus, P. M. B. 22, Oleh, Nigeria

E. O. Ohimor
Industries Department, Ministry of Commerce and Industry, Agbor, Delta State, Nigeria

O. Olayebi
Department of Chemical Engineering, Faculty of Engineering, Federal University of Petroleum Resources, Ugbomoro, Delta State, Nigeria

Gulab Chand Shah
School of Biotechnology, Rajiv Gandhi Proudyogiki Vishwavidyalaya, Bhopal, Madhya Pradesh (University of Technology of Madhya Pradesh), Airport Bypass Road, Gandhi Nagar, Bhopal-462 033, India

Mahavir Yadav
School of Biotechnology, Rajiv Gandhi Proudyogiki Vishwavidyalaya, Bhopal, Madhya Pradesh (University of Technology of Madhya Pradesh), Airport Bypass Road, Gandhi Nagar, Bhopal-462 033, India

Archana Tiwari
School of Biotechnology, Rajiv Gandhi Proudyogiki Vishwavidyalaya, Bhopal, Madhya Pradesh (University of Technology of Madhya Pradesh), Airport Bypass Road, Gandhi Nagar, Bhopal-462 033, India

Vijittra Chalatlon
School of Environment, Resources and Development, Asian Institute of Technology, P.O. Box 4, Klong Luang, Pathumthani 12120, Thailand

Murari Mohon Roy
Department of Engineering, Nova Scotia Agricultural College, 39 Cox Road, Banting Building, Truro, NS, Canada, B2N 5E3

Animesh Dutta
Mechanical Engineering Program, School of Engineering University of Guelph, Guelph, Ontario, Canada, N1G 2W1

Sivanappan Kumar
School of Environment, Resources and Development, Asian Institute of Technology, P.O. Box 4, Klong Luang, Pathumthani 12120, Thailand

Kambila Vijay Kumar
Centre for Nano Technology, FED, K. L University, Guntur-522 502(A.P), India

Krishna Bolla
Department of Microbiology, Kakatiya University, Warangal – 506009, A.P, India

S. V. S. S. S. L. Hima Bindu N
Department of Microbiology, Kakatiya University, Warangal – 506009, A.P, India

M. A. Singara Charya
Department of Microbiology, Kakatiya University, Warangal – 506009, A.P, India

P. V. Rao
Department of Mechanical Engineering, Andhra University, Visakhapatnam-530003, Andhra Pradesh, India

Ioana Stanciu
Faculty of Chemistry, University of Bucharest, 4-12 Regina Elisabeta Blvd., 030018 Bucharest, Romania

Ajay Chandak
SSVPS BSD College of Engineering, Deopur, Dhule: 424005, India.
Prince Suman Foundation, Dhule, India

Sunil Somani
Medicaps Institute of Technology and Management, Pigdambar, Indore - 453 331, MP, India

Fábio Viana de Abreu
Petrobras - Brazilian Oil S. A. Brazil

Mila Rosendal Avelino
Program of Pos Graduation in the Mechanical Engineering, Universidade do Estado do Rio de Janeiro (UERJ), Brazil

Mauro Carlos Lopes Souza
Program of Pos Graduation in the Mechanical Engineering, Universidade do Estado do Rio de Janeiro (UERJ), Brazil

Diego Preza Monaco
Program of Pos Graduation in the Mechanical Engineering, Universidade do Estado do Rio de Janeiro (UERJ), Brazil

Joevis J. Claveria
School of Engineering, Science and Health, Victoria University of Technology, Victoria, Australia

Akhtar Kalam
School of Engineering, Science and Health, Victoria University of Technology, Victoria, Australia

J. K. Tangka
Renewable Energy laboratory Department of Agricultural Engineering, University of Dschang, P.O. Box 373, Dschang, Cameroon

J. E. Berinyuy
Crop Processing laboratory, Department of Agricultural Engineering, University of Dschang , P.O. Box 373, Dschang, Cameroon

Tekounegnin
Department of Agricultural Engineering FASA University of Dschang Cameroon

A. N. Okale
Department of Agricultural Engineering FASA University of Dschang Cameroon

E. I. Bello
Department of Mechanical Engineering, The Federal University of Technology Akure, Ondo State, Nigeria

A. O. Fade-Aluko
Department of Mechanical Engineering, The Federal University of Technology Akure, Ondo State, Nigeria

S. A. Anjorin
Department of Mechanical Engineering, The Federal University of Technology Akure, Ondo State, Nigeria

T. S. Mogaji
Department of Mechanical Engineering, The Federal University of Technology Akure, Ondo State, Nigeria

Andrea Kleinová
Faculty of Chemical and Food Technology, Slovak University of Technology (STU), Bratislava, Slovak Republic

Ján Cvengroš
Faculty of Chemical and Food Technology, Slovak University of Technology (STU), Bratislava, Slovak Republic

B. K. Highina
Department of Chemical Engineering, University of Maiduguri, Borno State, Nigeria

I. M. Bugaje
Department of Chemical Engineering, University of Maiduguri, Borno State, Nigeria

B. Umar
Department of Agricultural and Water Resources Engineering, University of Maiduguri, Borno State, Nigeria

Moinuddin Sarker
Natural State Research Inc., 37 Brown House Road, Stamford, CT-06902, USA

Mohammad Mamunor Rashid
Natural State Research Inc., 37 Brown House Road, Stamford, CT-06902, USA

Md Sadikur Rahman
Natural State Research Inc., 37 Brown House Road, Stamford, CT-06902, USA

Sameh M. Metwalley
Faculty of Engineering, Helwan University, Cairo, Egypt

Shawki A. Abouel-seoud
Faculty of Engineering, Helwan University, Cairo, Egypt

Abdelfattah M. Farahat
Faculty of Engineering, Helwan University, Cairo, Egypt

C. S. Ezeanayanso
National Research Institute for Chemical Technology, P. M. B. 1052, Basawa, Zaria, Kaduna State, Nigeria

E. M. Okonkwo
National Research Institute for Chemical Technology, P. M. B. 1052, Basawa, Zaria, Kaduna State, Nigeria

E. B. Agbaji
Department of Chemistry, Ahmadu Bello University, Samara, Zaria, Kaduna State, Nigeria

V. I. Ajibola
Department of Chemistry, Ahmadu Bello University, Samara, Zaria, Kaduna State, Nigeria

O. J. Okunola
National Research Institute for Chemical Technology, P. M. B. 1052, Basawa, Zaria, Kaduna State, Nigeria